Current and Future Developments in Nanomaterials and Carbon Nanotubes:

(Volume 3)

Applications of Nanomaterials in Energy Storage and Electronics

Edited by

Gaurav Manik
Department of Polymer and Process Engineering
Indian Institute of Technology Roorkee (IITR)
Roorkee, Uttarakhand, India

&

Sushanta Kumar Sahoo
Materials Science and Technology Division
CSIR-National Institute for Interdisciplinary Science and
Technology (CSIR-NIIST)
Kerala, India

Current and Future Developments in Nanomaterials and Carbon Nanotubes

Volume # 3

Applications of Nanomaterials in Energy Storage and Electronics

Editors: Gaurav Manik and Sushanta Kumar Sahoo

ISSN (Online): 2589-2207

ISSN (Print): 2589-2193

ISBN (Online): 978-981-5050-71-4

ISBN (Print): 978-981-5050-72-1

ISBN (Paperback): 978-981-5050-73-8

need for a court order if at any point you breach any terms of this License Agreement. In no event will any delay or failure by Bentham Science Publishers in enforcing your compliance with this License Agreement constitute a waiver of any of its rights.

3. You acknowledge that you have read this License Agreement, and agree to be bound by its terms and conditions. To the extent that any other terms and conditions presented on any website of Bentham Science Publishers conflict with, or are inconsistent with, the terms and conditions set out in this License Agreement, you acknowledge that the terms and conditions set out in this License Agreement shall prevail.

Bentham Science Publishers Pte. Ltd.
80 Robinson Road #02-00
Singapore 068898
Singapore
Email: subscriptions@benthamscience.net

CONTENTS

FOREWORD

In the 21st century, depleting fossil fuel resources and environmental concerns have created huge challenges in meeting adequate energy production and storage. Undoubtedly, the ground-breaking nanotechnology may be the only solution to meet the demands of mankind in the future by producing a secure, green and sustainable energy. Carbon-based and other advanced functional nanomaterials have enormous potential to produce and save energy through effective and sustainable approach, and also to support suitable applications in the field of opto- and bio-electronics.

This unique book by the editors (Dr. Manik and Dr. Sahoo), comprising about fifteen chapters presents an excellent overview and current state-of-the-art future generation nanostructured materials like carbon nanotubes, carbon nano onions, nanowires, graphene, 2D boron nitride, metal oxides, quantum dots, metal organic frameworks (MoFs) *etc.* and their respective nanocomposites. Such materials find great applications with regard to efficient storage of energy like solar, hydrogen and electrochemical energy. These materials find enormous applications in fuel cells, supercapacitors, bio-sensors and opto-electronic nano-devices, ferroelectric liquid crystal nanocomposites for optical memory, future display devices and electronics. Considering these aspects, I believe that the book shall find great use in the scientific and engineering community.

It is notable that the book chapters have been contributed by several researchers from premier institutes who carry extensive experience and knowledge in the relevant fields. Further, the book is well edited, quite consistent and focused on enrichment of the knowledge of the readers. This is an excellent masterpiece for students, academics and industry researchers.

This book shall, therefore, be an impactful addition in the area of application of next generation nanomaterials in the field of energy storage and electronics. The reported work will be important to many groups of researchers and industry experts across the globe, including, but not limited to, those working in materials science, chemical engineering, biotechnology, polymer science, physics, chemistry and renewable energy.

Sabu Thomas
Mahatma Gandhi University
Kottayam, Kerala
India

PREFACE

Nanotechnology is one of the research areas most focused on the development of a new class of nanomaterials that offer enhanced material or product performance by exploiting synergism. Due to the growing demand in the area of energy and electronics, nanomaterials and their hybrids have shaped the research in the relevant fields with remarkable attention as advanced multifunctional materials with desired efficiency, higher durability and improved stability.

Currently, the applications of nanomaterials in the field of energy harvesting, storage and opto-electronics are booming, which has resulted in an upsurge in the number of publications, patents and technology transfers worldwide. Despite this, relatively only a few books have focused on nanomaterials applied to energy and electronics. Therefore, the editors found it an opportunity to edit a book on *"Current and Future Developments in Nanomaterials: Applications in Energy Storage and Electronics"*. We hope that the present book will immensely benefit scientists, engineers, academic researchers, research scholars and post graduate students working in the area of nanomaterials. This book consists of 15 chapters that describe the recent advancements in synthesis and applications of nanomaterials in energy harvesting and storage, and also technology in the field of opto-electronics for next generation devices. Some of the chapters summarize the recent progress in applications of nanomaterials like Carbon Nanotubes, Metal Oxides, and Graphene oxides-based hybrids in solar energy harvesting using recent Photovoltaic Technologies. Similarly, some of the chapters review the fundamentals and state-of-the-art developments in Nanowires, Graphene Quantum Dots, Boron nitrides, Carbon Nano Onions and Metal Organic Frameworks leading to the fabrication of Supercapacitors, Bio-sensors, Lithium ion batteries and Hydrogen storage applications. Further, a few chapters discuss the next generation fuel cells using Polymer Nanocomposites, Ferroelectric Liquid Crystal Nanocomposite and Opto-electronic Nanomaterials for optical memory and displays devices.

The editors are extremely thankful to the esteemed and experienced contributors of all chapters and also the Bentham Science Publishing team for their kind support. We hope that the content presented herewith in a simple and concise form shall serve as a comprehensive guide to benefit the readers and elevate their knowledge in this increasingly advancing area.

Gaurav Manik
Department of Polymer and Process Engineering
Indian Institute of Technology Roorkee (IITR)
Roorkee, Uttarakhand, India

&

Sushanta Kumar Sahoo
Materials Science and Technology Division
CSIR-National Institute of Interdisciplinary Science and
Technology (NIIST)
Kerala, India

List of Contributors

Achu Chandran	Materials Science and Technologyy Division, CSIR-National Institute for Interdisciplinary Science and Technologyy (NIIST), Thiruvananthapuram-695019, India Academy of Scientific and Innovative Research (AcSIR), Ghaziabad-201002, India
Ananthakumar Ramadoss	School for Advanced Research in Petrochemicals: Laboratory for Advanced Research in Polymeric Materials, Central Institute of Petrochemicals Engineering & Technologyy, Bhubaneswar-75102, India
Ankita Mohanty	School for Advanced Research in Petrochemicals: Laboratory for Advanced Research in Polymeric Materials, Central Institute of Petrochemicals Engineering & Technologyy, Bhubaneswar-75102, India Department of Physics, Utkal University, Bhubaneswar, 751004, India
Arijit Mitra	Institute of Physics, Sachivalaya Marg, Bhubaneswar 751005, India
Arpita Adhikari	Department of Polymer Science and Technologyy, University of Calcutta, 92 A.P.C. Road, Kolkata, 700009, India
Ashish Kalkal	NanobioTechnologyy Laboratory, Department of Biosciences and Bioengineering, Indian Institute of Technologyy Roorkee, Roorkee-24766, Uttarakhand, India
B. Bindhu	Department of Physics, Noorul Islam Centre for Higher Education, Kumaracoil, Tamil Nadu, 629180, India
D. Harimurugan	Department of Electrical Engineering, Dr. B. R. Ambedkar National Institute of Technologyy, Jalandhar-140011, Jalandhar-140011
Dipankar Chattopadhyay	Department of Polymer Science and Technology, University of Calcutta, 92 A.P.C. Road, Kolkata, 700009, India
Debabrata Panda	Department of Chemical Engineering, National Institute of Technologyy Rourkela, Rourkela-769008, Odisha, India
Gaurav Manik	Department of Polymer and Process Engineering, Indian Institute of Technologyy Roorkee, Roorkee, Uttarakhand, 247667, India
Gopinath Packirisamy	NanobioTechnologyy Laboratory, Department of Biosciences and Bioengineering, Indian Institute of Technologyy Roorkee, Roorkee-24766, Uttarakhand, India Centre for NanoTechnologyy, Indian Institute of Technologyy Roorkee, Roorkee-247667, Uttarakhand, India
H. N. Nagendra	Centre for Cryogenic Technologyy, Indian Institute of Science, Bangalore 560012, India
Harilal	School of Chemistry, University of Hyderabad, Hyderabad 500046, India
Harris Varghese	Materials Science and Technologyy Division, CSIR-National Institute for Interdisciplinary Science and Technologyy (NIIST), Thiruvananthapuram-695019, India Academy of Scientific and Innovative Research (AcSIR), Ghaziabad-201002, India

J. Ping Liu	Department of Physics, University of Texas at Arlington, Arlington, TX 76019, USA
Jeotikanta Mohapatra	Department of Physics, University of Texas at Arlington, Arlington, TX 76019, USA
Jingting Luo	Shenzhen Key Laboratory of Advanced Thin Films and Applications, College of Physics and Optoelectronic Engineering, Shenzhen University, 518060, Shenzhen, PR China
Jitendra Kumar	Advanced Research in Electrochemical Impedance Spectroscopy, Indian Institute of Technologyy Roorkee, Roorkee 247667, India
K. Prabakaran	Shenzhen Key Laboratory of Advanced Thin Films and Applications, College of Physics and Optoelectronic Engineering, Shenzhen University, 518060, Shenzhen, PR, China Department of Physics, KPR Institute of Engineering and Technology, Coimbatore, 641407, Tamil, Nadu, India
Krunal M. Gangawane	Department of Chemical Engineering, National Institute of Technologyy Rourkela, Rourkela-769008, Odisha, India
Manjinder Singh	Department of Polymer and Process Engineering, Indian Institute of Technologyy Roorkee, Roorkee, Uttarakhand, 247667, India
Monojit Bag	Advanced Research in Electrochemical Impedance Spectroscopy, Indian Institute of Technologyy Roorkee, Roorkee 247667, India
N.C. Shivaprakash	Instrumentation and Applied Physics, Indian Institute of Science, Bangalore 560012, India
Pralay Maiti	School of Materials Science and Technologyy, Indian Institute of Technologyy (BHU), Varanasi 221005, India
Prakash Chandra Ghosh	Indian Institute Technologyy Bombay, Mumbai, 400076, India
P.J. Jandas	Shenzhen Key Laboratory of Advanced Thin Films and Applications, College of Physics and Optoelectronic Engineering, Shenzhen University, 518060, Shenzhen, PR China
Ramesh Kumar	Advanced Research in Electrochemical Impedance Spectroscopy, Indian Institute of Technologyy Roorkee, Roorkee 247667, India
Ratikanta Nayak	Orson Resins & Coatings Pvt Ltd, Mumbai,400101, India NIST (AUTONOMOUS), Berhampur, Odisha, 761008, India
Ravi Prakash	School of Materials Science and Technologyy, Indian Institute of Technologyy (BHU), Varanasi 221005, India
Ravi Verma	Department of Control and Instrumentation Engineering, Dr. B. R. Ambedkar National Institute of Technologyy, Jalandhar-140011, India
Sachin Kadian	Department of Polymer and Process Engineering, Indian Institute of Technologyy Roorkee, Roorkee, Uttarakhand, 247667, India Department of Electrical and Computer Engineering, University of Alberta, Edmonton, AB T6G 1H9, Canada
Shamsiya Shams	Department of Physic, Noorul Islam Centre for Higher Education, Kumaracoil, Tamil Nadu, 629180, India

Shanky Jha   Department of Electrical Engineering, Dr. B. R. Ambedkar National Institute of Technology, Jalandhar-140011, India

Sukhila Krishnan   Sahrdaya College of Engineering and Technologyy, Department of Applied Science and Humanities, Kodakara, Thrissur-680684, Kerala, India

Srinivasan Kasthurirengan   Centre for Cryogenic Technologyy, Indian Institute of Science, Bangalore 560012, India

Sriparna De   Department of Allied Health Sciences, Brainware University, Kolkata, West Bengal 700125, India

Sudheer Kumar   School for Advanced Research in Petrochemicals (SARP), Laboratory for Advanced Research in Polymeric Materials (LARPM), Central Institute of Petrochemicals Engineering & Technologyy (CIPET), B/25, CNI Complex, Patia, Bhubaneswar 751024, Odisha, India

Sunil Kumar   School of Materials Science and Technologyy, Indian Institute of Technologyy (BHU), Varanasi 221005, India
Department of Chemistry, L.N.T. College (B.R.A. Bihar University), Muzaffarpur-842002, India

T.K. Abhilash   Materials Science and Technologyy Division, CSIR-National Institute for Interdisciplinary Science and Technologyy (NIIST), Thiruvananthapuram-695019, India
Academy of Scientific and Innovative Research (AcSIR), Ghaziabad-201002, India

Upendra Behera   Centre for Cryogenic Technologyy, Indian Institute of Science, Bangalore 560012, India

CHAPTER 1

Carbon Nanotube Based Nanomaterials for Solar Energy Storage Devices

Ravi Prakash[1], Sunil Kumar[1, 2] and Pralay Maiti[1,*]

[1] *School of Materials Science and Technology, Indian Institute of Technology (BHU), Varanasi 221005, India*

[2] *Department of Chemistry, L.N.T. College (B.R.A. Bihar University), Muzaffarpur-842002, India*

Abstract: Carbon nanotubes (CNTs) and their nanocomposites are used in various products and technologies due to their unique characteristics. For their future implementation, the manufacturing of CNTs with appropriate specifications has gained momentum in the area of nanoscience and technology. Conventional phase change materials used in solar thermal energy storage have low thermal conductivity. CNTs are used to prepare phase change materials with high thermal conductivity to solve this issue. This chapter addresses the synthesis, structure, and properties of CNTs. The different varieties of solar energy storage systems used to store solar radiation are also discussed. Further, we explain the phase change materials (PCMs) as suitable solar thermal energy storage systems and discuss the methods to prepare CNT-based nanomaterials for use as a heat transfer fluid (HTF) after using the CNTs based PCMs in solar storage systems. CNT based nanomaterials as a heat transfer fluid significantly increase the effective receiving efficiency, thermal conductivity, and absorption coefficient of such storage systems.

Keywords: Arc discharge, Carbon nanotubes, Chemical vapour deposition, Electrolysis, Graphene, Heat transfer fluid, Laser ablation, Multi-walled carbon nanotubes, Nanomaterials, Nanotechnology, Phase change materials, Photovoltaic, Single walled carbon nanotubes, Solar cells, Solar energy storage devices, Solar radiation, Thermal conductivity, Sonochemical, Specific heat capacity.

INTRODUCTION

The production of renewable energy, consumption, and storage are major global challenges for researchers [1 - 3]. Solar energy devices that convert solar energy directly into electrical energy are called solar cells [4, 5]. There are significant

---
[*] **Corresponding author Pralay Maiti:** School of Materials Science and Technology, Indian Institute of Technology (BHU), Varanasi 221005, India; E-mail: pmaiti.mst@itbhu.ac.in

Gaurav Manik and Sushanta Kumar Sahoo (Eds.)

research and development efforts underway to improve the device efficiency and lower the fabrication cost [3, 4, 6 - 8]. Photovoltaic devices are already used in the current era, but the devices suffer from insufficient durability and higher expense for fabrication. Further, solar energy production is mainly dependent on weather conditions; that is why solar power generation is irregular and unpredictable. Moreover, the energy requirement is considerably high in the daytime, and solar energy is available only for a small number of hours, creating the problem of maintaining a balance between the requirement and supply [9]. These two factors are the driving force behind the development of efficient solar energy storage devices, which may help reduce the fluctuation arising from the generation side and provide the possibility of performing auxiliary services. Energy storage systems are thus increasingly reducing the mismatch between need and supply and improving the work capability and reliability of energy systems which play a crucial role in conserving the produced energy [10, 11]. The conventional mechanism of the solar energy storage device is to convert the solar energy into electrical energy through solar panels and then store it in batteries, but it suffers from the issue of high manufacturing costs. In recent years, some research groups have introduced phase change materials (PCMs) as an alternative method for solar energy storage [12]. Since then, the developments of phase-change materials have become a hot research topic. The solar thermal energy storage devices are fabricated using PCMs due to their excellent stable form during phase transition.

In this method, the solar energy is converted into thermal energy using the PCMs and stored in a storage tank, which acts as a thermal battery [12 - 14]. Nanotechnology is an important field in the development of modern technology and attracts researchers in all fields. Carbon nanotubes are prime members of the research and development in nanotechnology. CNTs are classified into two categories (Fig. **1**) based on a number of layers present in the structure; (1) single-walled carbon nanotubes (SWCNTs) and (2) multi-walled carbon nanotubes (MWCNTs). The physicochemical properties are shown in Table **1**. CNTs possess high thermal, mechanical, and electrical characteristics, making them suitable for developing smart composite materials used in energy storage devices, field emitters, sensors, and so on [15]. This chapter deals with CNT-based nanomaterials, their synthesis, properties, and applications in solar energy storage devices. CNTs have been used in various technologies, depending on their attractive electrical, mechanical, and thermal properties [16]. They are primarily used in electronics [19], transistors, and display technologies, owing to their electrical properties [20, 21].

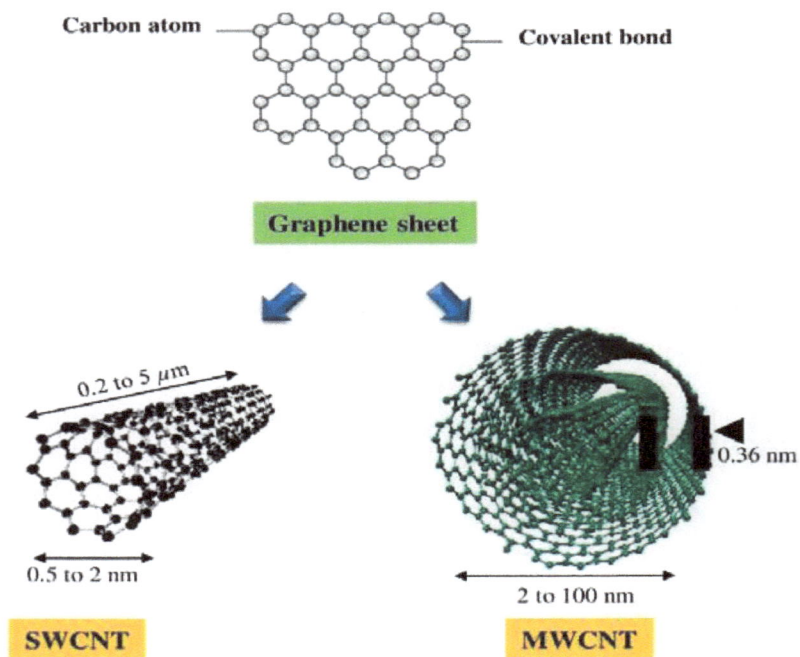

Fig. (1). Conceptual diagram of SWCNTs and MWCNTs; Reprinted with copyright permission from Ref [17, 18].

CARBON NANOTUBES

Carbon nanotubes contain sp^2 hybridisation and assume different structures with graphite as a well-known example. Graphene is a 2-dimensional (2D) single layer of graphite in the list of carbon nanomaterials. Graphene is stronger than diamond because it contains sp^2 hybridisation, which is stronger than sp^3 hybridisation in a diamond.

The sp^2 hybridized carbon can form open and closed cages with honeycomb structures [22] and Kroto *et al.* [23] discovered such kinds of structures. Carbon nanotubes are large molecules of pure carbon that are long, thin and tube-like, about 1-3 nanometres in diameter, and hundreds to thousands of nanometres long.

Table 1. Physiochemical properties of SWCNTs and MWCNTs.

SWCNTs	MWCNTs
1. Thermal conductivity range 6000 W/m.K	1. Thermal conductivity range 3000 W/m.K
2. Low purity	2. High purity
3. More defection during the functionalization	3. Less defection, but hard to improve
4. Catalyst needed for synthesis	4. No need of catalyst for manufacture
5. Synthesis is difficult in bulk	5. Synthesis is easy in bulk
6. Expensive	6. Cheaper
7. Easily twisted	7. Difficult to twist
8. Less accumulation body	8. Greater accumulation

The various types of carbon cage structures were studied long back when Iijima [24] first observed the tubular carbon (tube-like structure) structure in 1991. In carbon nanotubes, the carbon molecules are cylindrical and have unique electrical, thermal and mechanical properties that make carbon nanotubes suitable for use in different areas. The carbon nanotubes consist of up to several tens of graphite shells collectively known as multi-walled carbon nanotubes (MWCNTs) with a large length/diameter ratio. The diameter of carbon nanotubes being ~ 1.0 nm and the separation between the two adjacent shells being ~ 0.34 nm. After two years, Iijima, Ichihashi [25] and Bethune *et al.* [26] synthesised the single-walled carbon nanotubes (SWNTs). The CNTs are classified into three types on the basis of chirality; (1) armchair structure, (2) zigzag structure, and (3) chiral structure (Fig. **2**).

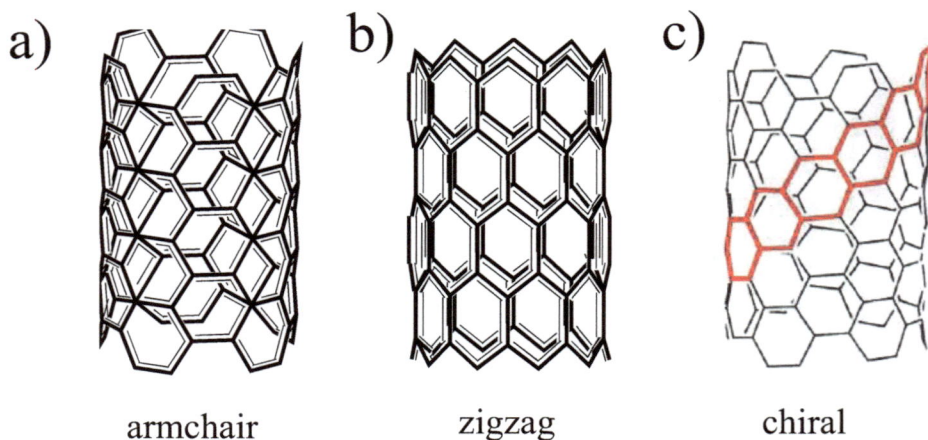

Fig. (2). Schematics diagrams of CNTs. **(a)** armchair structure. **(b)** zigzag structure, and **(c)** chiral structure. Reprinted with copyright permission from Ref [27].

The important structure of SWCNTs and MWCNTs are shown in Fig. **(3)**. The SWCNTs are cylindrical-shaped, and MWCNTs consist of several concentric SWCNTs. The structures of both CNTs are different, and hence, their properties are also different [28, 29].

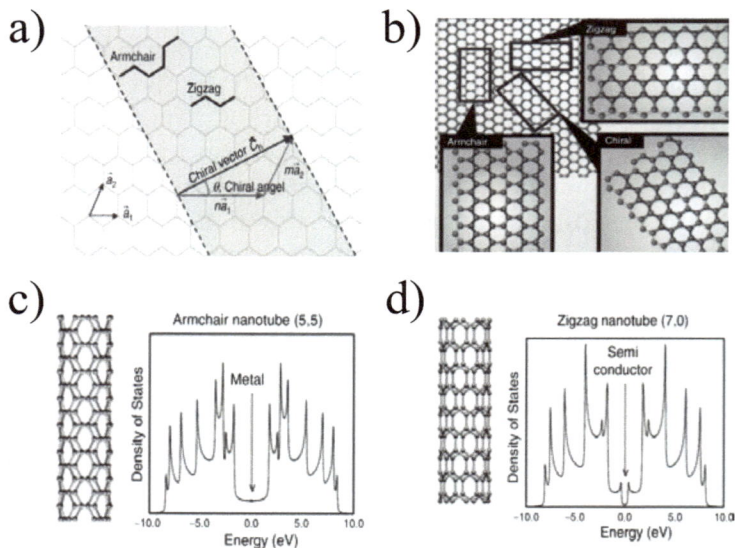

Fig. (3). (a) Schematic diagram of how grapheme sheets are rolled up to form CNTs; **(b)** representation of the three types of SWNTs structure obtained with the pair (*n, m*) from the chiral vector. Reprinted with copyright permission from Ref [28]. **(c)** Electronic structure of armchair (metallic); and **(d)** Electronic structure of zigzag (semiconductor). Reprinted with copyright permission from Ref [29].

PROPERTIES OF CARBON NANOTUBES

The properties of CNTs depend upon their atomic arrangement, tube length, diameter, and morphology [28 - 30]. SWCNT properties are governed by structure formation by the bonds between the carbon atoms of graphene sheets [28, 29]. The structure of SWCNTs depends upon the chirality of the tube, which refers to a chirality vector ($\overrightarrow{C_h}$) and chirality angle (Θ), which are shown in Fig. (3). The chirality vector $\overrightarrow{C_h}$ is the linear combination of a_1 and a_2 of the simple hexagonal through the following relationship:

$$\vec{C_h} = \vec{n}a_1 + \vec{m}a_2 \qquad (1)$$

where, n and m are integers and the single-cell vectors of two-dimensional matrix formed by the graphene sheet in which the direction of the nanotubes axis is perpendicular to the chiral vectors shown in Fig. (**3a**). The graphene sheets are rolled in the direction indicated by the chiral vector pair (n, m), and the values of the chiral vectors' pair allow three types of arrangement for SWCNTs, (1) zig-zag, (2) arm chair, and (3) chiral, which are shown in Fig. (**3b**). The mechanical, electrical, and optical properties and the nanotube's chirality are determined by the chiral vector pair (n, m) [28, 29]. If the value of the pair (n, m) is a multiple of 3, then the structure of nanotubes looks like an arm chair, has metallic behaviour, and its Fermi level is partially filled (Fig. **3c**). When the value of the pair (n, m) is a multiple of 3, the structure is zig-zag and has the behaviour of a semiconductor (Fig. **3d**). Carbon nanotubes have unique structures and properties like high aspect ratio, which gives them good electrical, thermo-mechanical properties [31], high tensile strength [28, 31, 32] (~50-500 GPa), very low density (~1.3 g/cm^3) and very high Young's modulus (~1500 GPa) [32, 33]. Due to these unique and excellent properties, CNTs are stronger and, at the same time, lighter than steel [31]. The perfect arrangement of carbon-carbon covalent bonds along the axis of nanotubes makes them very strong, with an excellent strength-to-weight ratio [31]. The MWCNTs have a broad range [34, 35] of UV-vis light absorbance [36, 37], which yields good light-thermal [38, 39] conversion capability [40, 41]. CNTs have very high thermal conductivity and are stable up to 2800 °C in vacuum [42, 43].

SYNTHESIS OF CARBON NANOTUBES

The MWNTs and SWNTs are synthesised by various methods: arc-discharge, electrolysis, laser-ablation, sonochemical or hydrothermal and chemical vapour deposition, *etc*. The various synthetic methods are shown in Fig. (**4**), which can be used to produce the CNTs in large quantities. The first production method of

CNTs was high temperature preparation techniques like arch discharge or laser ablation were used, but recently, such methods have been replaced by low chemical vapour technique deposition technique (< 800 ^{0}C), since the alignment, orientation, length diameter, purity and density of CNTs can precisely be controlled in the latter method [44].

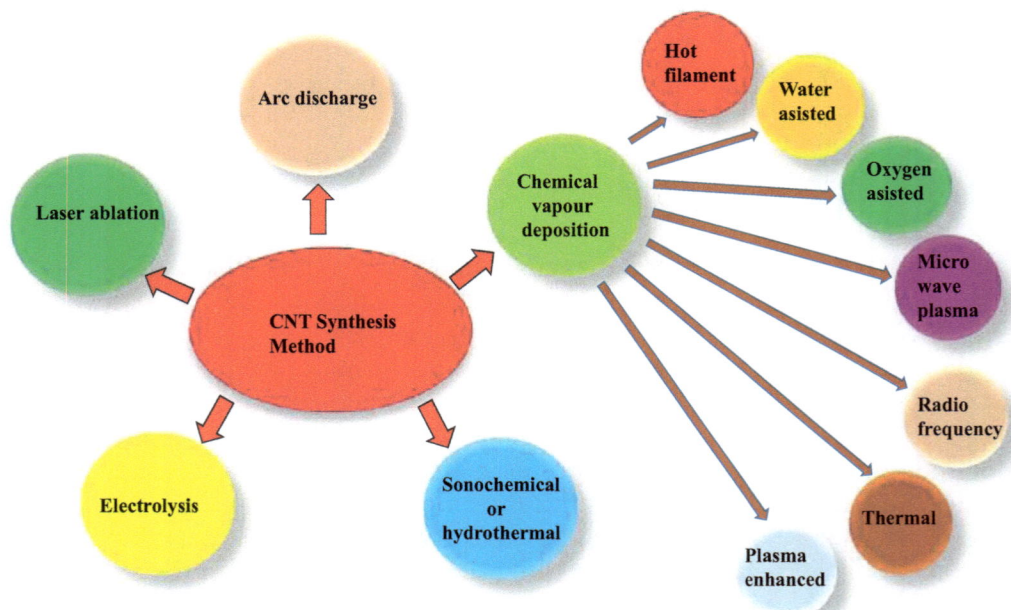

Fig. (4). Schematic diagram of various synthesis methods conventionally used for the preparation of CNTs.

Arc Discharge Method

In the arc discharge method using a high temperature, preferably above 1700 °C, CNTs are synthesised with a lower structural defects in comparison to other methods. Arc discharge method was used to synthesise the SWCNTs, MWCNTs, and double walled carbon nanotubes (DWCNTs).

Bethune *et al.* produced CNTs with a small diameter (1.2 nm) through the co-evaporation of carbon with cobalt in an arc generator [26]. Likewise, Ajayan *et al.* also synthesised the SWCNTs (1-2 nm diameter) through the arc discharge method with cobalt used in the helium atmosphere [45]. The most utilized catalyst in the synthesis of SWCNTs is nickel and Seraphin *et al.* have performed detailed studies on the catalytic activity of various catalysts like Ni, Pd, and Pt in the synthesis of carbon nanoclusters using the DC arc discharge method with operating conditions of 28 V and 70 A under the 550 Torr helium atmospheric pressure. They observed that the Ni filled anode stimulated the growth of

SWCNTs [46]. Similarly, Saitio *et al.* also reported the rapid growth of SWCNTs by using the fine Ni particles and Zhou *et al.* reported radial growth of SWCNTs by using the Yttrium carbide deposited anode.

The MWCNTs synthesis through the arc discharge method is very simple. Shimotani *et al.* [47] synthesized MWCNTs by using the arc discharge method under helium, acetone, ethanol and hexane atmosphere at different pressures from 150 to 500 Torr, and concluded that the arc discharge method produces MWCNTs three times more in ethanol, acetone and hexane as compared to helium atmospheric condition. This can be explained as follows: acetone, ethanol, and hexane may be ionized. The molecules may decompose to carbon and hydrogen atoms. These ionized species may contribute to the synthesis of MWCNTs, resulting in higher yields MWCNTs. MWCNTs yield increases in the presence of an organic atmosphere with increasing pressures up to 400 Torr [47]. The arc discharge deposition is usually promising with DC arc discharge and pulsed technique. Jung *et al.* synthesised high yield MWCNTs by arc discharge method using liquid nitrogen, and they concluded that the arc discharge method can be used to synthesize MWCNTs at a large scale with high purity. Montoro *et al.* [47, 48] synthesised high-quality SWCNTs and MWCNTs through arc discharge method using pure graphite electrode using the H_3VO_3 aqueous solution. The DWCNTs have been synthesized through the arc discharge method, but the process is not easy though some research groups have reported the successful synthesis of DWCNTs. Hutchison *et al.* first reported the successful synthesis of DWCNTs through arc discharge method under the mixture of hydrogen and argon atmosphere and SWCNTs were obtained as a by-product during the synthesis of the DWCNTs using arc discharge methods. Sugai *et al.* reported a novel synthesis of DWCNTs with high temperature pulsed arc discharge method using the Y/Ni alloy catalyst [49]. Qiu *et al.* [50] also reported their successful synthesis from coal through arc discharge method in hydrogen free atmosphere.

Laser Ablation

This method has been used to synthesize high-quality and high-purity SWCNTs. In 1995 [51], Smalley's group first used this method, which has similar principles and mechanism as of arc discharge, but laser hitting of a graphite pellet containing nickel or cobalt [52] catalyst was used to produce energy. Almost all lasers used for the ablation has been Nd:YAG and CO_2. Zhang *et al.* used continuous CO_2 laser ablation without applying any additional heat to the target for the preparation of SWCNTs, and they observed that the average diameter of prepared SWCNTs through CO_2 laser was increased with the increase of laser power [53 - 55].

Chemical Vapour Deposition

In this process, CNTs are prepared by heating the catalyst materials in a furnace (relatively lower temperature 500–1000 °C) under the flowing hydrocarbon through the tube reactor for a certain period of time. The catalysts are transition-metal nanoparticles having a high surface area and the catalysts serve as a seed to the growth of nanotubes.

SOLAR ENERGY STORAGE DEVICES

Solar energy storage systems are classified in different ways. The solar energy storage systems are classified as: (1) thermal storage, (2) electrical storage, (3) chemical storage, (4) mechanical storage, and (5) electromagnetic storage. The different solar storage systems used to store the solar energy are shown in Fig. (**5**). The solar thermal energy storage devices are traditionally classified into three major categories [56] depending on the physical principles used for energy conversion and storage [11]. The first method relates with general property of matter to experience bulk heating, and the energy storage is proportional to the specific heat capacity of the energy absorbing materials, which is often called sensible heat [11, 57]. The second method involves the heat absorption/release (latent heat) properties of the materials during the phase transition [58, 59]. In the third method, the chemical reaction process requires to produce chemical compounds having high energy chemical bonds, and which release their energy upon disruption [11, 60].

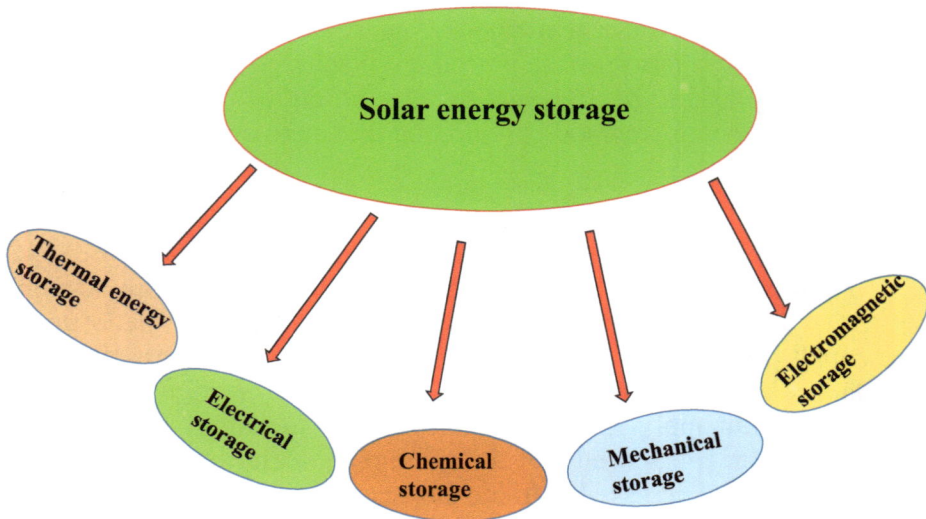

Fig. (5). Schematic diagrams of various solar energy storage systems.

Another method was also added in this series which is based on the absorption of a photon and the generation of electron hole pairs in energy storage devices such as lithium ion rechargeable batteries [11, 61] which is also combined with the generation of hydrogen. The above method can also be classified in the high energy (the light is utilized in the visible range) domain and another in low energy (IR light) domain of solar energy, which is converted as stored energy [11]. The electronic excitation in absorber materials involves absorbing the photons in the visible region of the solar radiation, and the electrical energy is produced by the photovoltaic cell, which may store energy in the form of chemical energy *via* a photochemical reaction. This method is referred to as the photonics conversion method and these are electrochemical batteries fed by photovoltaic cells. Photosynthesis and electrochemical energy storage devices directly absorb the photons by photo-electrode and produce a chemical fuel. The materials absorb the IR light of solar radiation, which excites the photons and produces heat that can be stored, and the method is conventionally called the thermal conversion method. The method is based on heat-induced chemical reactions and is classified into sensible and latent heat-based.

CNT NANOCOMPOSITES IN SOLAR ENERGY STORAGE DEVICES

CNTs are a good candidate as a filler in composite materials to enhance thermal transport due to their high thermal conductivity [62, 63]. Many research groups have reported enhanced thermal conductivity of the PCMs with the addition of CNTs [64 - 68]. Ji *et al.* reported the functionalised MWNTs/plasmatic acid composite materials to prepare PCMs in thermal storage systems [62]. Solar energy collector is an important part of solar energy storage device, which converts the solar energy directly into heat [69]. In 1970, Minradi *et al.* proposed an idea about the direct absorption solar collectors (DASC) [70]. The DASC system directly absorbs the solar energy radiation through heat transfer fluid and decreases the radiation heat loss [71] by avoiding the heat resistance between medium and absorption surface. The photo-absorption and thermo-physical properties of heat transfer fluids have affected the receiving efficiency of the DASC system [72]. Hence, to increase the receiving efficiency, it has been proposed to use novel slurry with fine optical and thermo-physical properties. Recently, some research groups have enhanced the optical absorption properties of working heat transfer fluid with the dispersion of the metal nanoparticles [73] and carbon-based composites [74]. However, the working fluids have very low specific heat capacity, which makes it impossible for them to store more energy, and hence, the reception efficiency is insignificant. Since PCMs have large specific heat capacity, it has been found that when the PCMs are dispersed with heat transfer carrying fluid, the latent functional heat of fluid can enhance the thermal energy storage capacity of the working fluid [75, 76]. Ma *et al.* used

CNTs effectively to prepare PCMs for utilization in solar energy storage materials [69].

The optimized amount of CNTs has been previously incorporated in PCMs, which enhanced the thermal conductivity, solar radiation absorption efficiency, solar heat storage capacity and specific heat capacity of the heat transfer working fluid. The effective receiving efficiency of the heat transfer fluid is determined by using the following equation (2) [77]:

$$\eta = \frac{m \int C_p\,(T)\,dT}{G_s\,A\,t} \times 100\,\%$$ (2)

where, η is the receiver efficiency, mass of the heat transfer slurry is m, C_p is the specific heat capacity, G_s is the intensity of solar radiation, T represents the real time temperature of the slurry, A is the area of energy received by the light source, the radiation time of the sample is t. The $G_s A\ t$ in the equation (2) represents the heat received by the simulated light source/collector from the sun, and the $m \int C_p$ $(T)\ dT$ represents the heat actually absorbed by heat transfer fluid through the irradiation to increase its temperature. The ratio of these two provides the so called receiving efficiency [69]. Ma *et al.* calculated the effective receiving efficiency from equation (2) for water, 5 wt.% paraffin@SnO$_2$ and paraffin@SnO$_2$/CNTs, which are shown in Fig. (**6a**) as a function of temperature. The CNTs mixed slurry was observed to exhibit higher effective receiving efficiency as compared to water and 5 wt.% paraffin@SnO$_2$, and also high photo-conversion efficiency or absorption as compared to the base fluid. UV-vis spectroscopy was used to determine the optical properties of prepared slurry, and from such measurements, it has been observed that the absorption of paraffin@SnO$_2$/CNTs composites exhibits three times more efficiency as compared to paraffin@SnO$_2$ slurry. Tong *et al.* [78] also used CNTs in solar radiation collectors, using different concentrations of MWNTs in MWNTs/water nano-fluid employed in u-tube solar collectors. When MWNTs with 0.24% volume concentration were employed in the nano-fluid, they enhanced the heat transfer coefficient by 8% between the tube and working fluid as compared to the base nano-fluid (without MWNTs).

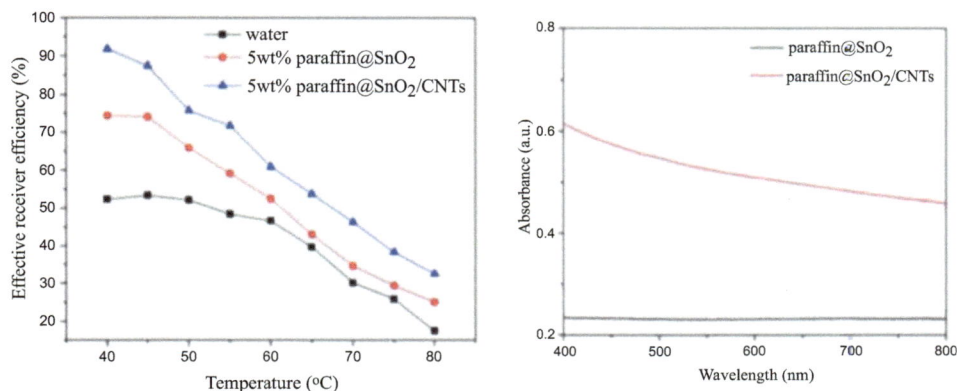

Fig. (6). (a) Effective receiving efficiency of 5 wt.% paraffin@SnO$_2$ microcapsules and paraffin@SnO$_2$/CNTs dispersion slurry; and **(b)** UV-vis reflection spectra of paraffin@SnO$_2$ microcapsules and paraffin@SnO$_2$/CNTs dispersion slurry. Reprinted with copyright permission from Ref [69].

CONCLUSION

CNTs are an important class of nanomaterials derived from carbon with unique properties, which makes them more attractive for use in various technologies. Although their scope of application is extremely wide and exhaustive, in this chapter, we have specifically dealt with the synthesis and application of CNTs based nanomaterials in solar energy storage devices. An optimum amount of CNTs needs to be added, for instance, in PCMs, to enhance the thermal conductivity and solar heat storage capacity of the materials (heat transfer working fluid in such a case) resulting in an increase in the efficiency of the storage device. A good number of important recent research studies are available in this field. Therefore in this chapter, we have focused on the important synthesis methods (laser ablation, arc discharge, CVD, *etc.*) and various types of CNTs, namely SWCNTs, DWCNTs and MWCNTs, characteristic properties and application of the CNTs, which are responsible for the boosting of nanotechnology applied to solar energy storage.

CONSENT FOR PUBLICATION

Not applicable.

CONFLICT OF INTERESTS

The authors declare no conflict of interest, financial or otherwise.

ACKNOWLEDGEMENTS

The author (Ravi Prakash) acknowledges the Institutes (IIT-BHU) internship for carrying out this work.

REFERENCE

[1]   Tang, B.; Wang, Y.; Qiu, M.; Zhang, S. A full-band sunlight-driven carbon nanotube/PEG/SiO2 composites for solar energy storage. *Sol. Energy Mater. Sol. Cells,* **2014**, *123*, 7-12.
[http://dx.doi.org/10.1016/j.solmat.2013.12.022]

[2]   Lafdi, K.; Mesalhy, O.; Elgafy, A. Graphite foams infiltrated with phase change materials as alternative materials for space and terrestrial thermal energy storage applications. *Carbon,* **2008**, *46*(1), 159-168.
[http://dx.doi.org/10.1016/j.carbon.2007.11.003]

[3]   Nozik, A.J. Quantum dot solar cells. *Physica E,* **2002**, *14*(1-2), 115-120.
[http://dx.doi.org/10.1016/S1386-9477(02)00374-0]

[4]   Kamat, P.V. Quantum dot solar cells. Semiconductor nanocrystals as light harvesters. *J. Phys. Chem. C,* **2008**, *112*(48), 18737-18753.
[http://dx.doi.org/10.1021/jp806791s]

[5]   Choi, H.; Radich, J.G.; Kamat, P.V. Sequentially layered CdSe/CdS nanowire architecture for improved nanowire solar cell performance. *J. Phys. Chem. C,* **2014**, *118*(1), 206-213.
[http://dx.doi.org/10.1021/jp410235s]

[6]   Prakash, R.; Maiti, P. Functionalized thermoplastic polyurethane Gel electrolytes for cosensitized TiO_2/CdS/CdSe photoanode solar cells with high efficiency. *Energy Fuels,* **2020**, *34*(12), 16847-16857.
[http://dx.doi.org/10.1021/acs.energyfuels.0c03250]

[7]   Kongkanand, A.; Tvrdy, K.; Takechi, K.; Kuno, M.; Kamat, P.V. Quantum dot solar cells. Tuning photoresponse through size and shape control of CdSe-TiO_2 architecture. *J. Am. Chem. Soc.,* **2008**, *130*(12), 4007-4015.
[http://dx.doi.org/10.1021/ja0782706] [PMID: 18311974]

[8]   Ghoreishi, F.S.; Ahmadi, V.; Samadpour, M. Improved performance of CdS/CdSe quantum dots sensitized solar cell by incorporation of ZnO nanoparticles/reduced graphene oxide nanocomposite as photoelectrode. *J. Power Sources,* **2014**, *271*, 195-202.
[http://dx.doi.org/10.1016/j.jpowsour.2014.07.165]

[9]   Vega☐Garita, V.; Ramirez☐Elizondo, L.; Narayan, N.; Bauer, P. Integrating a photovoltaic storage system in one device: A critical review. *Prog. Photovolt. Res. Appl.,* **2019**, *27*(4), 346-370.
[http://dx.doi.org/10.1002/pip.3093]

[10]   Deng, Y.Y.; Blok, K.; van der Leun, K. Transition to a fully sustainable global energy system. *Energy Strategy Reviews,* **2012**, *1*(2), 109-121.
[http://dx.doi.org/10.1016/j.esr.2012.07.003]

[11]   Dimitriev, O.; Yoshida, T.; Sun, H. Principles of solar energy storage. *Energy Storage,* **2020**, *2*(1), e96.
[http://dx.doi.org/10.1002/est2.96]

[12]   Nazir, H.; Batool, M.; Bolivar Osorio, F.J.; Isaza-Ruiz, M.; Xu, X.; Vignarooban, K.; Phelan, P.; Inamuddin, ; Kannan, A.M. Recent developments in phase change materials for energy storage applications: A review. *Int. J. Heat Mass Transf.,* **2019**, *129*, 491-523.
[http://dx.doi.org/10.1016/j.ijheatmasstransfer.2018.09.126]

[13]   Pelay, U.; Luo, L.; Fan, Y.; Stitou, D.; Rood, M. Thermal energy storage systems for concentrated solar power plants. *Renew. Sustain. Energy Rev.,* **2017**, *79*, 82-100.

[http://dx.doi.org/10.1016/j.rser.2017.03.139]

[14]   Esteves, L.; Magalhães, A.; Ferreira, V.; Pinho, C. Test of two phase change materials for thermal energy storage: Determination of the global heat transfer coefficient. *ChemEngineering,* **2018,** *2*(1), 10.
       [http://dx.doi.org/10.3390/chemengineering2010010]

[15]   aughman, R. H.; Zakhidov, A. A.; De Heer, W. A. Carbon nanotubes--the route toward applications. *science,* **2002,** , 792-787.*297*

[16]   Lau, C.H.; Cervini, R.; Clarke, S.R.; Markovic, M.G.; Matisons, J.G.; Hawkins, S.C.; Huynh, C.P.; Simon, G.P. The effect of functionalization on structure and electrical conductivity of multi-walledcarbon nanotubes. *J. Nanopart. Res.,* **2008,** *10*(S1), 77-88.
       [http://dx.doi.org/10.1007/s11051-008-9376-1]

[17]   Patel, D. K.; Kim, H.-B.; Dutta, S. D.; Ganguly, K.; Lim, K.-T. Carbon nanotubes-based nanomaterials and their agricultural and biotechnological applications. *materials 13, 1679***2020,**

[18]   He, H.; Pham-Huy, L.A.; Dramou, P.; Xiao, D.; Zuo, P.; Pham-Huy, C. Carbon nanotubes: applications in pharmacy and medicine. *BioMed Res. Int.,* **2013,** *2013*, 578290.
       [http://dx.doi.org/10.1155/2013/578290] [PMID: 24195076]

[19]   Ravindran, S.; Chaudhary, S.; Colburn, B.; Ozkan, M.; Ozkan, C.S. Covalent coupling of quantum dots to multiwalled carbon nanotubes for electronic device applications. *Nano Lett.,* **2003,** *3*(4), 447-453.
       [http://dx.doi.org/10.1021/nl0259683]

[20]   Wang, Q.; Yan, M.; Chang, R.P. Flat panel display prototype using gated carbon nanotube field emitters. *Appl. Phys. Lett.,* **2001,** *78*(9), 1294-1296.
       [http://dx.doi.org/10.1063/1.1351847]

[21]   Bachtold, A.; Hadley, P.; Nakanishi, T.; Dekker, C. Logic circuits with carbon nanotube transistors. *Science,* **2001,** *294*(5545), 1317-1320.
       [http://dx.doi.org/10.1126/science.1065824] [PMID: 11588220]

[22]   Popov, V.N. Carbon nanotubes: properties and application. *Mater. Sci. Eng. Rep.,* **2004,** *43*(3), 61-102.
       [http://dx.doi.org/10.1016/j.mser.2003.10.001]

[23]   Kroto, H. W.; Heath, J. R.; O'Brien, S. C.; Curl, R. F.; Smalley, R. E. **1985.**

[24]   Iijima, S. Helical microtubules of graphitic carbon. *Nature,* **1991,** *354*(6348), 56-58.
       [http://dx.doi.org/10.1038/354056a0]

[25]   Iijima, S.; Ichihashi, T. Single-shell carbon nanotubes of 1-nm diameter. *Nature,* **1993,** *363*(6430), 603-605.
       [http://dx.doi.org/10.1038/363603a0]

[26]   Bethune, D.S.; Kiang, C.H.; de Vries, M.S.; Gorman, G.; Savoy, R.; Vazquez, J.; Beyers, R. Cobalt-catalysed growth of carbon nanotubes with single-atomic-layer walls. *Nature,* **1993,** *363*(6430), 605-607.
       [http://dx.doi.org/10.1038/363605a0]

[27]   Sisto, T.J.; Zakharov, L.N.; White, B.M.; Jasti, R. Towards pi-extended cycloparaphenylenes as seeds for CNT growth: investigating strain relieving ring-openings and rearrangements. *Chem. Sci. (Camb.),* **2016,** *7*(6), 3681-3688.
       [http://dx.doi.org/10.1039/C5SC04218F] [PMID: 29997859]

[28]   Thostenson, E.T.; Ren, Z.; Chou, T-W. Advances in the science and technology of carbon nanotubes and their composites: a review. *Compos. Sci. Technol.,* **2001,** *61*(13), 1899-1912.
       [http://dx.doi.org/10.1016/S0266-3538(01)00094-X]

[29]   Charlier, J-C. Defects in carbon nanotubes. *Acc. Chem. Res.,* **2002,** *35*(12), 1063-1069.
       [http://dx.doi.org/10.1021/ar010166k] [PMID: 12484794]

[30]   Gooding, J.J. Nanostructuring electrodes with carbon nanotubes: A review on electrochemistry and applications for sensing. *Electrochim. Acta,* **2005**, *50*(15), 3049-3060.
[http://dx.doi.org/10.1016/j.electacta.2004.08.052]

[31]   Basheer, B.V.; George, J.J.; Siengchin, S.; Parameswaranpillai, J. Polymer grafted carbon nanotubes—Synthesis, properties, and applications: A review. *Nano-Structures & Nano-Objects,* **2020**, *22*, 100429.
[http://dx.doi.org/10.1016/j.nanoso.2020.100429]

[32]   Lau, A.K-T.; Hui, D. The revolutionary creation of new advanced materials—carbon nanotube composites. *Compos., Part B Eng.,* **2002**, *33*(4), 263-277.
[http://dx.doi.org/10.1016/S1359-8368(02)00012-4]

[33]   Al-Saleh, M.H.; Sundararaj, U. Electromagnetic interference shielding mechanisms of CNT/polymer composites. *Carbon,* **2009**, *47*(7), 1738-1746.
[http://dx.doi.org/10.1016/j.carbon.2009.02.030]

[34]   Arepalli, S.; Nikolaev, P.; Gorelik, O.; Hadjiev, V.G.; Holmes, W.; Files, B.; Yowell, L. Protocol for the characterization of single-wall carbon nanotube material quality. *Carbon,* **2004**, *42*(8-9), 1783-1791.
[http://dx.doi.org/10.1016/j.carbon.2004.03.038]

[35]   Arjmand, M.; Apperley, T.; Okoniewski, M.; Sundararaj, U. Comparative study of electromagnetic interference shielding properties of injection molded *versus* compression molded multi-walled carbon nanotube/polystyrene composites. *Carbon,* **2012**, *50*(14), 5126-5134.
[http://dx.doi.org/10.1016/j.carbon.2012.06.053]

[36]   Peng, H.; Alemany, L.B.; Margrave, J.L.; Khabashesku, V.N. Sidewall carboxylic acid functionalization of single-walled carbon nanotubes. *J. Am. Chem. Soc.,* **2003**, *125*(49), 15174-15182.
[http://dx.doi.org/10.1021/ja037746s] [PMID: 14653752]

[37]   Zhao, B.; Hu, H.; Haddon, R.C. Synthesis and properties of a water-soluble single-walled carbon nanotube–poly (m-aminobenzene sulfonic acid) graft copolymer. *Adv. Funct. Mater.,* **2004**, *14*(1), 71-76.
[http://dx.doi.org/10.1002/adfm.200304440]

[38]   Fan, L-W.; Fang, X.; Wang, X.; Zeng, Y.; Xiao, Y-Q.; Yu, Z-T.; Xu, X.; Hu, Y-C.; Cen, K-F. Effects of various carbon nanofillers on the thermal conductivity and energy storage properties of paraffin-based nanocomposite phase change materials. *Appl. Energy,* **2013**, *110*, 163-172.
[http://dx.doi.org/10.1016/j.apenergy.2013.04.043]

[39]   O'Connell, M.J.; Bachilo, S.M.; Huffman, C.B.; Moore, V.C.; Strano, M.S.; Haroz, E.H.; Rialon, K.L.; Boul, P.J.; Noon, W.H.; Kittrell, C.; Ma, J.; Hauge, R.H.; Weisman, R.B.; Smalley, R.E. Band gap fluorescence from individual single-walled carbon nanotubes. *Science,* **2002**, *297*(5581), 593-596.
[http://dx.doi.org/10.1126/science.1072631] [PMID: 12142535]

[40]   Moon, H.K.; Lee, S.H.; Choi, H.C. In vivo near-infrared mediated tumor destruction by photothermal effect of carbon nanotubes. *ACS Nano,* **2009**, *3*(11), 3707-3713.
[http://dx.doi.org/10.1021/nn900904h] [PMID: 19877694]

[41]   Kataura, H.; Kumazawa, Y.; Maniwa, Y.; Umezu, I.; Suzuki, S.; Ohtsuka, Y.; Achiba, Y. Optical properties of single-wall carbon nanotubes. *Synth. Met.,* **1999**, *103*(1-3), 2555-2558.
[http://dx.doi.org/10.1016/S0379-6779(98)00278-1]

[42]   Thostenson, E.T.; Li, C.; Chou, T-W. Nanocomposites in context. *Compos. Sci. Technol.,* **2005**, *65*(3-4), 491-516.
[http://dx.doi.org/10.1016/j.compscitech.2004.11.003]

[43]   Pop, E.; Mann, D.; Wang, Q.; Goodson, K.; Dai, H. Thermal conductance of an individual single-wall carbon nanotube above room temperature. *Nano Lett.,* **2006**, *6*(1), 96-100.
[http://dx.doi.org/10.1021/nl052145f] [PMID: 16402794]

[44]   He, Z.; Maurice, J-L.; Seok Lee, C.; Cojocaru, C.S.; Pribat, D. Growth mechanisms of carbon nanostructures with branched carbon nanofibers synthesized by plasma-enhanced chemical vapour deposition. *CrystEngComm,* **2014**, *16*(14), 2990-2995.
[http://dx.doi.org/10.1039/c3ce42241k]

[45]   Ajayan, P.; Lambert, J.M.; Bernier, P.; Barbedette, L.; Colliex, C.; Planeix, J.M. Growth morphologies during cobalt-catalyzed single-shell carbon nanotube synthesis. *Chem. Phys. Lett.,* **1993**, *215*(5), 509-517.
[http://dx.doi.org/10.1016/0009-2614(93)85711-V]

[46]   Seraphin, S.; Zhou, D.; Jiao, J.; Minke, M.; Wang, S.T. Yadav, and J. C. Withers. *Chem. Phys. Lett.,* **1994**, *217*, 1.

[47]   Shimotani, K.; Anazawa, K.; Watanabe, H.; Shimizu, M. New synthesis of multi-walled carbon nanotubes using an arc discharge technique under organic molecular atmospheres. *Appl. Phys., A Mater. Sci. Process.,* **2001**, *73*(4), 451-454.
[http://dx.doi.org/10.1007/s003390100821]

[48]   Lofrano, R.; Rosolen, J.; Montoro, L. Synthesis of single-walled and multi-walled carbon nanotubes by arc-water method. *Carbon,* **2005**, *43*(1), 200-203.
[http://dx.doi.org/10.1016/j.carbon.2004.08.014]

[49]   Sugai, T.; Yoshida, H.; Shimada, T.; Okazaki, T.; Shinohara, H.; Bandow, S. New synthesis of high-quality double-walled carbon nanotubes by high-temperature pulsed arc discharge. *Nano Lett.,* **2003**, *3*(6), 769-773.
[http://dx.doi.org/10.1021/nl034183+]

[50]   Qiu, J.; Wang, Z.; Zhao, Z.; Wang, T. Synthesis of double-walled carbon nanotubes from coal in hydrogen-free atmosphere. *Fuel,* **2007**, *86*(1-2), 282-286.
[http://dx.doi.org/10.1016/j.fuel.2006.05.024]

[51]   Guo, T.; Nikolaev, P.; Thess, A.; Colbert, D.T.; Smalley, R.E. Catalytic growth of single-walled manotubes by laser vaporization. *Chem. Phys. Lett.,* **1995**, *243*(1-2), 49-54.
[http://dx.doi.org/10.1016/0009-2614(95)00825-O]

[52]   Ikegami, T.; Nakanishi, F.; Uchiyama, M.; Ebihara, K. Optical measurement in carbon nanotubes formation by pulsed laser ablation. *Thin Solid Films,* **2004**, *457*(1), 7-11.
[http://dx.doi.org/10.1016/j.tsf.2003.12.033]

[53]   Notarianni, M.; Liu, J.; Vernon, K.; Motta, N. Synthesis and applications of carbon nanomaterials for energy generation and storage. *Beilstein J Nanotechnol 7,* **2016**, 196-149. http://europepmc.org/abstract/MED/26925363
[http://dx.doi.org/10.3762/bjnano.7.17]

[54]   Prasek, J.; Drbohlavova, J.; Chomoucka, J.; Hubalek, J.; Jasek, O.; Adam, V.; Kizek, R. Methods for carbon nanotubes synthesis. *J. Mater. Chem.,* **2011**, *21*(40), 15872-15884.
[http://dx.doi.org/10.1039/c1jm12254a]

[55]   Marchiori, R.; Braga, W.; Mantelli, M.; Lago, A. Analytical solution to predict laser ablation rate in a graphitic target. *J. Mater. Sci.,* **2010**, *45*(6), 1495-1502.
[http://dx.doi.org/10.1007/s10853-009-4112-5]

[56]   Gao, L.; Zhao, J.; Tang, Z. A review on borehole seasonal solar thermal energy storage. *Energy Procedia,* **2015**, *70*, 209-218.
[http://dx.doi.org/10.1016/j.egypro.2015.02.117]

[57]   Kuravi, S.; Trahan, J.; Goswami, D.Y.; Rahman, M.M.; Stefanakos, E.K. Thermal energy storage technologies and systems for concentrating solar power plants. *Pror. Energy Combust. Sci.,* **2013**, *39*(4), 285-319.
[http://dx.doi.org/10.1016/j.pecs.2013.02.001]

[58]   Hadiya, J.; Shukla, A.K.N. Thermal energy storage using phase change materials: a way forward. *Int.*

J. Glob. Energy Issues, **2018**, *41*(1/2/3/4), 108-127.
[http://dx.doi.org/10.1504/IJGEI.2018.092311]

[59]   Abhat, A. Low temperature latent heat thermal energy storage: heat storage materials. *Sol. Energy,* **1983**, *30*(4), 313-332.
[http://dx.doi.org/10.1016/0038-092X(83)90186-X]

[60]   Garg, H.; Mullick, S.; Bhargava, V.K. *Solar thermal energy storage*; Springer Science & Business Media, **2012**.

[61]   Chen, X.; Li, C.; Grätzel, M.; Kostecki, R.; Mao, S.S. Nanomaterials for renewable energy production and storage. *Chem. Soc. Rev.,* **2012**, *41*(23), 7909-7937.
[http://dx.doi.org/10.1039/c2cs35230c] [PMID: 22990530]

[62]   Ji, P.; Sun, H.; Zhong, Y.; Feng, W. Improvement of the thermal conductivity of a phase change material by the functionalized carbon nanotubes. *Chem. Eng. Sci.,* **2012**, *81*, 140-145.
[http://dx.doi.org/10.1016/j.ces.2012.07.002]

[63]   Choi, S.; Zhang, Z.; Yu, W.; Lockwood, F.; Grulke, E. Anomalous thermal conductivity enhancement in nanotube suspensions. *Appl. Phys. Lett.,* **2001**, *79*(14), 2252-2254.
[http://dx.doi.org/10.1063/1.1408272]

[64]   Fujii, M.; Zhang, X.; Xie, H.; Ago, H.; Takahashi, K.; Ikuta, T.; Abe, H.; Shimizu, T. Measuring the thermal conductivity of a single carbon nanotube. *Phys. Rev. Lett.,* **2005**, *95*(6), 065502.
[http://dx.doi.org/10.1103/PhysRevLett.95.065502] [PMID: 16090962]

[65]   Wang, H.; Feng, J.; Hu, X.; Ng, K.M. Reducing thermal contact resistance using a bilayer aligned CNT thermal interface material. *Chem. Eng. Sci.,* **2010**, *65*(3), 1101-1108.
[http://dx.doi.org/10.1016/j.ces.2009.09.064]

[66]   Wang, J.; Xie, H.; Xin, Z. Thermal properties of heat storage composites containing multiwalled carbon nanotubes. *J. Appl. Phys.,* **2008**, *104*(11), 113537.
[http://dx.doi.org/10.1063/1.3041495]

[67]   Clancy, T.C.; Gates, T.S. Modeling of interfacial modification effects on thermal conductivity of carbon nanotube composites. *Polymer (Guildf.),* **2006**, *47*(16), 5990-5996.
[http://dx.doi.org/10.1016/j.polymer.2006.05.062]

[68]   Xie, H.; Lee, H.; Youn, W.; Choi, M. Nanofluids containing multiwalled carbon nanotubes and their enhanced thermal conductivities. *J. Appl. Phys.,* **2003**, *94*(8), 4967-4971.
[http://dx.doi.org/10.1063/1.1613374]

[69]   Ma, X.; Liu, H.; Chen, C.; Liu, Y.; Zhang, L.; Xu, B.; Xiao, F. Synthesis of novel microencapsulated phase change material with SnO2/CNTs shell for solar energy storage and photo-thermal conversion. *Mater. Res. Express,* **2020**, *7*(1), 015513.
[http://dx.doi.org/10.1088/2053-1591/ab657e]

[70]   Minardi, J.E.; Chuang, H.N. Performance of a "black" liquid flat-plate solar collector. *Sol. Energy,* **1975**, *17*(3), 179-183.
[http://dx.doi.org/10.1016/0038-092X(75)90057-2]

[71]   Otanicar, T.P.; Golden, J.S. Comparative environmental and economic analysis of conventional and nanofluid solar hot water technologies. *Environ. Sci. Technol.,* **2009**, *43*(15), 6082-6087.
[http://dx.doi.org/10.1021/es900031j] [PMID: 19731722]

[72]   Delgado, M.; Lázaro, A.; Mazo, J.; Zalba, B. Review on phase change material emulsions and microencapsulated phase change material slurries: materials, heat transfer studies and applications. *Renew. Sustain. Energy Rev.,* **2012**, *16*(1), 253-273.
[http://dx.doi.org/10.1016/j.rser.2011.07.152]

[73]   Karami, M.; Akhavan-Bahabadi, M.; Delfani, S.; Raisee, M. Experimental investigation of CuO nanofluid-based direct absorption solar collector for residential applications. *Renew. Sustain. Energy Rev.,* **2015**, *52*, 793-801.

[http://dx.doi.org/10.1016/j.rser.2015.07.131]

[74]   Karami, M.; Bahabadi, M.A.; Delfani, S.; Ghozatloo, A. A new application of carbon nanotubes nanofluid as working fluid of low-temperature direct absorption solar collector. *Sol. Energy Mater. Sol. Cells,* **2014**, *121*, 114-118.
[http://dx.doi.org/10.1016/j.solmat.2013.11.004]

[75]   Jamekhorshid, A.; Sadrameli, S.; Farid, M. A review of microencapsulation methods of phase change materials (PCMs) as a thermal energy storage (TES) medium. *Renew. Sustain. Energy Rev.,* **2014**, *31*, 531-542.
[http://dx.doi.org/10.1016/j.rser.2013.12.033]

[76]   Liu, C.; Rao, Z.; Zhao, J.; Huo, Y.; Li, Y. Review on nanoencapsulated phase change materials: preparation, characterization and heat transfer enhancement. *Nano Energy,* **2015**, *13*, 814-826.
[http://dx.doi.org/10.1016/j.nanoen.2015.02.016]

[77]   Yuan, K.; Wang, H.; Liu, J.; Fang, X.; Zhang, Z. Novel slurry containing graphene oxide-grafted microencapsulated phase change material with enhanced thermo-physical properties and photo-thermal performance. *Sol. Energy Mater. Sol. Cells,* **2015**, *143*, 29-37.
[http://dx.doi.org/10.1016/j.solmat.2015.06.034]

[78]   Tong, Y.; Kim, J.; Cho, H. Effects of thermal performance of enclosed-type evacuated U-tube solar collector with multi-walled carbon nanotube/water nanofluid. *Renew. Energy,* **2015**, *83*, 463-473.
[http://dx.doi.org/10.1016/j.renene.2015.04.042]

CHAPTER 2

Recent Advances on Carbon Nanostructure-Based Biosensors

Ashish Kalkal[1] and **Gopinath Packirisamy**[1,2,*]

[1] *Nanobiotechnology Laboratory, Department of Biosciences and Bioengineering, Indian Institute of Technology Roorkee, Roorkee-247667, Uttarakhand, India*

[2] *Centre for Nanotechnology, Indian Institute of Technology Roorkee, Roorkee-247667, Uttarakhand, India*

Abstract: Carbon-based nanostructured materials have derived substantial attention as novel functional materials towards the fabrication of various biosensing platforms owing to their interesting physicochemical and optoelectronic properties, as well as desired surface functionalities. These nanomaterials provide increased and oriented immobilization of biomolecules along with maintaining their biological activity in view of their lower cytotoxicity and higher biocompatibility. The integration of carbon nanomaterials with biosensing platforms has provided new opportunities and paved the way for the efficient detection of various biomolecules and analytes. These nanostructured materials-based biosensors have improved biosensing characteristics, including broader linear detection range, lower detection limit, better selectivity, and higher sensitivity. This chapter summarizes the results of different electrochemical and fluorescent biosensors related to various nanostructured carbon materials, namely carbon nanotubes (CNTs), graphene and its derivatives (reduced graphene oxide (rGO), graphene oxide (GO), graphene quantum dots (GQDs) and carbon dots (CDs).

Keywords: Biosensor, Carbon dots, Carbon nanotubes, Electrochemical, Fluorescent, Graphene, Graphene oxide, Graphene quantum dots, Heteroatom doping, Nanomaterial, Nanostructured, Reduced graphene oxide, Surface functionalization.

INTRODUCTION

The interest in developing advanced biosensors accompanied by nanostructured materials is rapidly increasing over the last two decades. With the advent of nanoscience and nanotechnology, these biosensors can provide various desirable

---

[*] **Corresponding author Gopinath Packirisamy:** Department of Biosciences and Bioengineering, Centre for Nanotechnology, Indian Institute of Technology Roorkee, Roorkee-247667, Uttarakhand, India; Tel: +91-1332-285650; Fax: +91-1332-273560; E-mails: gopi@bt.iitr.ac.in, genegopi@gmail.com

Gaurav Manik and Sushanta Kumar Sahoo (Eds.)

characteristics, including higher sensitivity, good selectivity, low cost, faster response, miniaturization, portability, simple operation, and accurate detection [1 - 6]. Among the various types of nanostructured materials, carbon nanomaterials have aroused enormous scientific attention toward creating biosensing platforms because of their fascinating functional, optical, electronic, physiochemical, mechanical, catalytic, and biocompatible properties [7, 8]. Additionally, these nanomaterials possess strong adsorption ability, larger surface area, greater electron-transfer kinetics, ease of surface functionalization, enabling the oriented immobilization of biomolecules, providing enhanced analytical performance with desirable biosensing characteristics [1, 9, 10].

The family of nanostructured carbon materials includes fullerene, carbon nanotubes (CNTs), graphene and its derivatives (GO, rGO), nanodiamond, carbon dots (CDs), carbon nanofibers, and graphene quantum dots (GQDs), *etc.* [11, 12]. The graphite with sp^2 hybridization and diamond with sp^3 hybridization are the two commonly known allotropic forms of carbon [13]. These nanostructured materials are classified according to the geometrical structure of the particles (spheres, ellipsoids, horns, rods, sheets, foams, or tubes) and dimensionalities (OD, 1D, 2D, and 3D). For instance, Fullerenes 0D form of carbon possesses spherical or ellipsoidal nanoparticles [14], whereas CNTs 1D form of carbon contains tube-shaped particles [15]. Interestingly, carbon-based nanostructured systems can lead to the development of novel bioanalytical technologies, advantageous for detecting various infectious and non-infectious diseases [16, 17]. The surface properties of carbon nanomaterials can be easily tailored, facilitating them to conjugate with different diagnostic or imaging agents toward the development of next-generation biosensors [1].

As stated by the International Union of Pure and Applied Chemistry (IUPAC), the biosensor is "a device that utilizes specific biochemical reactions mediated by isolated organelles, whole cells, tissues, immunosystems or enzymes for detecting chemical compounds or analytes generally *via* optical, thermal or electrical signals" [18]. Basically, a biosensing system comprises three key components: a bio-recognition element, also known as a bioreceptor that selectively recognizes the target analyte, an immobilization matrix wherein biomolecules are immobilized, and the transducer that converts the input biological signal into a measurable output signal [19, 20]. The nanostructured materials can be used as an immobilization matrix in the electrochemical biosensors to integrate the biomolecules with a transducer surface. Owing to strong adsorption ability and a larger surface-to-volume ratio, these nanomaterials offer increased biomolecule loading and maintain immobilized biomolecules' functionality. Moreover, these materials provide faster electron transfer kinetics between the electrode surface and active sites of biomolecules, prevailing the fabrication of an efficient biosensing platform [21]. Alternatively, owing to the inherent photoluminescent and interesting optical (absorption in UV or NIR regions) properties, carbon nanomaterials (CDs and GQDs) are utilized as energy donor species for the fabrication of fluorescent biosensor [7]. On the other hand, graphene-based layered materials have been used as energy acceptor species or fluorescence quencher to detect various analytes [22].

In this chapter, efforts have been made to present the recent advancements over the last decade in carbon nanostructured-based biosensors. This chapter's application is restricted to electrochemical and fluorescent biosensors based on various nanomaterials, including CNTs, graphene, GO, rGO, CDs and GQDs. The first section focuses on CNTs based biosensors. The second section outlines graphene and its derivatives-based biosensors. The third section explores the recent advancements pertaining to CDs and GQDs based biosensors.

CNTS BASED BIOSENSORS

CNTs are the 1D allotropes of carbon and were discovered by S. Ijima in 1991 [15]. CNTs can be classified into two dichotomies: multi-walled carbon nanotubes (MWCNTs) and single-walled carbon nanotubes (SWCNTs). SWCNTs are the single graphene sheet rolled into a cylinder, whereas MWCNTs are the stacked concentric graphene sheets, having 0.34 nm interlayer spacing and rolled into a cylinder. The basic structure of CNTs can be chiral, armchair, or zigzag, described by a vector (n, m) that decides the chemical and physical characteristics of CNTs [1]. In general, CNTs possess a unique sp^2-hybridized carbon surface, higher thermal and electrical conductivity, high surface area, high mechanical strength, and excellent chemical stability [2]. Besides, CNTs provide the ease to functionalize with numerous targeting ligands and polymers (functional groups) through noncovalent or covalent interactions [1]. In line with this, Farias *et al.* reported the carboxylic functionalized MWCNT (MWCNT$_f$) based electrochemical biosensor toward the sensitive detection of flutamide (an anticancer drug) in urine and pharmaceuticals samples using the cyclic voltammetry (CV) technique. The presence of MWCNT$_f$ presented an improved catalytic ability to reduce or oxidize the flutamide [23]. In another work, Nasrabadi and the team demonstrated the electrochemical biosensor based on fullerene functionalized ionic liquid (1-butyl-3-methylimidazolium tetrafluoro borate) and CNT for the quantification of diazepam in real samples (tablets, urine, and serum). Enhanced electrocatalytic behavior toward the reduction of diazepam was observed with a limit of detection (LOD) of 87 ± 2 nM and a linear detection range of 0.3–700.0 μM [24]. Nishimura and co-workers reported the sulphuric acid (H_2SO_4) functionalized SWCNT-based flexible electrode for dopamine detection with an improved LOD of ~100 nM compared to the non-functionalized electrode with LOD of ~1 µM. It was mentioned that the acid functionalization of SWCNT introduced the defects providing the increased electron transfer kinetics [25]. Singh and co-workers designed an immunosensor for determining the food toxin named aflatoxin (AFB1) by electrophoretically depositing the carboxylic MWCNT on the working electrode. The as-designed sensing platform revealed lower LOD (0.08 ng mL^{-1}) and higher sensitivity (95.2 µA ng^{-1} mL cm^{-2}) owing to the excellent electrocatalytic activity of MWCNT [26]. Recently, Huang *et al.* reported MWCNT/polymeric nanoparticle nanocomposite (PAKB/NPs-CNTs) coated screen-printed carbon electrode (SPCE) based highly sensitive enzymatic biosensor (Fig. **1a**) toward hydrogen peroxide (H$_2$O$_2$) detection. The proposed bioelectrode exhibited good stability, short response time, excellent selectivity, and high sensitivity with a broader detection range (0.02 to 6.48 mM) and lower LOD (2.7 µM). The improved sensing characteristics were attributed to the synergistic effect of PAKB/NPs-CNTs that provides faster electron transfer, increased and stable enzyme immobilization, and higher specific surface area [27]. Further, Waqas *et al.* modified the glassy carbon electrode (GCE) with a composite of sulfur-doped MWCNT and Cu$_2$O microspheres (Cu$_2$OMSs/S-MWCNTs) for the non-enzymatic glucose quantification (Fig. **1b**). The Cu$_2$OMSs/S-MWCNTs composite offers highly defected structures having greater catalytic active sites, resulting in an improved electrocatalytic activity [28]. Bafrooei and co-workers fabricated an ultrasensitive biosensing platform utilizing rGO-MWCNT/AuNPs for the quantitative detection of prostate-specific antigen (PSA) over the detection range of 0.005–100 ng mL^{-1} and LOD of 1.0 pg mL^{-1} [29]. Kumar *et al.* utilized conducting paper for carcinoembryonic antigen (CEA) detection based on the nanocomposite of CNT and poly (3,4-ethylenedioxythiophene): poly(styrenesulfonate) (PEDOT:PSS). The modified paper was further processed with formic acid for improving the conductivity from ~6.5 × 10^{-4} to 2.2 × 10^{-2} S cm^{-1}. The fabricated conducting paper-based biosensor revealed higher sensitivity (7.8 µA ng^{-1} ml cm^{-2}) toward CEA detection [30]. Singh *et al.* used the GO rapped MWCNTs as transducer material toward the development of a microfluidic biochip. The developed biochip was utilized for the *Salmonella typhimurium* estimation with an LOD of 0.376 CFU/mL [31].

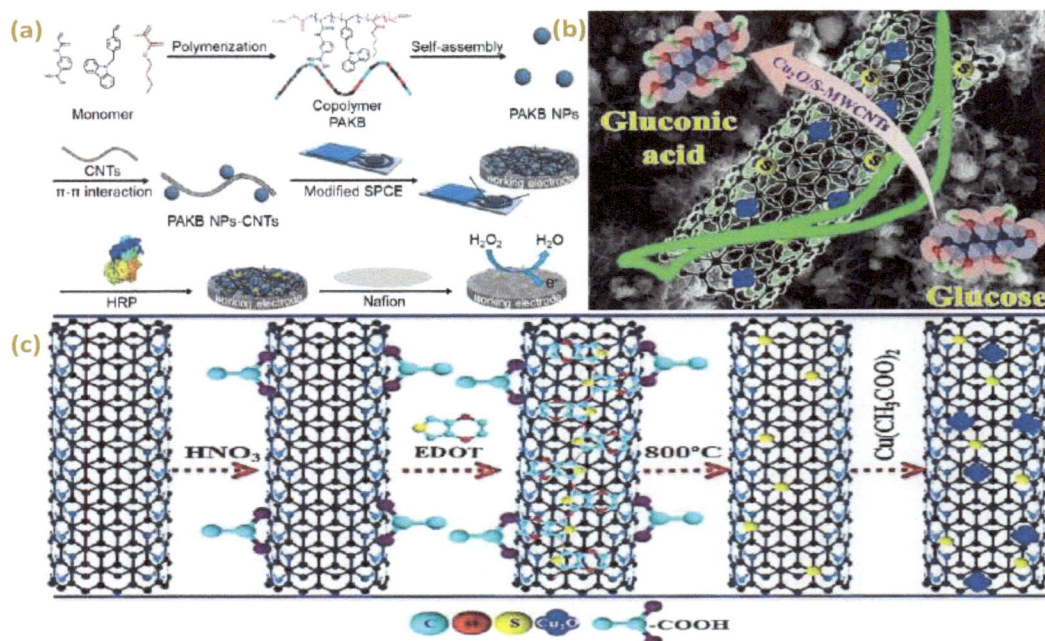

Fig. (1). (a) PAKB/NPs-CNTs coated SPE based enzymatic biosensor for H_2O_2 detection Reprinted with permission from reference [27] Copyright © 2020 American chemical society **(b)** Cu_2OMSs/S-MWCNTs based non-enzymatic glucose biosensor **(c)** process flow for the preparation of Cu_2OMSs/S-MWCNTs composite Reproduced with permission from reference [28] Copyright © 2020 American chemical society.

GRAPHENE AND ITS DERIVATIVES BASED BIOSENSORS

Graphene is the 2D carbon material representing a single graphite sheet wherein carbon atoms are present in the honeycomb structure. It was discovered by Novoselov *et al.* in 2004 utilizing the scotch tape method by mechanically exfoliating the graphite [32]. It is regarded as the building block of other carbon allotropes, namely graphite, CNTs, and fullerenes. When graphene is stacked up, it becomes 3D Graphite, whereas it forms fullerene upon wrapping and CNTs upon rolling up cylindrically [33]. Graphene depicts interesting thermal, physiochemical, and mechanical characteristics such as high surface area, good biocompatibility, better thermal and electrical conductivity, and high Young's modulus [34]. Owing to its exciting properties, this 2D material has derived enormous interest for various biomedical and environmental applications over the last decade [35]. Pristine graphene provides ease of conjugation to different biomolecules through hydrophobic interactions and π–π stacking [1]. However, lack of surface functional groups, low aqueous solubility are the main obstacles associated with graphene. Therefore, to bridge the gap, graphene derivatives, namely GO and rGO, were explored through the chemical method that opened up a new window to graphene research [36]. These graphene derivatives are rich in functional groups (epoxy, carboxylic, hydroxyl) and provide higher aqueous solubility, offering unparalleled opportunities to fabricate biosensing platforms [37]. In this context, Jaiswal *et al.* reported the electrophoretic deposition of carboxylated graphene nanoflakes (c-GNF) and aminopropyltrimethoxysilane (APTMS)-functionalized zinc oxide (ZnO) nanorods for detecting *E.coli* through the electrochemical impedance spectroscopy (EIS) technique. The deposited nanohybrid offers multiple functional groups for the efficient and increased loading of probe DNA, resulting in excellent biosensing characteristics including low detection limit, better sensitivity, and wider detection range [38]. Augustine *et al.* demonstrated the molybdenum trioxide and rGO nanohybrid (MoO_3@rGO) based ultrasensitive biosensor for the quantitative detection of breast cancer-specific biomarker named human epidermal growth factor receptor-2 (HER-2) (Fig. **2a**). Herein, 1D MoO_3 was anchored onto 2D rGO that provided improved surface area, faster heterogeneous electron transfer kinetics, increased mechanical stability and better charge transferability. The fabricated biosensor revealed remarkable biosensing parameters towards HER-2 detection with a broad detection range (0.001–500 ng mL^{-1}) improved sensitivity (13 uA mL ng^{-1} cm^{-2}) [39]. Sandil and co-workers utilized tungsten trioxide (WO_3) and rGO based nanocomposite as an immobilization matrix for the efficient detection of human cardiac biomarker Troponin I (cTnI). The WO_3@rGO nanohybrid offered enhanced heterogeneous electron transfer rate constant ($K_o = 2.4 \times 10^{-4}$ cms^{-1}) owing to the synergistic behavior between WO_3 nanorods and rGO that further results in enhanced biosensor efficiency [40]. Singh *et al.* demonstrated scalable biosensor fabrication based on chemical vapor deposition (CVD) grown graphene for CEA detection [41]. Kumar and the team utilized 2D nanosheets of rGO for the uniform decoration of nanostructured hafnia (nHfO$_2$) and avoided the agglomeration by trapping their Brownian motion. The reduced agglomeration provided the increased heterogeneous electron transfer, resulting in the efficient biosensing characteristics toward non-invasive CYFRA-21-1 (oral cancer biomarker) detection through differential pulse voltammetry (DPV) technique [3, 42].

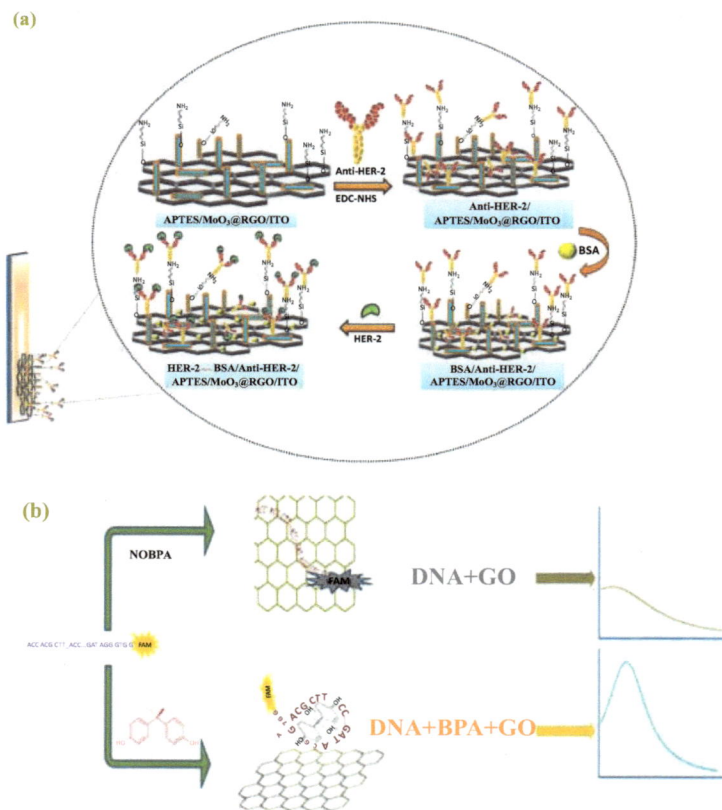

Fig. (2). (a) Schematic illustration of MoO_3@rGO nanohybrid based electrochemical biosensor for HER-2 detection Reproduced with permission from reference [39] Copyright © 2019, American Chemical Society **(b)** GO-based fluorescent biosensor for BPA detection Reprinted with permission from reference [43] Copyright © 2015, American Chemical Society.

Graphene and its derivatives have also been explored as energy acceptor species in the fabrication of energy transfer-based fluorescent biosensing platforms. Generally, fluorescent biosensors involve the energy transfer phenomenon from excited donor species to nearby acceptor species align="center"to study various biomolecular interactions. Based on the energy transfer phenomenon, several biosensors have been developed over the last decade. For example, Bhatnagar and co-workers reported the fluorescence resonance energy transfer (FRET) based immunosensor for myocardial infarction (early heart attack) detection by utilizing the graphene nanosheets as an energy acceptor. The fluorescence intensity of cTnI specific antibody conjugated amine-functionalized graphene quantum dots (anti-cTnI/afGQDs) was quenched upon the addition of graphene nanosheets. Further, the addition of target analyte expelled graphene apart from afGQD, leading to the linear recovery of quenched fluorescence [44]. Similar to graphene, GO has also been explored as an energy acceptor species in fluorescent biosensors. For example, Shi *et al.* utilized the green fluorescent protein (GFP) and GO as the donor-acceptor pair for the specific detection of Botulinum neurotoxins (BoNTs) [45]. Recently, Guo *et al.* reported the imidacloprid detection in environmental and food samples based on up-converting nanoparticles (UCNPs) and GO as FRET pair. Similarly, Zhu *et al.* designed an aptasensor for Bisphenol A (BPA) detection (Fig. **2b**) based on FRET between GO and fluorescently modified anti-BPA aptamer (FAM-ssDNA). The designed aptasensor revealed LOD of 0.05 ng/mL over the dynamic range of 0.1–10 ng/mL toward BPA detection [43].

CARBON DOTS AND GRAPHENE QUANTUM DOTS BASED BIOSENSORS

CDs and GQDs are both the zero-dimensional (0D) carbon-based fluorescent material in solution state with spectral features, having a size less than 10 nm [46, 47]. The major difference between them lies in their inside structure. GQDs possess an sp^2 hybridized graphene structure [48]. In contrast, CDs contain an sp^3 hybridized spherical-shaped amorphous carbon structure [49]. Owing to their eco-friendly nature and inherent biocompatibility, these materials have emerged as potential alternatives to toxic heavy metal-based quantum dots [50]. Besides, these materials endow distinct properties, such as strong quantum confinement and edge effects, presence of several active functional groups, higher aqueous solubility, tunable bandgap, broadband optical absorption, good conductivity, broad excitation spectra, and good photostability [7, 50]. Since the day of CDs origin in 2004 by Xu *et al.* [51], enormous efforts have been accomplished toward the exploration of various bottom-up and top-down strategies for the preparation of GQDs and CDs in the quest of uniform size, large-scale production, tunable surface properties, low cost, high yield, and environmentally benign material [52]. These strategies include microwave synthesis, electrochemical methods, hydrothermal synthesis, thermal pyrolysis, incomplete carbonization, physical grinding, *etc.* [53]. Interestingly, the size and intrinsic properties of these fluorescent materials can be regulated during or after synthesis according to the desired application, making them predominant material compared to other carbon allotropes. For instance, the size can be varied by controlling reactants' concentration, stabilizing agent, reaction time, and temperature [53]. Kwon and co-workers prepared different-sized GQDs (2 to 10 nm) by regulating amine concentration during amidative cutting of tattered graphite [54]. Besides, heteroatom doping has also been reported as an effective approach for tailoring the optical and electronic properties of these fluorescent carbon materials, including conductivity, catalytic properties, and fluorescence quantum yield (QY). For instance, Kalkal *et al.* demonstrated the *in-situ* nitrogen (N) doping of GQDs *via* diethylenetriamine (DETA) for improving the QY. The QY of doped GQDs increased from 14.57% to 78.83% compared to bare GQDs [7]. Similarly, Kadian *et al.* reported the sulphur doping of GQDs *via* thermal pyrolysis of citric acid (CA) and 3-mercaptopropionic acid (MPA). Utilizing both theoretical and experimental investigations, authors revealed the effect of sulphur doping in the enhancement of fluorescence intensity and QY of GQDs [8]. Apart from heteroatom doping, surface functionalization during or after synthesis is another important strategy for tailoring the properties of CDs and GQDs. The surface or edge of these materials can be facilitated with different functional groups such as hydroxyl, amine, and carboxyl, making them viable to interact with different biomolecules *via* covalent and electrostatic interactions.

Generally, GQDs and CDs have pretty similar optical properties, including up-conversion photoluminescence (PL), phosphorescence electrochemiluminescence (ECL), chemiluminescence (CL), fluorescence, and optical absorption. Among them, fluorescence is the most explored property in terms of practical applications and fundamental studies. Upon excitation, these materials can emit different colored (blue, yellow, green, red, *etc..*) fluorescence. Interestingly, these materials can exhibit both excitation-independent and excitation-dependent emission spectra wherein there is no shift in emission peak and emission peak shift towards higher wavelength with varying excitation wavelength, respectively. Apart from interesting optical properties, these 0D materials also exhibit remarkable electrochemical properties such as good conductivity, fast electron transferability, and high current density because of the quantum confinement and edge effect. CDs and GQDs have been widely explored in the fabrication of fluorescent and electrochemical biosensing platforms. These materials serve as an efficient energy donor species in fluorescent biosensors to detect cancer biomarkers, small molecules, heavy metal ions, bioflavonoids, nucleic acid, *etc.*. In this context, Kalkal *et al.* (Fig. **3**) reported the neuron-specific enolase (NSE) detection utilizing a fluorescent biosensing platform based on gold nanoparticles (AuNPs) and *in-situ* nitrogen-doped and amine-functionalized GQDs (amine-N-GQDs) as nano surface energy transfer (NSET) pair. The fluorescence intensity of NSE-specific antibody conjugated amine-N-GQDs (anti-NSE/amine-N-GQDs) was quenched upon the addition of AuNPs. Further, the addition of target analyte (NSE) expelled AuNPs apart from amine-N-GQDs, leading to the linear recovery of quenched fluorescence.

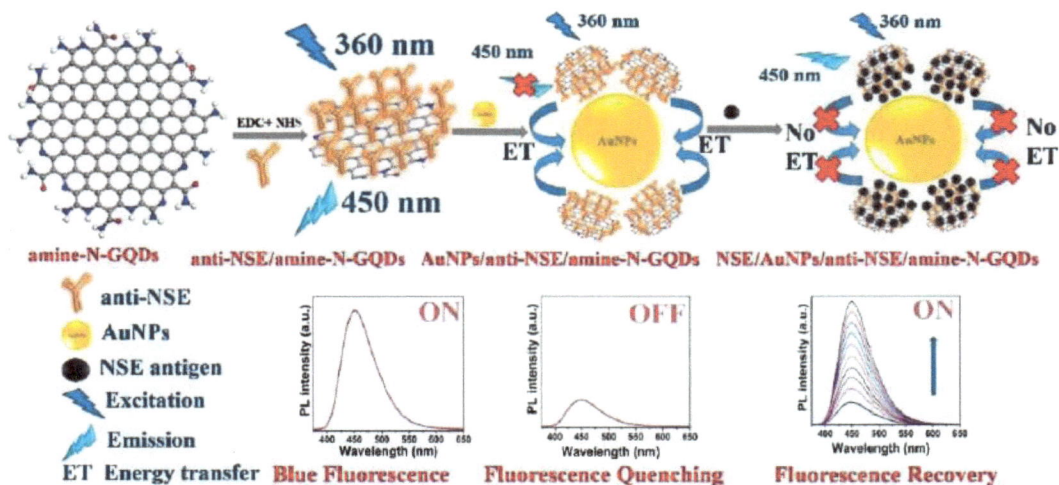

Fig. (3). Process flow for the NSE detection utilizing amine-N-GQDs and AuNPs as donor-acceptor pair Reprinted with permission from reference [7] Copyright © 2020, American Chemical Society.

Further, Kadian *et al.* reported the selective detection of picric acid and quercetin based on the inner filter effect (IFE) by using S-doped GQDs as the fluorescent sensing probes [55, 56]. Liu *et al.* demonstrated specific and sensitive fluorescent probes based on carbon quantum dots (CQDs) toward the ferric ion (Fe^{3+}) determination in the wider dynamic ranges of 10–200 μM [57]. Shi *et al.* reported fluorescent biosensing assay toward detecting gene sequence of *staphylococcus aureus* (*S. aureus*) using AuNPs and GQDs as FRET pair with a LOD of 1 nM [58]. Recently, Farahani *et al.* reported boric acid-functionalized S, N co-doped GQDs based fluorescent nanosensor for sensitive glucose detection [59]. Besides, these materials have also been explored towards the detection of trypsin [60], miRNA-34a [61], glutathione [62], ascorbic acid [63], tetracycline [64], methyl-paraoxon [65] *etc.*. (Table **1**).

Table 1. List of nanostructured carbon materials (CNTs, graphene, GO, rGO, and GQDs) based fluorescent and electrochemical biosensors.

Nanostructured Material	Type of Biosensor	Detection Technique	Analyte Detected	Detection Range	Sensitivity	LOD	Refs.
Carboxylic MWCNT	Electrochemical	CV	Flutamide	0.1 to 1000 μ mol L^{-1}	0.30 μA μmol^{-1} L	0.03 μmol L^{-1}	[23]
C_{60}-CNTs	Electrochemical	DPV	Diazepam	0.3–700.0 μM	0.173 μA μM^{-1}	87 ± 2 nM	[24]
SWCNTs	Electrochemical	CV	Dopamine	100 nM-10 μM	-	100 nM	[25]
c-MWCNTs	Electrochemical	EIS	Aflatoxin	0.25–1.375 ng mL^{-1}	95.2 μA ng^{-1} mL cm^{-2}	0.08 ng mL^{-1}	[26]
PAKB NPs-CNTs	Electrochemical	Amperometry	H_2O_2	0.02 to 6.48 mM	72.08 μA mM^{-1} cm^{-2}	2.7 μM	[27]

(Table 1) cont.....

Nanostructured Material	Type of Biosensor	Detection Technique	Analyte Detected	Detection Range	Sensitivity	LOD	Refs.
Cu$_2$O MSs/S-MWCNTs	Electrochemical	Amperometry	Glucose	4.95 µM to 7 mM	581.89 µA mM^{-1} cm^{-2}	1.46 µM	[28]
rGO-MWCNT/AuNPs	Electrochemical	EIS	PSA	0.005–100 ng mL−1	-	1.0 pg mL^{-1}	[29]
CNT-PEDOT:PSS	Amperometric	Chrono-amperometry	CEA	2–15 ng ml^{-1}	7.8 µA ng^{-1} ml cm^{-2}	-	[30]
c-MWCNTs/GO	Electrochemical	CV	*S. typhimurium*	10^1 to 10^7 CFU/mL	162.47 µA/CFU^{-1}/mLcm^{-2}	0.376 CFU/mL	[31]
c-GNF/APTMS-ZnO	Impedimetric	EIS	*E.coli*	10^{-16} M to 10^{-6} M	-	0.1 fM	[38]
MoO$_3$@rGO	Electrochemical	DPV	HER-2	0.001–500 ng mL^{-1}	13 uA mLng^{-1}cm^{-2}	0.001 ng mL^{-1}	[39]
WO$_3$@rGO	Electrochemical	DPV	cTnI	0.01–250 ng/mL	58.24 µA/cm2	0.01 ng/mL	[40]
CVD-Graphene	Electrochemical	EIS	CEA	1.0–25.0 ng mL−1	563.4 Ω ng^{-1} mL cm^{-2}	0.23 ng mL^{-1}	[41]
nHfO$_2$@rGO	Electrochemical	DPV	CYFRA-21-1	0 to 30 ng mL^{-1}	18.24 µA mL ng^{-1}	0.16 ng mL^{-1}	[3]
afGQD/Graphene	Fluorescent	FRET	cTnI	1 pg–1000 ng mL^{-1}	-	0.19 pg mL^{-1}	[44]
GFP-GO	Fluorescent	FRET	BoNTs	1 fg/mL to 1 pg/mL	-	1 fg/mL	[45]
FAM-ssDNA/GO	Fluorescent	FRET	BPA	0.1–10 ng/mL	-	0.05 ng/mL	[43]
Amine-N-GQDs/AuNPs	Fluorescent	NSET	NSE	0.1 pg mL^{-1} to 1000 ng mL^{-1}	-	0.09 Pg/mL	[7]
S-GQDs	Fluorescent	IFE	Picric acid	0.1–100 µM	-	0.093 µM	[55]
S-GQDs	Fluorescent	IFE	Quercetin	0–50.0 µM	2.009 × 104 M^{-1}	0.006 µg/mL	[56]
CQDs	Fluorescent	Quenching	Fe^{3+}	10–200 µM	-	1.8 µM	[57]
GQDs/AuNPs	Fluorescent	FRET	*S. aureus*	100 pM to 400 nM	-	1 nM	[58]
Boric acid-S,N GQDs	Fluorescent	PL	Glucose	5.5–66 mM	-	5.5 µM	[59]
GQDs/UCNP@SiO2	Fluorescent	FRET	Trypsin	0 to 23 µg/mL	-	0.7 µg/mL	[60]
AuNF@GQDs	Fluorescent	FRET	miRNA-34a	0.4–4 fM	-	0.1 fM	[61]
GQDs-AuNPs	Fluorescent	FRET	Glutathione	0.005–0.13 mU/mL	-	0.005 mU/mL	[62]
HRP-GQDs	Fluorescent	PL	Ascorbic acid	1.11 to 300 µM	-	0.32 µM	[63]
PdNPs/aptamer/GQDs	Fluorescent	PL	Tetracycline	40–90 ng mL^{-1}	-	18 ng mL^{-1}	[64]
GQDs/AChE/CHOx	Fluorescent	PL	methyl-paraoxon	0.40–4.05 µM	-	0.342 µM	[65]
Hb-GQDs-Chit	Electrochemical	EIS	H$_2$O$_2$	1.5–195 µM	0.53 µAµM^{-1} cm^{-2}	(0.68 µM)	[66]
N, S-GQDs@Au-PANI	Impedimetric	EIS	CEA	0.5 to 1000 ng mL^{-1}	-	0.01 ng mL^{-1}	[67]
MWCNT-GQDs	Electrochemical	DPV	DA	0.005 to 100.0 µM	-	0.87 nM	[68]
GQDs-MoS$_2$	Impedimetric	EIS	Aflatoxin	0.1 to 3.0 ng mL^{-1}	44.44Ω/(ng mL^{-1})/cm^2	0.09 ng mL^{-1}	[69]
GQDs-HRP	Amperometric	Amperometry	miRNA-155	1 fM to 100 pM	-	0.14 fM	[70]

(Table 1) cont.....

Nanostructured Material	Type of Biosensor	Detection Technique	Analyte Detected	Detection Range	Sensitivity	LOD	Refs.
GQDs-SPE	Electrochemical	EIS	Myoglobin	0.01–100 ng/mL	-	0.01 ng/mL	[71]
MoS_2-GQDs	Electrochemical	CV	Caffeic acid	0.38–100 μM	17.92 nA μM^{-1}	0.32 μM	[72]
AgNPs/thiol-GQDs	Electrochemical	DPV	Trinitrotoluene	1.00×10^{-6} –3.00×10^{4}	-	3.33×10^{-7}	[74]
PANI-GQDs	Electrochemical	DPV	Calycosin	1.1×10^{-5}- 3.52×10^{-4} mol L^{-1}	-	9.8×10^{-6} mol L^{-1}	[75]
GQDs/f-MWCNTs	Electrochemical	DPV	Interleukin-6	0.01–2.0 pg mL^{-1}	-	0.0030 pg mL^{-1}	[76]
GQDs-GCE	Impedimetric	EIS	HCV	10–400 pg mL^{-1}	-	3.3 pg mL^{-1}	[77]

On the other hand, in the case of electrochemical biosensors, CDs and GQDs provide either direct electron transfer or helps in increasing electron transfer rate among active sites of biomolecules and electrode surface. In line with this, Rezai and the team demonstrated an electrochemical biosensor for H_2O_2 quantification using hemoglobin immobilized GQDs-chitosan nanocomposite. The as-fabricated biosensor exhibited efficient biosensing parameters *viz.,* faster heterogeneous electron transfer rate, low LOD, better sensitivity, broad detection range owing to the high surface area, and unique electron transfer ability of GQDs [67]. Huang *et al.* reported the sensitive dopamine detection (Fig. **4**) based on the nanocomposite of MWCNT-GQDs and revealed a LOD of 0.87 nM. The anionic groups of GQDs interacted with the cationic groups of dopamine *via* π-π interaction, providing selective dopamine detection [68]. Bhardwaj *et al.* demonstrated the selective quantification of aflatoxin B1 (AFB1) based on GQDs decorated molybdenum disulfide (MoS_2) nanosheets over the detection range of 0.1-3.0 ng mL^{-1} and LOD of 0.09 ng mL^{-1} [69]. CDs and GQDs have also been employed in detecting miRNA-155 [70], myoglobin [71], caffeic acid [72], heavy metal ions [73], trinitrotoluene [74], calycosin [75], interleukin-6 [76], and hepatitis C virus (HCV) [77]. The sensing characteristics of these biosensing platforms have been provided in Table **1**.

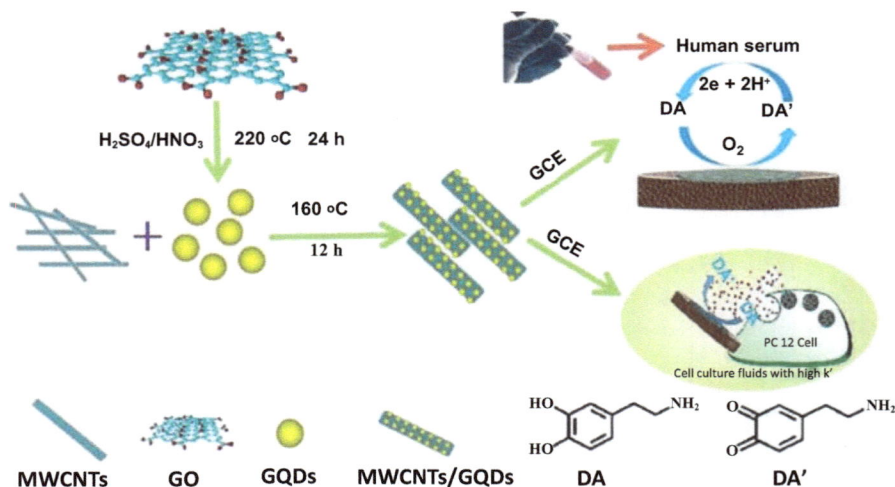

Fig. (4). MWCNT-GQDs based electrochemical biosensor for dopamine detection Reproduced with permission from reference [68] Copyright © 2020, American Chemical Society.

CONCLUSIONS AND PERSPECTIVES

In this chapter, we have highlighted the recent advancements in fluorescent and electrochemical biosensors based on carbon nanostructured materials, namely CNTs, graphene and its derivatives (GO, rGO), CDs, and GQDs. These interesting nanostructured materials have been extensively employed for detecting a number of analytes, comprising cancer biomarkers, small molecules, heavy metal ions, bioflavonoids, nucleic acid, food toxins, viruses, isoflavone, neurotransmitters, pathogens, enzymes, aptamers, antioxidants, and proteins. When it comes to fluorescent sensing, carbon nanomaterials such as graphene nanosheets GO act as efficient energy acceptor species or fluorescence quenchers. In contrast, carbon nanostructured materials such as CDs and GQDs act as prominent energy donor species, which provide photostable, and tunable optical properties. On the other hand, in electrochemical biosensors, these materials exhibit high electrical conductivity, increased electrocatalytic activity, and faster heterogeneous electron transfer kinetics. Besides, due to strong adsorption ability, larger surface area, and ample surface functionalities, these nanomaterials provide the oriented and increased loading of biomolecules, resulting in align="center"a superior analytical performance with desirable biosensing characteristics. Furthermore, the lower cytotoxicity and better biocompatibility of these materials help in maintaining the activity of loaded biomolecules. Despite the enormous exploration and significant advancement in this field, some challenges regarding the mass-scale production, the mechanism behind their inherent fluorescence and electron transfer capability, long-term cytotoxicity of these carbon-based nanostructured materials remain to be addressed before their application as a clinical point of care devices. An in-depth investigation of these challenges will provide new opportunities for their utilization in the sensing field. Conclusively, it can be anticipated that these carbon-based nanostructured materials align="center"can have enormous potential in developing next-generation point-of-care devices in the near future.

CONSENT FOR PUBLICATION

Not applicable.

CONFLICT OF INTERESTS

The authors declare no conflict of interest, financial or otherwise.

ACKNOWLEDGEMENTS

Ashish Kalkal is thankful to the Ministry of Education, Government of India, for the fellowship. Gopinath Packirisamy is thankful to the Department of Science and Technology (DST), Government of India, Technology Innovation Hub (TIH) for Devices Materials and Technology Foundation, IIT Roorkee [A Section-8 Company: Divyasampark IHub for Devices Materials and Technology Foundation].

REFERENCES

[1]   Augustine, S.; Singh, J.; Srivastava, M.; Sharma, M.; Das, A.; Malhotra, B.D. Recent advances in carbon based nanosystems for cancer theranostics. *Biomater. Sci.,* **2017**, *5*(5), 901-952.
      [http://dx.doi.org/10.1039/C7BM00008A] [PMID: 28401206]

[2]   Bezzon, V.D.N.; Montanheiro, T.L.A.; de Menezes, B.R.C.; Ribas, R.G.; Righetti, V.A.N.; Rodrigues, K.F.; Thim, G.P. Carbon Nanostructure-based Sensors: A brief review on recent advances. *Adv. Mater. Sci. Eng.,* **2019**, *2019*, 4293073.
      [http://dx.doi.org/10.1155/2019/4293073]

[3]   Kumar, S.; Ashish, ; Kumar, S.; Augustine, S.; Yadav, S.; Yadav, B.K.; Chauhan, R.P.; Dewan, A.K.; Malhotra, B.D. Effect of Brownian motion on reduced agglomeration of nanostructured metal oxide towards development of efficient cancer biosensor. *Biosens. Bioelectron.,* **2018**, *102*, 247-255.
      [http://dx.doi.org/10.1016/j.bios.2017.11.004] [PMID: 29153946]

[4]   Pradhan, R.; Kalkal, A.; Jindal, S.; Packirisamy, G.; Manhas, S. Four electrode-based impedimetric biosensors for evaluating cytotoxicity of tamoxifen on cervical cancer cells. *RSC Advances,* **2021**, *11*(2), 798-806.
      [http://dx.doi.org/10.1039/D0RA09155C]

[5]   Kumar, S.; Kumar, J.; Sharma, S.N. Nanostructured titania based electrochemical impedimetric biosensor for non-invasive cancer detection. *Mater. Res. Express,* **2018**, *5*(12), 125405.

[http://dx.doi.org/10.1088/2053-1591/aae1e2]

[6]   Pradhan, R.; Raisa, S.A.; Kumar, P.; Kalkal, A.; Kumar, N.; Packirisamy, G.; Manhas, S. Optimization, fabrication, and characterization of four electrode-based sensors for blood impedance measurement. *Biomed. Microdevices,* **2021**, *23*(1), 9.
[http://dx.doi.org/10.1007/s10544-021-00545-4] [PMID: 33449205]

[7]   Kalkal, A.; Pradhan, R.; Kadian, S.; Manik, G.; Packirisamy, G. Biofunctionalized graphene quantum dots based fluorescent biosensor toward efficient detection of small cell lung cancer. *ACS Appl. Bio Mater.,* **2020**, *3*(8), 4922-4932.
[http://dx.doi.org/10.1021/acsabm.0c00427] [PMID: 35021736]

[8]   Kadian, S.; Manik, G.; Ashish, K.; Singh, M.; Chauhan, R.P. Effect of sulfur doping on fluorescence and quantum yield of graphene quantum dots: an experimental and theoretical investigation. *Nanotechnology,* **2019**, *30*(43), 435704.
[http://dx.doi.org/10.1088/1361-6528/ab3566] [PMID: 31342919]

[9]   Iqbal, S.; Khatoon, H.; Hussain Pandit, A.; Ahmad, S. Recent development of carbon based materials for energy storage devices. *Mater. Sci. Energy Technol.,* **2019**, *2*(3), 417-428.
[http://dx.doi.org/10.1016/j.mset.2019.04.006]

[10]   Kumar, S.; Kumar, J.; Narayan Sharma, S.; Srivastava, S. rGO integrated MEHPPV and P3HT polymer blends for bulk hetero junction solar cells: A comparative insight. *Optik (Stuttg.),* **2019**, *178*, 411-421.
[http://dx.doi.org/10.1016/j.ijleo.2018.09.148]

[11]   Kumar, S.; Kumar, J.; Sharma, S.N. Investigation of charge transfer properties in MEHPVV and rGO-AA nanocomposites for Green organic photovoltaic application. *Optik (Stuttg.),* **2020**, *208*, 164540.
[http://dx.doi.org/10.1016/j.ijleo.2020.164540]

[12]   Hirsch, A. The era of carbon allotropes. *Nat. Mater.,* **2010**, *9*(11), 868-871.
[http://dx.doi.org/10.1038/nmat2885] [PMID: 20966925]

[13]   Kour, R.; Arya, S.; Young, S-J.; Gupta, V.; Bandhoria, P.; Khosla, A. Review—Recent Advances in Carbon Nanomaterials as Electrochemical Biosensors. *J. Electrochem. Soc.,* **2020**, *167*(3), 037555.
[http://dx.doi.org/10.1149/1945-7111/ab6bc4]

[14]   Kroto, H.W.; Heath, J.R.; O'Brien, S.C.; Curl, R.F.; Smalley, R.E. C 60: Buckminsterfullerene. *Nature,* **1985**, *318*(6042), 162-163.
[http://dx.doi.org/10.1038/318162a0]

[15]   Iijima, S. Helical microtubules of graphitic carbon. *Nature,* **1991**, *354*(6348), 56-58.
[http://dx.doi.org/10.1038/354056a0]

[16]   Kumar, S.; Kumar, S.; Srivastava, S.; Yadav, B.K.; Lee, S.H.; Sharma, J.G.; Doval, D.C.; Malhotra, B.D. Reduced graphene oxide modified smart conducting paper for cancer biosensor. *Biosens. Bioelectron.,* **2015**, *73*, 114-122.
[http://dx.doi.org/10.1016/j.bios.2015.05.040] [PMID: 26057732]

[17]   Goyal, R.N.; Chatterjee, S.; Rana, A.R.S. The effect of modifying an edge-plane pyrolytic graphite electrode with single-wall carbon nanotubes on its use for sensing diclofenac. *Carbon,* **2010**, *48*(14), 4136-4144.
[http://dx.doi.org/10.1016/j.carbon.2010.07.024]

[18]   McNaught, AD; Wilkinson, A *Compendium of chemical terminology.,* **1997**.

[19]   Pérez, J.A.C.; Sosa-Hernández, J.E.; Hussain, S.M.; Bilal, M.; Parra-Saldivar, R.; Iqbal, H.M.N. Bioinspired biomaterials and enzyme-based biosensors for point-of-care applications with reference to cancer and bio-imaging. *Biocatal. Agric. Biotechnol.,* **2019**, *17*, 168-176.
[http://dx.doi.org/10.1016/j.bcab.2018.11.015]

[20]   Kalkal, A.; Kumar, S.; Kumar, P.; Pradhan, R.; Willander, M.; Packirisamy, G.; Kumar, S.; Malhotra, B.D. Recent advances in 3D printing technologies for wearable (bio)sensors. *Addit. Manuf.,* **2021**, *46*,

102088.
[http://dx.doi.org/10.1016/j.addma.2021.102088]

[21]   Riley, P.R.; Narayan, R.J. Recent advances in carbon nanomaterials for biomedical applications: A review. *Curr. Opin. Biomed. Eng.,* **2021**, *17*, 100262.
[http://dx.doi.org/10.1016/j.cobme.2021.100262] [PMID: 33786405]

[22]   Shi, J.; Tian, F.; Lyu, J.; Yang, M. Nanoparticle based fluorescence resonance energy transfer (FRET) for biosensing applications. *J. Mater. Chem. B Mater. Biol. Med.,* **2015**, *3*(35), 6989-7005.
[http://dx.doi.org/10.1039/C5TB00885A] [PMID: 32262700]

[23]   Farias, J.S.; Zanin, H.; Caldas, A.S.; dos Santos, C.C.; Damos, F.S.; de Cássia Silva Luz, R. Functionalized multiwalled carbon nanotube electrochemical sensor for determination of anticancer drug flutamide. *J. Electron. Mater.,* **2017**, *46*(10), 5619-5628.
[http://dx.doi.org/10.1007/s11664-017-5630-6]

[24]   Rahimi-Nasrabadi, M.; Khoshroo, A.; Mazloum-Ardakani, M. Electrochemical determination of diazepam in real samples based on fullerene-functionalized carbon nanctubes/ionic liquid nanocomposite. *Sens. Actuators B Chem.,* **2017**, *240*, 125-131.
[http://dx.doi.org/10.1016/j.snb.2016.08.144]

[25]   Nishimura, K.; Ushiyama, T.; Viet, N.X.; Inaba, M.; Kishimoto, S.; Ohno, Y. Enhancement of the electron transfer rate in carbon nanotube flexible electrochemical sensors by surface functionalization. *Electrochim. Acta,* **2019**, *295*, 157-163.
[http://dx.doi.org/10.1016/j.electacta.2018.10.147]

[26]   Singh, C.; Srivastava, S.; Ali, M.A.; Gupta, T.K.; Sumana, G.; Srivastava, A.; Mathur, R.B.; Malhotra, B.D. Carboxylated multiwalled carbon nanotubes based biosensor for aflatoxin detection. *Sens. Actuators B Chem.,* **2013**, *185*, 258-264.
[http://dx.doi.org/10.1016/j.snb.2013.04.040]

[27]   Huang, X.; Xu, S.; Zhao, W.; Xu, M.; Wei, W.; Luo, J.; Li, X.; Liu, X. Screen-printed carbon electrodes modified with polymeric nanoparticle-carbon nanotube composites for enzymatic biosensing. *ACS Appl. Nano Mater.,* **2020**, *3*(9), 9158-9166.
[http://dx.doi.org/10.1021/acsanm.0c01800]

[28]   Waqas, M.; Wu, L.; Tang, H.; Liu, C.; Fan, Y.; Jiang, Z.; Wang, X.; Zhong, J.; Chen, W. Cu_2O microspheres supported on sulfur-doped carbon nanotubes for glucose sensing. *ACS Appl. Nano Mater.,* **2020**, *3*(5), 4788-4798.
[http://dx.doi.org/10.1021/acsanm.0c00847]

[29]   Heydari-Bafrooei, E.; Shamszadeh, N.S. Electrochemical bioassay development for ultrasensitive aptasensing of prostate specific antigen. *Biosens. Bioelectron.,* **2017**, *91*, 284-292.
[http://dx.doi.org/10.1016/j.bios.2016.12.048] [PMID: 28033557]

[30]   Kumar, S.; Willander, M.; Sharma, J.G.; Malhotra, B.D. A solution processed carbon nanotube modified conducting paper sensor for cancer detection. *J. Mater. Chem. B Mater. Biol. Med.,* **2015**, *3*(48), 9305-9314.
[http://dx.doi.org/10.1039/C5TB02169C] [PMID: 32262929]

[31]   Singh, C.; Ali, M.A.; Reddy, V.; Singh, D.; Kim, C.G.; Sumana, G.; Malhotra, B.D. Biofunctionalized graphene oxide wrapped carbon nanotubes enabled microfluidic immunochip for bacterial cells detection. *Sens. Actuators B Chem.,* **2018**, *255*, 2495-2503.
[http://dx.doi.org/10.1016/j.snb.2017.09.054]

[32]   Novoselov, K.S.; Geim, A.K.; Morozov, S.V.; Jiang, D.; Zhang, Y.; Dubonos, S.V.; Grigorieva, I.V.; Firsov, A.A. Electric field effect in atomically thin carbon films. *Science,* **2004**, *306*(5696), 666-669.
[http://dx.doi.org/10.1126/science.1102896] [PMID: 15499015]

[33]   Lui, C.H.; Liu, L.; Mak, K.F.; Flynn, G.W.; Heinz, T.F. Ultraflat graphene. *Nature,* **2009**, *462*(7271), 339-341.
[http://dx.doi.org/10.1038/nature08569] [PMID: 19924211]

[34]   Soldano, C.; Mahmood, A.; Dujardin, E. Production, properties and potential of graphene. *Carbon,* **2010**, *48*(8), 2127-2150.
        [http://dx.doi.org/10.1016/j.carbon.2010.01.058]

[35]   Shao, Y.; Wang, J.; Wu, H.; Liu, J.; Aksay, I.A.; Lin, Y. Graphene based electrochemical sensors and biosensors: A review *Electroanal,* **2010**, *22*(10), 1027-1036.

[36]   Sturala, J.; Luxa, J.; Pumera, M.; Sofer, Z. *Chemistry of Graphene Derivatives: Synthesis, Applications, and Perspectives.,* **2018**, *24*(23), 5992-6006.

[37]   Bai, Y.; Xu, T.; Zhang, X. *Graphene-Based Biosensors for Detection of Biomarkers.,* **2020**, *11*(1), 60.

[38]   Jaiswal, N.; Pandey, C.M.; Solanki, S.; Tiwari, I.; Malhotra, B.D. An impedimetric biosensor based on electrophoretically assembled ZnO nanorods and carboxylated graphene nanoflakes on an indium tin oxide electrode for detection of the DNA of Escherichia coli O157:H7. *Mikrochim. Acta,* **2019**, *187*(1), 1.
        [http://dx.doi.org/10.1007/s00604-019-3921-8] [PMID: 31797052]

[39]   Augustine, S.; Kumar, P.; Malhotra, B.D. Amine-Functionalized MoO$_3$@RGO Nanohybrid-Based Biosensor for Breast Cancer Detection. *ACS Appl. Bio Mater.,* **2019**, *2*(12), 5366-5378.
        [http://dx.doi.org/10.1021/acsabm.9b00659] [PMID: 35021536]

[40]   Sandil, D.; Srivastava, S.; Malhotra, B.D.; Sharma, S.C.; Puri, N.K. Biofunctionalized tungsten trioxide-reduced graphene oxide nanocomposites for sensitive electrochemical immunosensing of cardiac biomarker. *J. Alloys Compd.,* **2018**, *763*, 102-110.
        [http://dx.doi.org/10.1016/j.jallcom.2018.04.293]

[41]   Singh, V.K.; Kumar, S.; Pandey, S.K.; Srivastava, S.; Mishra, M.; Gupta, G.; Malhotra, B.D.; Tiwari, R.S.; Srivastava, A. Fabrication of sensitive bioelectrode based on atomically thin CVD grown graphene for cancer biomarker detection. *Biosens. Bioelectron.,* **2018**, *105*, 173-181.
        [http://dx.doi.org/10.1016/j.bios.2018.01.014] [PMID: 29412942]

[42]   Kumar, S.; Kalkal, A. 3 - Electrochemical detection: Cyclic voltammetry/differential pulse voltammetry/impedance spectroscopy. In: *Nanotechnology in Cancer Management*; Khondakar, K.R.; Kaushik, A.K., Eds.; Elsevier, **2021**; pp. 43-71.
        [http://dx.doi.org/10.1016/B978-0-12-818154-6.00008-1]

[43]   Zhu, Y.; Cai, Y.; Xu, L.; Zheng, L.; Wang, L.; Qi, B.; Xu, C. Building an aptamer/graphene oxide FRET biosensor for one-step detection of bisphenol A. *ACS Appl. Mater. Interfaces,* **2015**, *7*(14), 7492-7496.
        [http://dx.doi.org/10.1021/acsami.5b00199] [PMID: 25799081]

[44]   Bhatnagar, D.; Kumar, V.; Kumar, A.; Kaur, I. Graphene quantum dots FRET based sensor for early detection of heart attack in human. *Biosens. Bioelectron.,* **2016**, *79*, 495-499.
        [http://dx.doi.org/10.1016/j.bios.2015.12.083] [PMID: 26748366]

[45]   Shi, J.; Guo, J.; Bai, G.; Chan, C.; Liu, X.; Ye, W.; Hao, J.; Chen, S.; Yang, M. A graphene oxide based fluorescence resonance energy transfer (FRET) biosensor for ultrasensitive detection of botulinum neurotoxin A (BoNT/A) enzymatic activity. *Biosens. Bioelectron.,* **2015**, *65*, 238-244.
        [http://dx.doi.org/10.1016/j.bios.2014.10.050] [PMID: 25461164]

[46]   Li, M.; Chen, T.; Gooding, J.J.; Liu, J. Review of carbon and graphene quantum dots for sensing. *ACS Sens.,* **2019**, *4*(7), 1732-1748.
        [http://dx.doi.org/10.1021/acssensors.9b00514] [PMID: 31267734]

[47]   Kalkal, A.; Allawadhi, P.; Pradhan, R.; Khurana, A.; Bharani, K.K.; Packirisamy, G. *Allium sativum* derived carbon dots as a potential theranostic agent to combat the COVID-19 crisis. *Sensors International,* **2021**, *2*, 100102.
        [http://dx.doi.org/10.1016/j.sintl.2021.100102] [PMID: 34766058]

[48]   Pan, D.; Zhang, J.; Li, Z.; Wu, M. Hydrothermal Route for Cutting Graphene Sheets into Blue-Luminescent Graphene Quantum Dots. *Adv Mats,* **2010**, *22*(6), 734-738.

[49] Lim, S.Y.; Shen, W.; Gao, Z. Carbon quantum dots and their applications. *Chem. Soc. Rev.,* **2015**, *44*(1), 362-381.
[http://dx.doi.org/10.1039/C4CS00269E] [PMID: 25316556]

[50] Sachdev, A.; Gopinath, P. Green synthesis of multifunctional carbon dots from coriander leaves and their potential application as antioxidants, sensors and bioimaging agents. *Analyst (Lond.),* **2015**, *140*(12), 4260-4269.
[http://dx.doi.org/10.1039/C5AN00454C] [PMID: 25927267]

[51] Xu, X.; Ray, R.; Gu, Y.; Ploehn, H.J.; Gearheart, L.; Raker, K.; Scrivens, W.A. Electrophoretic analysis and purification of fluorescent single-walled carbon nanotube fragments. *J. Am. Chem. Soc.,* **2004**, *126*(40), 12736-12737.
[http://dx.doi.org/10.1021/ja040082h] [PMID: 15469243]

[52] Kalkal, A.; Kadian, S.; Pradhan, R.; Manik, G.; Gopinath, P. Recent advances in graphene quantum dots based optical and electrochemical (bio)analytical sensors. In: *Mater. Adv*; , **2021**; 2, pp. 5541-5513.(7)
[http://dx.doi.org/10.1039/D1MA00251A]

[53] Kadian, S.; Sethi, S.K.; Manik, G. Recent advancements in synthesis and property control of graphene quantum dots for biomedical and optoelectronic applications. *Mater. Chem. Front.,* **2021**, *5*(2), 627-658.
[http://dx.doi.org/10.1039/D0QM00550A]

[54] Kwon, W.; Kim, Y-H.; Lee, C-L.; Lee, M.; Choi, H.C.; Lee, T-W.; Rhee, S.W. Electroluminescence from graphene quantum dots prepared by amidative cutting of tattered graphite. *Nano Lett.,* **2014**, *14*(3), 1306-1311.
[http://dx.doi.org/10.1021/nl404281h] [PMID: 24490804]

[55] Kadian, S; Manik, G. A highly sensitive and selective detection of picric acid using fluorescent sulfur-doped graphene quantum dots. *Luminescence,* **2020**, , 772-763.(5)*35*
[http://dx.doi.org/10.1002/bio.3782]

[56] Kadian, S.; Manik, G. Sulfur doped graphene quantum dots as a potential sensitive fluorescent probe for the detection of quercetin. *Food Chem.,* **2020**, *317*, 126457.
[http://dx.doi.org/10.1016/j.foodchem.2020.126457] [PMID: 32106009]

[57] Liu, M.; Xu, Y.; Niu, F.; Gooding, J.J.; Liu, J. Carbon quantum dots directly generated from electrochemical oxidation of graphite electrodes in alkaline alcohols and the applications for specific ferric ion detection and cell imaging. *Analyst (Lond.),* **2016**, *141*(9), 2657-2664.
[http://dx.doi.org/10.1039/C5AN02231B] [PMID: 26878217]

[58] Shi, J.; Chan, C.; Pang, Y.; Ye, W.; Tian, F.; Lyu, J.; Zhang, Y.; Yang, M. A fluorescence resonance energy transfer (FRET) biosensor based on graphene quantum dots (GQDs) and gold nanoparticles (AuNPs) for the detection of mecA gene sequence of Staphylococcus aureus. *Biosens. Bioelectron.,* **2015**, *67*, 595-600.
[http://dx.doi.org/10.1016/j.bios.2014.09.059] [PMID: 25288044]

[59] Masteri-Farahani, M.; Ghorbani, F.; Mosleh, N. Boric acid modified S and N co-doped graphene quantum dots as simple and inexpensive turn-on fluorescent nanosensor for quantification of glucose. *Spectrochim. Acta A Mol. Biomol. Spectrosc.,* **2021**, *245*, 118892.
[http://dx.doi.org/10.1016/j.saa.2020.118892] [PMID: 32916423]

[60] Poon, C-Y.; Li, Q.; Zhang, J.; Li, Z.; Dong, C.; Lee, A.W-M.; Chan, W.H.; Li, H.W. FRET-based modified graphene quantum dots for direct trypsin quantification in urine. *Anal. Chim. Acta,* **2016**, *917*, 64-70.
[http://dx.doi.org/10.1016/j.aca.2016.02.032] [PMID: 27026601]

[61] Sun, J.; Cui, F.; Zhang, R.; Gao, Z.; Ji, J.; Ren, Y.; Pi, F.; Zhang, Y.; Sun, X. Comet-like heterodimers "Gold Nanoflower @Graphene Quantum Dots" Probe with FRET "Off" to DNA circuit signal "On" for sensing and imaging microRNA *In Vitro* and *In Vivo. Anal. Chem.,* **2018**, *90*(19), 11538-11547.

[http://dx.doi.org/10.1021/acs.analchem.8b02854] [PMID: 30182713]

[62] Yan, X.; Zhao, X-E.; Sun, J.; Zhu, S.; Lei, C.; Li, R.; Gong, P.; Ling, B.; Wang, R.; Wang, H. Probing glutathione reductase activity with graphene quantum dots and gold nanoparticles system. *Sens. Actuators B Chem.,* **2018**, *263*, 27-35.
[http://dx.doi.org/10.1016/j.snb.2018.02.096]

[63] Liu, H.; Na, W.; Liu, Z.; Chen, X.; Su, X. A novel turn-on fluorescent strategy for sensing ascorbic acid using graphene quantum dots as fluorescent probe. *Biosens. Bioelectron.,* **2017**, *92*, 229-233.
[http://dx.doi.org/10.1016/j.bios.2017.02.005] [PMID: 28222367]

[64] Ahmed, S.R.; Kumar, S.; Ortega, G.A.; Srinivasan, S.; Rajabzadeh, A.R. Target specific aptamer-induced self-assembly of fluorescent graphene quantum dots on palladium nanoparticles for sensitive detection of tetracycline in raw milk. *Food Chem.,* **2021**, *346*, 128893.
[http://dx.doi.org/10.1016/j.foodchem.2020.128893] [PMID: 33387835]

[65] Sahub, C.; Tuntulani, T.; Nhujak, T.; Tomapatanaget, B. Effective biosensor based on graphene quantum dots *via* enzymatic reaction for directly photoluminescence detection of organophosphate pesticide. *Sens. Actuators B Chem.,* **2018**, *258*, 88-97.
[http://dx.doi.org/10.1016/j.snb.2017.11.072]

[66] Mohammad-Rezei, R.; Razmi, H. Preparation and characterization of hemoglobin immobilized on graphene quantum dots-chitosan nanocomposite as a sensitive and stable hydrogen peroxide biosensor. *Sens. Lett.,* **2016**, *14*(7), 685-691.
[http://dx.doi.org/10.1166/sl.2016.3691]

[67] Ganganboina, A.B.; Doong, R-A. Graphene quantum dots decorated gold-polyaniline nanowire for impedimetric detection of carcinoembryonic antigen. *Sci. Rep.,* **2019**, *9*(1), 7214.
[http://dx.doi.org/10.1038/s41598-019-43740-3] [PMID: 31076624]

[68] Huang, Q.; Lin, X.; Tong, L.; Tong, Q-X. Graphene quantum dots/multiwalled carbon nanotubes composite-based electrochemical sensor for detecting dopamine release from living cells. *ACS Sustain. Chem.& Eng.,* **2020**, *8*(3), 1644-1650.
[http://dx.doi.org/10.1021/acssuschemeng.9b06623]

[69] Bhardwaj, H.; Marquette, C.A.; Dutta, P.; Rajesh, ; Sumana, G. Integrated graphene quantum dot decorated functionalized nanosheet biosensor for mycotoxin detection. *Anal. Bioanal. Chem.,* **2020**, *412*(25), 7029-7041.
[http://dx.doi.org/10.1007/s00216-020-02840-0] [PMID: 32797305]

[70] Hu, T.; Zhang, L.; Wen, W.; Zhang, X.; Wang, S. Enzyme catalytic amplification of miRNA-155 detection with graphene quantum dot-based electrochemical biosensor. *Biosens. Bioelectron.,* **2016**, *77*, 451-456.
[http://dx.doi.org/10.1016/j.bios.2015.09.068] [PMID: 26453906]

[71] Tuteja, S.K.; Chen, R.; Kukkar, M.; Song, C.K.; Mutreja, R.; Singh, S.; Paul, A.K.; Lee, H.; Kim, K.H.; Deep, A.; Suri, C.R. A label-free electrochemical immunosensor for the detection of cardiac marker using graphene quantum dots (GQDs). *Biosens. Bioelectron.,* **2016**, *86*, 548-556.
[http://dx.doi.org/10.1016/j.bios.2016.07.052] [PMID: 27448545]

[72] Vasilescu, I.; Eremia, S.A.V.; Kusko, M.; Radoi, A.; Vasile, E.; Radu, G-L. Molybdenum disulphide and graphene quantum dots as electrode modifiers for laccase biosensor. *Biosens. Bioelectron.,* **2016**, *75*, 232-237.
[http://dx.doi.org/10.1016/j.bios.2015.08.051] [PMID: 26319166]

[73] Lu, L.; Zhou, L.; Chen, J.; Yan, F.; Liu, J.; Dong, X.; Xi, F.; Chen, P. Nanochannel-confined graphene quantum dots for ultrasensitive electrochemical analysis of complex samples. *ACS Nano,* **2018**, *12*(12), 12673-12681.
[http://dx.doi.org/10.1021/acsnano.8b07564] [PMID: 30485066]

[74] Shahdost-Fard, F.; Roushani, M. Designing an ultra-sensitive aptasensor based on an AgNPs/thiol-GQD nanocomposite for TNT detection at femtomolar levels using the electrochemical oxidation of

Rutin as a redox probe. *Biosens. Bioelectron.,* **2017**, *87*, 724-731.
[http://dx.doi.org/10.1016/j.bios.2016.09.048] [PMID: 27649328]

[75]   Cai, J.; Sun, B.; Gou, X.; Gou, Y.; Li, W.; Hu, F. A novel way for analysis of calycosin *via* polyaniline functionalized graphene quantum dots fabricated electrochemical sensor. *J. Electroanal. Chem. (Lausanne),* **2018**, *816*, 123-131.
[http://dx.doi.org/10.1016/j.jelechem.2018.03.035]

[76]   Özcan, N.; Karaman, C.; Atar, N.; Karaman, O.; Yola, M.L. A novel molecularly imprinting biosensor including graphene quantum dots/multi-walled carbon nanotubes composite for interleukin-6 detection and electrochemical biosensor validation. *ECS J. Solid State Sci. Technol.,* **2020**, *9*(12), 121010.
[http://dx.doi.org/10.1149/2162-8777/abd149]

[77]   Ghanbari, K.; Roushani, M.; Azadbakht, A. Ultra-sensitive aptasensor based on a GQD nanocomposite for detection of hepatitis C virus core antigen. *Anal. Biochem.,* **2017**, *534*, 64-69.
[http://dx.doi.org/10.1016/j.ab.2017.07.016] [PMID: 28728900]

CHAPTER 3

Carbon Nano-Onions: Synthesis, Properties and Electrochemical Applications

Jeotikanta Mohapatra[1,*], **Arijit Mitra**[2] and **J. Ping Liu**[1]

[1] Department of Physics, University of Texas at Arlington, Arlington, TX 76019, USA

[2] Institute of Physics, Sachivalaya Marg, Bhubaneswar 751005, India

Abstract: Carbon nano-onions (CNOs) or multilayered fullerenes have received considerable attention in diversified research areas such as supercapacitors, fuel cells, batteries, photovoltaics, and biosensors due to their unique physicochemical, optical, catalytic, and electronic properties. These structures were first observed in 1992, and ever since, a considerable amount of research on their physical properties and development of CNOs based supercapacitors and sensors has been successfully witnessed. CNOs are prepared *via* different experimental techniques, and their structural and physical properties often rely upon the fabrication process or parameters. This chapter presents an overview of different methods that have been adapted to prepare CNOs and their novel properties with a focus on the fundamental curvature morphology effects. A comprehensive discussion on the potential applications, citing recent research, is provided. The challenges and the potential directions of CNOs-based materials with an eye to develop highly efficient and long-term stable CNOs-based energy storage devices and sensors are also addressed.

Keywords: Anode materials, Carbon nano-onions, Carbon nanomaterials, Fullerenes, Flame-assisted synthesis, Glucose sensors, Immunosensors, Lithium-ion batteries, Raman spectra of CNOs, Supercapacitor electrodes.

INTRODUCTION

Nanostructured carbon materials have been a subject of interdisciplinary research since Curl, Kroto, and Smalley discovered the fullerene 'C_{60}' in 1985 [1]. Gradually, several other carbon-based nanostructures have been found, including carbon nanotubes (CNTs) [2], carbon nanohorns [3], nanodiamonds [4], graphene [5], and graphene quantum dots (GQDs) [6]. In 1992, Ugarte [7] discovered multi-shell fullerenes composed of concentric shells of carbon atoms known as carbon nano-onions (CNOs). He noticed that an irradiation process causes *in-situ*

---

* **Corresponding author Jeotikanta Mohapatra:** Department of Physics, University of Texas at Arlington, Arlington, TX 76019, USA; E-mail: jeotikanta.mohapatra@uta.edu

Gaurav Manik and Sushanta Kumar Sahoo (Eds.)

structural modification of amorphous carbon into spherical-shaped CNOs with a diameter of less than 50 nm. In this process, an intense electron beam causes graphitization and curling of the amorphous carbon. Typically, the distance between each fullerene layer is found to be 0.335 nm, which is almost the same as the distance between two graphitic planes [7].

Recently, the CNOs have received enormous interest due to their potential industrial applications in catalysis, supercapacitors, sensors, etc [8 - 14]. Subsequently, synthesizing CNOs of the desired size with a simple and economical approach becomes the core research theme for numerous researchers. To date, several approaches for producing CNOs have been identified, including thermal annealing of nanodiamonds, arc-discharge in water between two graphite electrodes, chemical vapor deposition (CVD) and plasma radiation, etc [14 - 18]. However, most approaches are limited to solid-state strategies that necessitate relatively high energy-assisted conditions. Recently, a flame-assisted pyrolysis method has been reported to synthesize CNOs using vegetable oils and clarified butter as carbon precursors [19, 20]. Furthermore, the flame-based pyrolysis method renders CNOs in gram scale yields with well-controlled chemical, structural and morphological properties. This chapter encompasses the recent advances in the synthesis and characterization of CNOs, and their structural, physical, and chemical properties and potential applications.

PRODUCTION AND STRUCTURAL PROPERTIES OF CNOS

The most widely used method to produce CNOs is align="center"annealing nanodiamond (ND) particles at high temperatures under high vacuum conditions [21, 22]. Fig. (1) depicts a summary of the selected structural and chemical properties of CNOs derived from nanodiamonds at different annealing temperatures [23]. In brief, the graphitization of nanodiamonds began on the surface at 600 °C and eventually proceeded to the core of the NDs. The conversion of sp^3 to sp^2 occurred in the temperature range from 900 °C to 1100 °C and was followed by the graphitization of amorphous carbon. However, it has been discovered that the NDs are completely transformed into CNOs at temperatures above 1600 °C. Further increase in annealing temperature and annealing time of NDs resulted in polygonal-shaped CNOs. This research also revealed that increasing the annealing temperature of NDs from 1100 °C to 1900 °C increases the ratio of sp^3 carbon to sp^2 carbon atoms in the as-produced CNOs from 0.39 to 1. Simultaneously, the increase in the annealing temperature, particularly above 1700 °C, causes an increase in electrical conductivity up to 4 S cm^1 due to more sp^2 carbon formation [23, 24]. The carbon shells of CNOs become more graphitic with increased sp^2 carbon content, leading to an increase in electrical conductivity. Even though the CNOs have a lower conductivity than graphite and graphene, it is comparable to carbon black (1–2 S cm⁻¹) and activated carbon (0.5 S cm⁻¹) [25]. A recent study found that annealing in the argon atmosphere rather than vacuum condition results in few-layer graphene flakes between the carbon onions, which improves electrical conductivity [26].

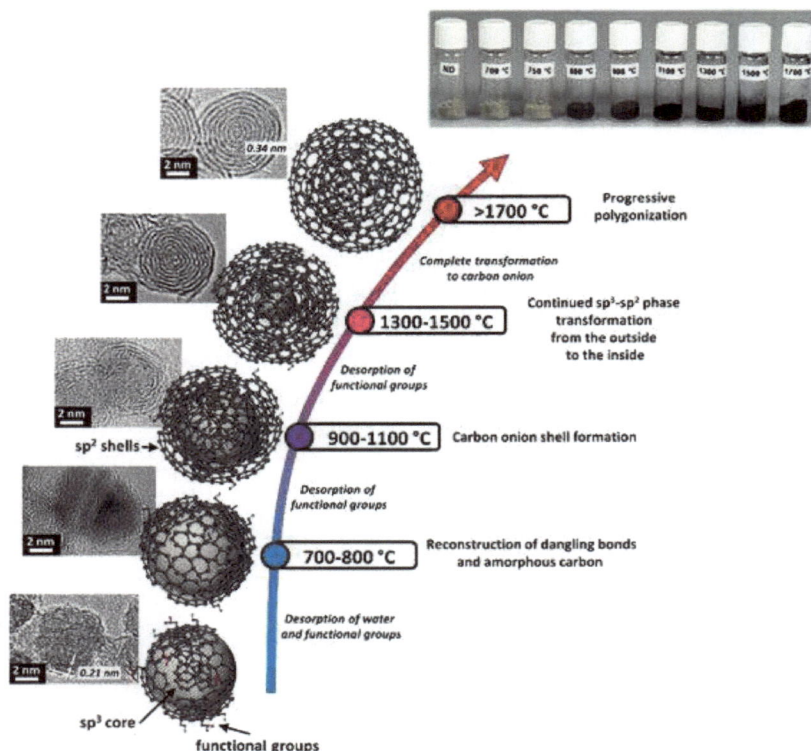

Fig. (1). Structural, and chemical properties of CNOs obtained *via* high-temperature annealing of NDs. Transmission electron micrographs (TEM), digital photographs, and schematic representation of intermediate stages of NDs depending on the annealing temperature during the transformation from NDs to CNOs [23]. *(Reprinted with or reproduced from the permission from Royal Society of Chemistry ref.23).*

CNOs are also synthesized using the arc-discharge method [27], which involves passing a high voltage and current through two graphite electrodes immersed in water. This process produces hollow-CNOs with 20–30 graphitic layers with particle diameters ranging from 15 to 25 nm. Chemical vapor deposition (CVD) was also deployed to synthesize spherical-shaped CNOs with diameters ranging from 5 to 50 nm [28, 29]. Furthermore, CNOs were efficiently and cost-effectively prepared by the pyrolysis of propane [30] and plastic wastes [31].

Recently, the CNOs have been produced *via* a flame-assisted pyrolysis of vegetable oils and clarified butter [20]. The flame naturally creates a high-temperature environment, required for CNO growth. This flame-assisted pyrolysis is a scalable and continuous-flow process with a much higher potential of producing high-purity CNOs at a significantly lower cost than the high-temperature annealing method discussed above. Recently, a facile flame-assisted synthesis approach has been reported to produce CNOs in the gram-scale by burning clarified butter or vegetable oils under ambient conditions. In this method, a polished copper foil is placed 1-3.5 cm above the flame, and the complete combustion of the oil results in the formation of a thick layer of carbon particles on the Cu foil. The role of the Cu plate and the reaction mechanism is discussed below.

Clarified butter and vegetable oils are primarily composed of hydrocarbons and aliphatic long-chain fatty acids [32]. The incomplete combustion of these fatty acids results in the formation of acetylene and polyacetylene soot on the metal substrate [33]. The metal (Cu, Ni, and Fe) plates exposed to flame get heated to high temperatures leading to thermal oxidation to form M(I) (metal in the oxidation state of +1), which further reacts with soot and forms metal acetylides and cyclic or linear chain α-carbynes (poly-ynes) in the presence of atmospheric oxygen [34]. This reaction is favorable with a copper substrate as it is known to readily form metal acetylides, whereas aluminium, nickel, and iron may get deactivated due to metal carbide formation at high-temperature conditions [35]. The cyclic α-carbynes are thermally annealed and undergo a [2+2] cycloaddition reaction followed by a cyclobutadiene tautomerisation and a Bergmann enediyne cyclisation to form a highly reactive parabenzyne diradical. The benzyne diradical undergoes further radical cyclisation followed by a retro [2+2] cycloaddition and forms the fullerene ring [36]. Once the fullerene rings are formed, these fullerene rings act as the seed material for the nucleation and growth of larger fullerenes. Gradually, smaller fullerenes are encapsulated in the larger fullerenes which ultimately leads to the formation of CNOs. The catalyzed mechanism may incorporate additional steps in certain reaction conditions, as indicated in Fig. (2).

The crystalline structure, size, and morphology of CNOs can be manipulated by varying the height of the copper substrate from the lamp. Fig. (3) shows TEM micrographs of CNOs synthesized by the combustion of various natural oils. The prepared CNOs are crystalline with concentric spherical graphite shells (see Fig. 3f) and are not mixed with other carbon materials or unreacted precursor materials; thus, no additional post-synthesis purification is required.

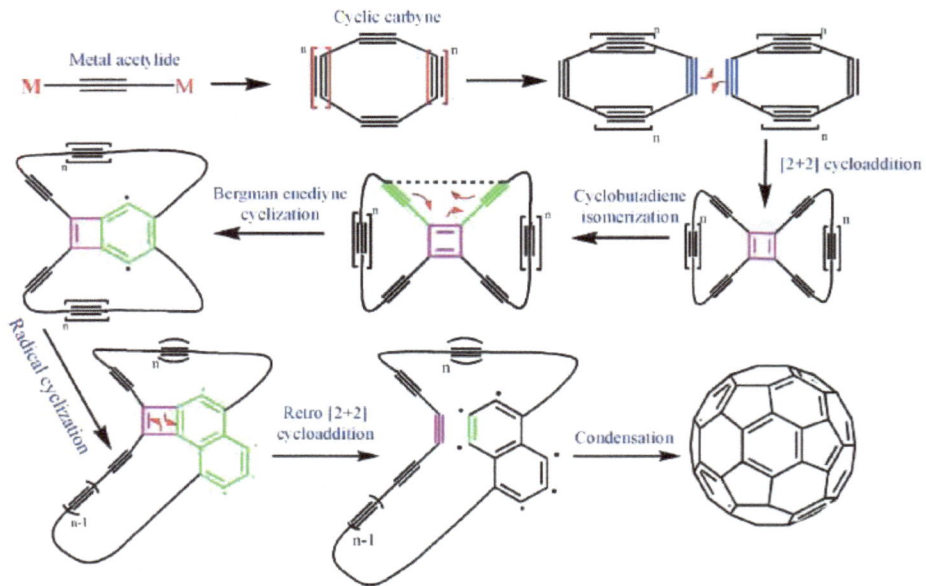

Fig. (2). Mechanism for the growth of CNOs from acetylenic and poly acetylenic soot, formed during combustion of vegetable oils on a metal substrate through [2+2] cycloaddition and radical cyclization reaction [37].

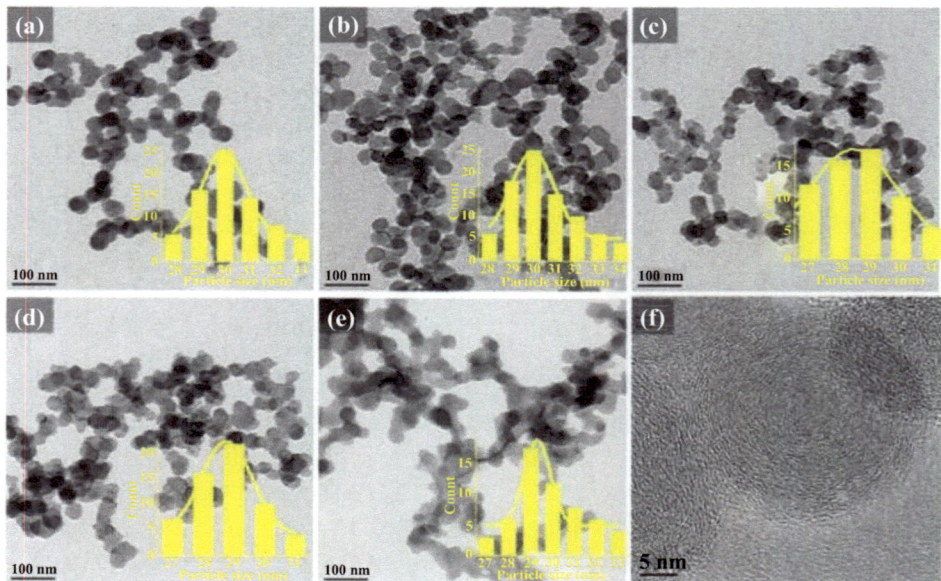

Fig. (3). TEM micrographs of CNOs produced from the combustion of different oils: (**a**) clarified butter, (**b**) mustard oil, (**c**) sunflower oil, (**d**) paraffin and (**e**) kerosene, where the Cu plate is placed at 2.5 cm above the lamp (**f**) HRTEM image of CNO displays an onion-like structure with an interlayer spacing of ~0.35 nm [20]. *(Reprinted with permission from ref.20).*

Raman spectroscopy reveals the chemical and crystalline structures of CNOs as vibrational frequencies are specific to the chemical bonds and symmetry of carbon nanomaterials. For instance, the Raman spectrum of the bulk diamond usually shows a single prominent Raman peak in the vicinity of 1332 cm^{-1}, which is a characteristic peak of the sp^3 structured diamond [38]. However, in case of nanodiamonds, this peak shifts to a higher frequency due to quantum confinement effects. At around 1603 cm^{-1}, an additional Raman wide peak related to the surface defects is also found primarily due to the sp^2 graphitate shell on the NDs. When nanodiamonds are annealed for 1 hour at a temperature of 1150 °C, a graphitic (G) mode around 1587 cm$^{-1,}$ which corresponds to the phonon modes of E_{2g} symmetry in graphite, termed as a tangential mode for the carbon nano-onions, appears in the Raman spectra [39]. The broadening of the G-band and its position are strongly related to the degree of crystallinity of the CNOs obtained through the annealing process. Another peak at around 1338 cm^{-1}, known as D-band has also been observed. The ratio of the intensities of the D and G peaks provides information on the degree of crystallinity and chemical structure of CNOs [40, 41]. A weak D-band and a low I_D / I_G ratio value reveal that the CNOs contained a very small amount of amorphous carbon, implying the formation of high-purity CNOs. Fig. (**4a**) shows the Raman spectra of the CNOs synthesized *via* flame-assisted pyrolysis of different oils. The CNOs show two broad Raman peaks of the disordered carbon, D peak at around 1340 cm^{-1} and the graphitic carbon, G peak centered at 1578 cm^{-1} [42]. In contrast to the CNOs obtained by thermal annealing of NDs, the CNOs derived from the combustion of oils showed a shift in the G peak to the lower wavenumber. The redshift of the G peak is related to the curvature-like structure of CNOs, which produces tensile strain in the graphite planes [42]. CNOs have a large I_D/I_G ratio (~0.8) and a broadening in Raman peaks owing to the presence of sp^3/sp^2 fraction of their graphitic planes and relatively poor degree of crystallinity [42, 43].

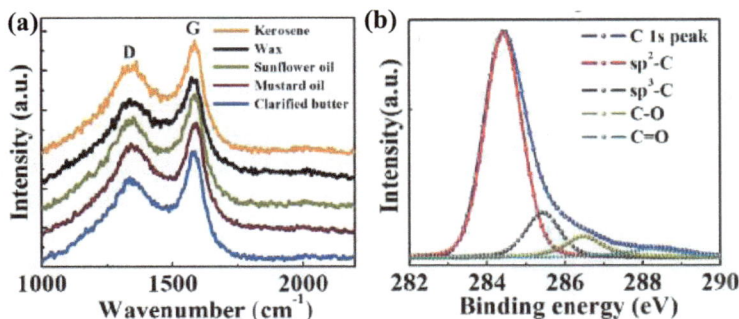

Fig. (**4**). (**a**) Raman spectra of CNOs obtained by combustion of various natural and vegetable oils. (**b**) C1s XPS spectrum and the deconvoluted spectra for CNOs derived from combustion of clarified butter [20]. *(Reprinted with permission from ref.20).*

X-ray photoelectron spectroscopy (XPS) is used to investigate the chemical composition and electronic structures of CNOs. Fig. (**4b**) shows the deconvoluted C-1s spectrum of CNOs synthesized *via* the combustion of clarified butter [20]. The peaks around 284.5 eV and 285.4 eV are corresponding to sp^2 graphitic carbon and sp^3-C/damaged alternant hydrocarbon structures, respectively. The XPS peaks at ~286.3 eV and ~288 eV are ascribed to the functional groups attached on the CNOs surface, specifically C–O and C=O groups, respectively [44, 45]. The XPS data accurately quantifies non-graphitic carbon, which includes amorphous carbon or damaged alternant hydrocarbon, and other chemical functionalities. The ratio of the total intensity of non-graphitic to that of the graphitic carbon (sp^2-C) peak in Fig. (**4b**) was found to be 0.4, which indicates that the flame assisted pyrolysis approach yields CNOs with a fractional content of sp^3 amorphous carbon [46].

Similar to the activated carbon materials used for electrochemical energy storage, CNOs also exhibit a relatively large specific surface area for catalytic activity, which is considered to be entirely related to their external surface and small size. CNOs have specific surface area values in the range of 300–600 m^2 g^{-1}, with inter-particle pore volumes exceeding 1 cm^3 g^{-1} [25, 47]. In comparison, the activated carbons typically have a specific surface area and pore volumes of approximately ~ 1500 m^2 g^{-1} and 1 cm^3 g^{-1}, receptively [47]. Importantly, the external surface area of CNOs is fully accessible for ion transport during charge and discharge in electrochemical energy storage devices. To increase the surface area of CNOs beyond 600 m^2 g^{-1}, chemical activation or physical activation can be performed. For example, KOH activation of CNOs produced by laser-assisted methane combustion resulted in higher specific surface area values above 800 m^2 g^{-1} [48].

SUPERCAPACITOR ELECTRODES BASED ON CNOS

A supercapacitor or ultracapacitor is an electrochemical capacitor in which charges accumulate at the interface of electrode–electrolyte. The electrode composition, which is typically made of nanomaterials with high specific surface areas and porosities, is the most significant component of an electrochemical capacitor. Theoretically, when the electrode surface area and electrolyte concentrations increase, the capacitance values increase. At the electrode surface/electrolyte interface, charges are stored and separated, and the corresponding electrical double-layer capacitance can be determined using the equation below [49],

$$C = \left(\frac{\varepsilon_r \varepsilon_0}{d}\right) A \qquad (1)$$

where, ε_r, ε_0, A and d are the relative permittivity, permittivity of vacuum, the specific surface area of the electrode and effective thickness of the electrical double layer, respectively.

Improved supercapacitor performance necessitates the use of appropriate electrode materials with a large conductive surface area and well-defined pore size, long electrochemical cycling stability, fast charge-discharge processes, low self-discharging and good stability at high temperatures.

Recent research have shown that CNOs are promising materials to develop high-power supercapacitors owing to their high specific surface area, which facilitates electrolyte ions to easily pass through [50]. However, the nonporous and inaccessible inner shell limits the energy density of the CNOs. The porous outer shells of CNOs raise the specific surface area significantly and expose the inner shells to electrolytes. Gao *et al.* [48] demonstrated that chemical activation of CNOs results in pores on the outer shells, which improves the capacitance of the CNOs-based supercapacitor electrodes. The capacitance of activated CNOs is found to be five times higher than that of non-modified CNOs, with an experimentally obtained power density of 153 kW/kg and an energy density of 8.5 W.h/kg in a 2 mol/L potassium nitrate electrolyte. At current densities of up to 25 A/g, activated CNOs retain approximately 71% of their initial capacitance (at 0.75 A/g). Pech *et al.* [51], fabricated supercapacitors by depositing colloidal CNO dispersion onto interdigital gold electrodes patterned on an Si substrate. The supercapacitors have powers per volume comparable to commercial electrolytic capacitors, four orders of magnitude higher capacitances, and one order of magnitude higher energies per volume.

To further improve the performance of the electrode, the CNOs are functionalized with transition metal oxides. Borgohain *et al.* [52] functionalized the CNOs with 67.5 wt. % RuO_2 loading. The obtained nanocomposites showed an energy density of 11.6 W.h.kg^{-1} at a maximum power density of 242.8 kW.kg^{-1}. Along with high specific capacitance, the nanocomposites displayed a significantly enhanced cyclic stability and power density. Fig. (**5**) shows the Ragone plot of a flexible RuO_2/CNO nanocomposite-based electrode in an acid electrolyte (0.5 M sulfuric acid, H_2SO_4) and PVA/H_2SO_4 gel electrolyte [53]. The gel electrolyte renders a maximum energy density of 10.59 W.h.kg^{-1} and power density of 4.475 kW.kg^{-1} whereas, the acid electrolyte renders the maximum energy density and power density of 19.78 W.h.kg^{-1} and 4.782 kW.kg^{-1}, respectively. Furthermore, even after 4000 cycles, the RuO_2/CNOs nanocomposite-based solid-state device demonstrated remarkable electrochemical activity with 94.47% cycling stability. Efforts have also been made in the past to produce composites of CNOs and other materials with superior electrochemical properties by combining the large pseudo-capacitance of redox-capacitive materials including MnO_2, $Ni(OH)_2$, and conducting polymers with the robustness of CNOs [53 - 55]. However, significant progress must be made in this direction to realize CNOs-based supercapacitors. The charge storage performance of nanocomposites can be improved considerably by regulating the degree of hydration and crystallinity of transition metal oxides, as well as the surface area of the composites.

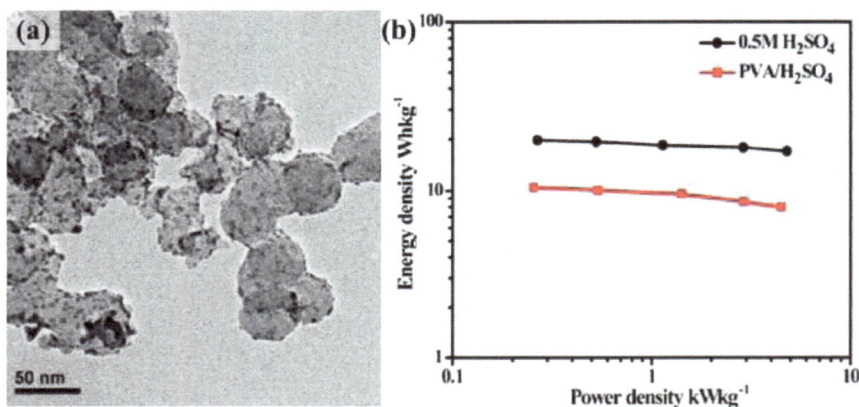

Fig. (5). (a) TEM image of RuO_2/CNO nanocomposites. **(b)** Ragone plot of the electrodes based on RuO_2/CNO nanocomposites, the experiments are conducted in an aqueous electrolyte and a solid-state device with gel electrolyte as indicated in the figure [53]. *(Reprinted with permission from ref.53)*

To improve charge storage efficiency, heteroatom doping with nitrogen, boron, or sulfur into sp^2-hybridized carbon frameworks has been adopted [56 - 58]. This allows the improvement of the electronic structures and surface chemistry of CNOs, resulting in improved capacitive performance of CNO-based electrodes. For instance, sulfur inclusion into the carbon matrix of the CNO has led to a highly polarized surface with uniformly distributed reversible pseudo-sites, which improves the overall conductivity of the electrode and electrode-electrolyte interactions [56]. In a symmetrical two-electrode cell design, the S-doped CNOs showed a specific capacitance of ~ 305 F/g, a power density of ~ 1004 W/kg and an energy density of ~ 10.6 Wh/kg at an applied current density of 2 A/g. The realized energy density is nearly three times greater than that of the pure CNO-based electrode with a similar electrochemical measurement condition.

CNOS AS ANODE MATERIAL FOR LITHIUM-ION BATTERIES

CNO-based electrodes work exceptionally well as anodes in lithium-ion batteries [59 - 61]. The benefits of using CNO-based anodes for lithium ion batteries over graphite electrodes can be described as follows: (i) CNOs provide a much higher contact area between the active material and the electrolyte due to its quasi-spherical shape and high specific surface area, which allow lithium (Li) ions to freely pass through the electrolyte from all directions. (ii) CNO has an appropriate electrical conductivity, which promotes fast transport of Li ions and electrons, (iii) The structural stability of spherical nanostructures is higher than that of sheet-like 2D materials [62, 63]. These fundamental properties and the spherical morphology of CNOs make them more stable than layered graphite and accommodate much greater volume expansion while also decreasing mechanical strain during Li-ion insertion, thus slowing the pulverization of the anode materials.

To improve the performance and stability of Li-ion batteries, Prasanna *et al.* [64] synthesized nitrogen-containing CNOs and a composite of them with Si nanoparticles. The developed CNO/Si nanocomposite-based electrodes outperformed pure Si electrodes in terms of cycling performance. In the beginning, the CNO/Si electrodes displayed decreased capacity and columbic efficiency when the charge-discharge cycles were tested between the voltage range of 0.01 V and 2 V at a current rate of 0.1 C, however, the results became stable after a few cycles at a slow Columbic rate. Furthermore, after 100 cycles at 1 C, the CNO/Si electrode produced a reversible capacity of ~852 mAh g^{-1} and an areal discharge capacity of ~2.9 mAh cm^{-2} (Fig. **6**) [64]. The presence of nitrogen functional groups in CNOs facilitates electron transport between the electrodes, which is another factor contributing to the enhanced capacity. The first cycle exhibited a high irreversible capacity due to the formation of a solid electrolyte interphase layer and the use of lithium ions. A strong interaction and a stable network between the particles and the current collector result in a higher columbic efficiency and power for the nanocomposite electrode. The nanocomposite electrode retained 62% of the capacity that was first incurred (at 0.2 C°). The presence of nitrogen in the carbon structure contributes to the formation of multiple active sites, which aid in the acquisition of additional Li-ions and result in an increased Li-ion storage capacity.

BIOSENSING CHARACTERISTICS OF CNOS

The physical and electrochemical properties of CNOs have prompted the development of highly sensitive and low-cost biosensors. Particularly, CNOs have been proven very efficient in glucose sensing [65, 66]. Before we discuss the glucose sensing characteristics of CNOs, let us just take a look at the present issues concerning glucose sensors.

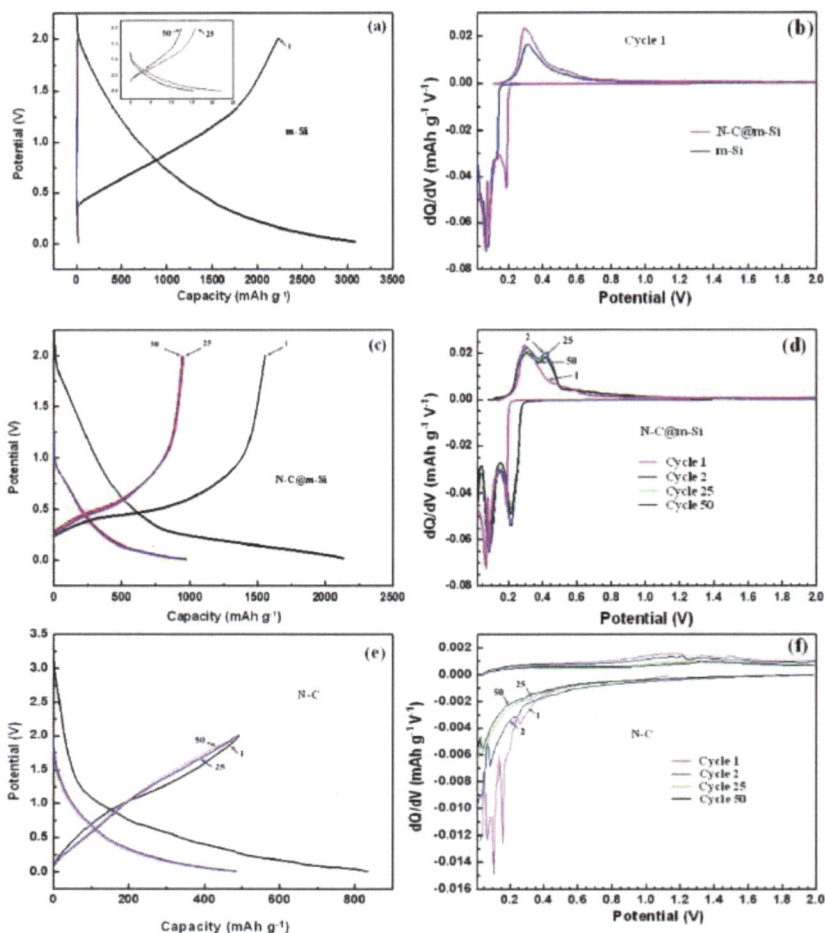

Fig. (6). Charge-discharge graphs of **(a)** CNO/Si nanocomposites; **(b)** pristine CNOs; **(c)** discharge capacity retention (%) for CNO/Si nanocomposites and CNOs electrodes at different Coulombic rates. **(d)** Columbic efficiency and specific capacity for Si and CNO/Si nanocomposites at repeated cycles, **(e)** CNOs electrodes, and **(f)** specific efficiency and areal capacity [64]. *(Reprinted with permission from ref. 64)*

Glucose detection and quantification have received a lot of attention in the last decade due to an increase in demand for inexpensive and reliable sensors, especially for use in clinical diagnosis, the food industry, and biotechnology [67]. One of the most promising approaches for developing highly sensitive sensors, which are selective, easy to operate and cost-effective, is electrochemical sensing [68]. Typical glucose sensors use the glucose oxidase enzyme (GOx) to selectively catalyze the conversion of glucose to hydrogen peroxide (H_2O_2), resulting in a sensitive and selective sensor [69, 70]. At present, a range of nanomaterials, including metals, metal oxides, and carbon, are utilized for glucose sensing due to their high specific surface area and excellent electrical conductivity

[71 - 75]. Recently, an electrochemical glucose sensor has been fabricated where CNOs are used as supports to immobilize the glucose oxidase enzyme. The sensor responds linearly over a glucose concentration of 1–10 mM with a sensitivity of 26.5 $\mu A\,mM^{-1}\,cm^{-2}$. Despite substantial advances in the development of enzymatic glucose sensors, there are still various challenges in determining their practical use. The bio-electrocatalytic activity of GOx enzyme on the sensor electrode, for example, is affected by temperature, pH, humidity, and other environmental factors. Furthermore, the sensor's efficiency is harmed by poor enzymatic stability in operating conditions. As a result, developing a glucose sensor that does not require the use of enzymes is crucial.

Many attempts have been made to improve non-enzymatic glucose sensors. Platinum (Pt) electrodes have been widely used for their strong electrical and catalytic capabilities and corrosion resistance [76 - 79]. To reduce material costs without compromising the sensor performance, it is convenient to decorate Pt on an electrically conducting support material that possesses a very high specific surface area [80 - 82]. The support materials facilitate the transport of reactants to the catalysts by increasing the surface area of the electrode for electron transfer [83, 84]. Furthermore, the conductive support helps to collect and transfer the electrons to the electrodes [84 - 87]. Graphene and carbon nanotubes (CNTs) have recently become popular as electrocatalyst supports, which result in an increased electrochemical sensing performance [86, 88]. CNOs are more electroactive than CNTs and graphene owing to their curvature morphology, smaller size, and higher surface area [89 - 91]. Being a low-cost material and having highly sensitive surfaces, particularly when they are synthesized *via* the flame-assisted pyrolysis process, the Pt nanoparticles functionalized CNOs deliver an affordable and accurate detection of glucose.

Recently, a nonenzymatic glucose sensor has been designed by functionalizing CNOs with 2.5 nm Pt nanoparticles (Fig. 7). The sensor based on Pt/CNOs displayed outstanding catalytic performance for glucose oxidation over 2–28 mM of glucose concentration using hydrodynamic amperometry at 0.45 V. The glucose sensor developed by us has a linear response with a sensitivity of 21.6 $\mu A\,mM^{-1}\,cm^{-2}$ and a minimum detection limit of 0.09 mM, indicating that it can be used to detect glucose in physiological conditions. This research successfully establishes the potential of CNOs as effective glucose sensors. Shing *et al.* combined CNOs with multi-walled carbon nanotubes (MWCNT) to create an amperometric glucose sensor. As compared to free CNOs and nanotube-based electrodes under alkaline measurement conditions, the CNO/MWCNT composites were found to be 10 and 2 times more efficient in sensing glucose. The large size of the CNOs, which reduced the surface to volume ratio for successful electrostatic attraction of glucose molecules, and its non-metallic character (which is particularly related to the CNOs synthesis process) could explain the low sensitivity in glucose sensing from the pristine CNOs electrode surface.

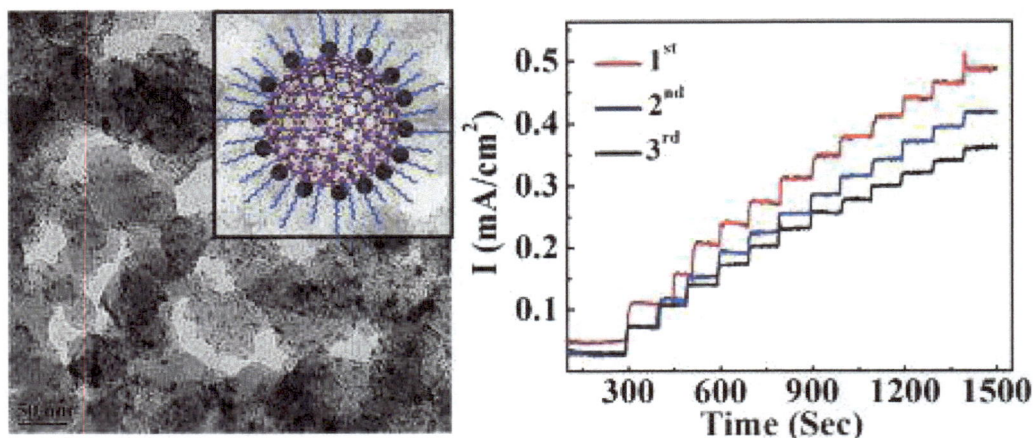

Fig. (7). (a) TEM images show the uniform coating of Pt nanoparticles on the CNOs. **(b)** Under the hydrodynamic condition, the amperometric response of Pt/CNOs electrode for glucose sensing (0-28 mM) at an applied potential 0.45 V *vs.* Ag/AgCl. The glucose sensing characteristics of the Pt/CNOs electrodes are reproducible, with a linear response for three cycles [20]. *(Reprinted with permission from ref.20).*

CNOs have been employed to electrochemically detect dopamine (DA) by interference studies in the presence of ascorbic and uric acids. Electrochemical experiments show that the theoretical detection limit and sensitivity for DA are 10 μM and 175 mA/μM^{-1}, respectively [92]. Moreover, Yang *et al.* [93] used the CNOs to identify biologically important molecules, including dopamine, epinephrine, and norepinephrine. The electrochemical performance of sensors incorporating CNOs for the detection of redox-active biomolecules demonstrated greater sensitivity, selectivity, and very stable electrode responses. In a similar approach, the CNOs-based electrode has been used to detect model DNA of the human papillomavirus oncogene [94]. The development of many low-cost immunosensors based on CNOs has been investigated; however, significant optimizations in material synthesis and sensor design are still needed to detect these biomarkers.

We believe CNOs can also be used as electrode materials for the rapid detection of SARS-CoV-2. The CNO-based electrode can be functionalized with the SARS-CoV-2 spike antibody through the cross-linking with EDC (1-Ethyl-3-(3-dimethylaminopropyl) carbodiimide) & NHS (N-Hydroxysuccinimide) (Fig. **8**) [95, 96]. As the antibody is specific to the SARS-CoV-2 spike protein, the bio-recognition events can occur at the electrode by measuring a change in the current due to interfacial electron transfer resistance caused by the biomolecular interaction or electrostatic interaction between the antibody and antigen. The same approach can be extended for the fabrication of other immunosensors, where antibodies serve as the recognition element and the sensor can convert the antibody–antigen binding event into a measurable electrical signal. Then, the developed sensor can be integrated with a microfluidic chip for continuous flow of analytes to the electrodes working area for continuous monitoring of biomarkers.

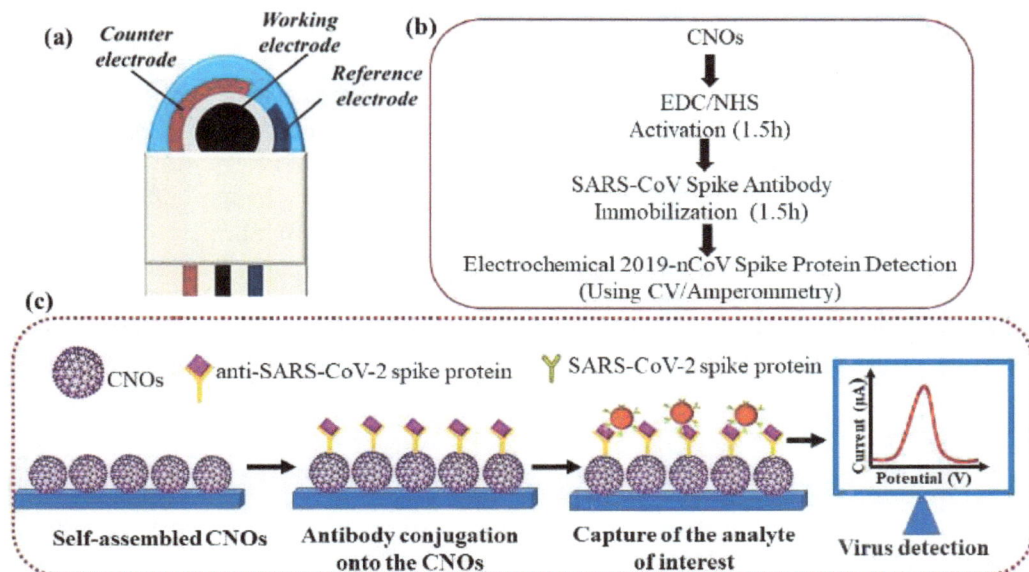

Fig. (8). Analytical principle and fabrication process of an electrochemical biosensor. **(a)** Schematic representation of amperometric glucose sensor. **(b)** Presents the steps involved in the electrode fabrication for 2019-nCoV Spike protein detection. After each step electrode can be cleaned with DI water and enzyme-modified electrode can be stored at 4 °C before use. Carbodiimide (EDC) and N-hydroxysuccinimide (NHS) activation steps can be found in reference [95, 96]. **(c)** Stepwise construction of CNOs-based electrochemical immunosensor.

CONCLUSION AND FUTURE OUTLOOK

In this chapter, the, methods for high yield CNOs synthesis, including both conventional and newly developed flame-assisted pyrolysis techniques have been described. The effects of different process parameters on the structural properties of CNOs are comprehensively discussed. The synthesized CNOs and CNO-based nanocomposites have been used in a wide variety of technological applications, such as anode materials in lithium-ion batteries, electrode materials in biosensors and electrode materials in capacitors, owing to their curvature morphology, high specific surface area, and novel physical and chemical properties. Despite numerous attempts,the large-scale synthesis of CNOs with high purity, regulated size and structure align="center"remain a challenging task. Therefore, further research in optimizing the properties of CNOs for the development of clean energy and electrochemical immunosensors is essential.

CNO-based hybrid materials have outperformed expensive noble metal-based catalysts and traditional capacitors in terms of commercial potential. However, the supercapacitors and Li-ion batteries based on CNO electrode materials have a lot of room for improvement in terms of overall performance, scalability, low-cost production, and long-term stability.

Further, CNO-based sensors have received significant attention in healthcare applications. Therefore, it is critical to examine the effects of CNOs on human health, such as biocompatibility, toxicity, and potential environmental risks before CNOs are integrated with human skin, especially when transplanted into the human body. Also, mechanical durability and long-term stability need to be assessed for the fabrication of flexible electrochemical immunosensors.

CONSENT FOR PUBLICATION

Not applicable.

CONFLICT OF INTERESTS

The authors declare no conflict of interest, financial or otherwise.

ACKNOWLEDGEMENTS

Declared none.

REFERENCES

[1]   Kroto, H.W.; Heath, J.R.; O'Brien, S.C.; Curl, R.F.; Smalley, R.E. C60: Buckminsterfullerene. *Nature,* **1985**, *318*(6042), 162-163.
        [http://dx.doi.org/10.1038/318162a0]

[2]   Iijima, S. Helical microtubules of graphitic carbon. *Nature,* **1991**, *354*(6348), 56-58.
        [http://dx.doi.org/10.1038/354056a0]

[3]   Iijima, S.; Yudasaka, M.; Yamada, R.; Bandow, S.; Suenaga, K.; Kokai, F.; Takahashi, K. Nano-aggregates of single-walled graphitic carbon nano-horns. *Chem. Phys. Lett.,* **1999**, *309*(3), 165-170.
        [http://dx.doi.org/10.1016/S0009-2614(99)00642-9]

[4]   Danilenko, V.V. On the history of the discovery of nanodiamond synthesis. *Phys. Solid State,* **2004**, *46*(4), 595-599.
        [http://dx.doi.org/10.1134/1.1711431]

[5]   Novoselov, K.S.; Geim, A.K.; Morozov, S.V.; Jiang, D.; Zhang, Y.; Dubonos, S.V.; Grigorieva, I.V.; Firsov, A.A. Electric field effect in atomically thin carbon films. *Science,* **2004**, *306*(5696), 666-669.
        [http://dx.doi.org/10.1126/science.1102896] [PMID: 15499015]

[6]   Kalita, H.; Mohapatra, J.; Pradhan, L.; Mitra, A.; Bahadur, D.; Aslam, M. Efficient synthesis of rice

based graphene quantum dots and their fluorescent properties. *RSC Advances,* **2016**, *6*(28), 23518-23524.
[http://dx.doi.org/10.1039/C5RA25706A]

[7]   Ugarte, D. Curling and closure of graphitic networks under electron-beam irradiation. *Nature,* **1992**, *359*(6397), 707-709.
[http://dx.doi.org/10.1038/359707a0] [PMID: 11536508]

[8]   Henstridge, M. C.; Shao, L.; Wildgoose, G. G.; Compton, R. G.; Tobias, G.; Green, M. L. H. The Electrocatalytic Properties of Arc-MWCNTs and Associated 'Carbon Onions'. *Electroanalysis,* *498*(5), 20.**2008**,

[9]   Sano, N.; Wang, H.; Alexandrou, I.; Chhowalla, M.; Teo, K.B.K.; Amaratunga, G.A.J.; Iimura, K. Properties of carbon onions produced by an arc discharge in water. *J. Appl. Phys.,* **2002**, *92*(5), 2783-2788.
[http://dx.doi.org/10.1063/1.1498884]

[10]   Wang, M-S.; Golberg, D.; Bando, Y. Carbon "onions" as point electron sources. *ACS Nano,* **2010**, *4*(8), 4396-4402.
[http://dx.doi.org/10.1021/nn1013353] [PMID: 20731425]

[11]   Jin, H.; Wu, S.; Li, T.; Bai, Y.; Wang, X.; Zhang, H.; Xu, H.; Kong, C.; Wang, H. Synthesis of porous carbon nano-onions derived from rice husk for high-performance supercapacitors. *Appl. Surf. Sci.,* **2019**, *488*, 593-599.
[http://dx.doi.org/10.1016/j.apsusc.2019.05.308]

[12]   Mykhailiv, O.; Zubyk, H.; Plonska-Brzezinska, M.E. Carbon nano-onions: Unique carbon nanostructures with fascinating properties and their potential applications. *Inorg. Chim. Acta,* **2017**, *468*, 49-66.
[http://dx.doi.org/10.1016/j.ica.2017.07.021]

[13]   Candelaria, S.L.; Shao, Y.; Zhou, W.; Li, X.; Xiao, J.; Zhang, J-G.; Wang, Y.; Liu, J.; Li, J.; Cao, G. Nanostructured carbon for energy storage and conversion. *Nano Energy,* **2012**, *1*(2), 195-220.
[http://dx.doi.org/10.1016/j.nanoen.2011.11.006]

[14]   Shenderova, O.; Grishko, V.; Cunningham, G.; Moseenkov, S.; McGuire, G.; Kuznetsov, V. Onion-like carbon for terahertz electromagnetic shielding. *Diamond Related Materials,* **2008**, *17*(4–5), 462-466.
[http://dx.doi.org/10.1016/j.diamond.2007.08.023]

[15]   Hou, S-S.; Chung, D-H.; Lin, T-H. High-yield synthesis of carbon nano-onions in counterflow diffusion flames. *Carbon,* **2009**, *47*(4), 938-947.
[http://dx.doi.org/10.1016/j.carbon.2008.11.054]

[16]   Lange, H.; Sioda, M.; Huczko, A.; Zhu, Y.Q.; Kroto, H.W.; Walton, D.R.M. Nanocarbon production by arc discharge in water. *Carbon,* **2003**, *41*(8), 1617-1623.
[http://dx.doi.org/10.1016/S0008-6223(03)00111-8]

[17]   Radhakrishnan, G.; Adams, P.M.; Bernstein, L.S. Plasma characterization and room temperature growth of carbon nanotubes and nano-onions by excimer laser ablation. *Appl. Surf. Sci.,* **2007**, *253*(19), 7651-7655.
[http://dx.doi.org/10.1016/j.apsusc.2007.02.033]

[18]   Fan, J-C.; Sung, H-H.; Lin, C-R.; Lai, M-H. The production of onion-like carbon nanoparticles by heating carbon in a liquid alcohol. *J. Mater. Chem.,* **2012**, *22*(19), 9794.
[http://dx.doi.org/10.1039/c2jm13273g]

[19]   Mongwe, T.H.; Matsoso, B.J.; Mutuma, B.K.; Coville, N.J.; Maubane, M.S. Synthesis of chain-like carbon nano-onions by a flame assisted pyrolysis technique using different collecting plates. *Diamond Related Materials,* **2018**, *90*, 135-143.
[http://dx.doi.org/10.1016/j.diamond.2018.10.002]

[20]   Mohapatra, J.; Ananthoju, B.; Nair, V.; Mitra, A.; Bahadur, D.; Medhekar, N.V.; Aslam, M. Enzymatic and non-enzymatic electrochemical glucose sensor based on carbon nano-onions. *Appl. Surf. Sci.,* **2018**, *442*, 332-341.
[http://dx.doi.org/10.1016/j.apsusc.2018.02.124]

[21]   Kuznetsov, V.L.; Chuvilin, A.L.; Butenko, Y.V.; Mal'kov, I.Y.; Titov, V.M. Onion-like carbon from ultra-disperse diamond. *Chem. Phys. Lett.,* **1994**, *222*(4), 343-348.
[http://dx.doi.org/10.1016/0009-2614(94)87072-1]

[22]   Tomita, S.; Sakurai, T.; Ohta, H.; Fujii, M.; Hayashi, S. Structure and electronic properties of carbon onions. *J. Chem. Phys.,* **2001**, *114*(17), 7477-7482.
[http://dx.doi.org/10.1063/1.1360197]

[23]   Zeiger, M.; Jäckel, N.; Mochalin, V.N.; Presser, V. Review: carbon onions for electrochemical energy storage. *J. Mater. Chem. A Mater. Energy Sustain.,* **2016**, *4*(9), 3172-3196.
[http://dx.doi.org/10.1039/C5TA08295A]

[24]   Yadav, R.; Tirumali, M.; Wang, X.; Naebe, M.; Kandasubramanian, B. Polymer composite for antistatic application in aerospace. *Defence Technology,* **2020**, *16*(1), 107-118.
[http://dx.doi.org/10.1016/j.dt.2019.04.008]

[25]   Portet, C.; Yushin, G.; Gogotsi, Y. Electrochemical performance of carbon onions, nanodiamonds, carbon black and multiwalled nanotubes in electrical double layer capacitors. *Carbon,* **2007**, *45*(13), 2511-2518.
[http://dx.doi.org/10.1016/j.carbon.2007.08.024]

[26]   Zeiger, M.; Jäckel, N.; Weingarth, D.; Presser, V. Vacuum or flowing argon: What is the best synthesis atmosphere for nanodiamond-derived carbon onions for supercapacitor electrodes? *Carbon,* **2015**, *94*, 507-517.
[http://dx.doi.org/10.1016/j.carbon.2015.07.028]

[27]   Xing, G.; Jia, S.; Shi, Z. The production of carbon nano-materials by arc discharge under water or liquid nitrogen. *N. Carbon Mater.,* **2007**, *22*(4), 337-341.
[http://dx.doi.org/10.1016/S1872-5805(08)60005-0]

[28]   Chen, X.H.; Deng, F.M.; Wang, J.X.; Yang, H.S.; Wu, G.T.; Zhang, X.B.; Peng, J.C.; Li, W.Z. New method of carbon onion growth by radio-frequency plasma-enhanced chemical vapor deposition. *Chem. Phys. Lett.,* **2001**, *336*(3), 201-204.
[http://dx.doi.org/10.1016/S0009-2614(01)00085-9]

[29]   Du, J.; Zhao, R.; Zhu, Z. facile approach for synthesis and *In situ* modification of onion-like carbon with molybdenum carbide. *Physica Status Solidi,* **2011**, *208*(4), 878.

[30]   Garcia-Martin, T.; Rincon-Arevalo, P.; Campos-Martin, G. Method to obtain carbon nano-onions by pyrolisys of propane. *Cent. Eur. J. Phys.,* **2013**, *11*(11), 1548.

[31]   Sawant, S.Y.; Somani, R.S.; Panda, A.B.; Bajaj, H.C. Formation and characterization of onions shaped carbon soot from plastic wastes. *Mater. Lett.,* **2013**, *94*, 132-135.
[http://dx.doi.org/10.1016/j.matlet.2012.12.035]

[32]   Moreda, W.; Pérez-Camino, M.C.; Cert, A. Gas and liquid chromatography of hydrocarbons in edible vegetable oils. *J. Chromatogr. A,* **2001**, *936*(1-2), 159-171.
[http://dx.doi.org/10.1016/S0021-9673(01)01222-5] [PMID: 11760997]

[33]   Boubel, R. W.; Fox, D. L.; Turner, D. B.; Stern, A. C. *Fundamentals of air pollution.,* **1994**.

[34]   Cataldo, F. From dicopper acetylide to carbyne. *Polym. Int.,* **1999**, *48*(1), 15-22.
[http://dx.doi.org/10.1002/(SICI)1097-0126(199901)48:1<15::AID-PI85>3.0.CO;2-#]

[35]   Wiltner, A.; Linsmeier, C. Formation of endothermic carbides on iron and nickel. *Phys. Status Solidi, A Appl. Res.,* **2004**, *201*(5), 881-887.
[http://dx.doi.org/10.1002/pssa.200304362]

[36]   Hunter, J.M.; Fye, J.L.; Roskamp, E.J.; Jarrold, M.F. Annealing Carbon Cluster Ions: A Mechanism for Fullerene Synthesis. *J. Phys. Chem.,* **1994**, *98*(7), 1810-1818.
       [http://dx.doi.org/10.1021/j100058a015]

[37]   Esselman, B.J.; McMahon, R.J. Effects of ethynyl substitution on cyclobutadiene. *J. Phys. Chem. A,* **2012**, *116*(1), 483-490.
       [http://dx.doi.org/10.1021/jp206478q] [PMID: 22185351]

[38]   Zou, Q.; Wang, M.; Li, Y. Onion-like carbon synthesis by annealing nanodiamond at lower temperature and vacuum. *J. Exp. Nanosci.,* **2010**, *5*(5), 375-382.
       [http://dx.doi.org/10.1080/17458080903583899]

[39]   Codorniu Pujals, D.; Arias de Fuentes, O.; Desdín García, L.F.; Cazzanelli, E.; Caputi, L.S. Raman spectroscopy of polyhedral carbon nano-onions. *Appl. Phys., A Mater. Sci. Process.,* **2015**, *120*(4), 1339-1345.
       [http://dx.doi.org/10.1007/s00339-015-9315-9]

[40]   Singh, G.; Botcha, V.D.; Sutar, D.S.; Narayanam, P.K.; Talwar, S.S.; Srinivasa, R.S.; Major, S.S. Near room temperature reduction of graphene oxide Langmuir-Blodgett monolayers by hydrogen plasma. *Phys. Chem. Chem. Phys.,* **2014**, *16*(23), 11708-11718.
       [http://dx.doi.org/10.1039/c4cp00875h] [PMID: 24810932]

[41]   Swain, A.K.; Li, D.; Bahadur, D. UV-assisted production of ferromagnetic graphitic quantum dots from graphite. *Carbon,* **2013**, *57*(0), 346-356.
       [http://dx.doi.org/10.1016/j.carbon.2013.01.082]

[42]   Roy, D.; Chhowalla, M.; Wang, H.; Sano, N.; Alexandrou, I.; Clyne, T.W.; Amaratunga, G.A.J. Characterisation of carbon nano-onions using Raman spectroscopy. *Chem. Phys. Lett.,* **2003**, *373*(1–2), 52-56.
       [http://dx.doi.org/10.1016/S0009-2614(03)00523-2]

[43]   Jorio, A.; Ferreira, E.H.M.; Moutinho, M.V.O.; Stavale, F.; Achete, C.A.; Capaz, R.B. Measuring disorder in graphene with the G and D bands. *Phys. Status Solidi, B Basic Res.,* **2010**, *247*(11-12), 2980-2982.
       [http://dx.doi.org/10.1002/pssb.201000247]

[44]   Swain, A.K.; Bahadur, D. Enhanced stability of reduced graphene oxide colloid using cross-linking polymers. *J. Phys. Chem. C,* **2014**, *118*(18), 9450-9457.
       [http://dx.doi.org/10.1021/jp500205n]

[45]   Shinde, D.B.; Debgupta, J.; Kushwaha, A.; Aslam, M.; Pillai, V.K. Electrochemical unzipping of multi-walled carbon nanotubes for facile synthesis of high-quality graphene nanoribbons. *J. Am. Chem. Soc.,* **2011**, *133*(12), 4168-4171.
       [http://dx.doi.org/10.1021/ja1101739] [PMID: 21388198]

[46]   Hu, S.; Bai, P.; Tian, F.; Cao, S.; Sun, J. Hydrophilic carbon onions synthesized by millisecond pulsed laser irradiation. *Carbon,* **2009**, *47*(3), 876-883.
       [http://dx.doi.org/10.1016/j.carbon.2008.11.041]

[47]   Jäckel, N.; Weingarth, D.; Zeiger, M.; Aslan, M.; Grobelsek, I.; Presser, V. Comparison of carbon onions and carbon blacks as conductive additives for carbon supercapacitors in organic electrolytes. *J. Power Sources,* **2014**, *272*, 1122-1133.
       [http://dx.doi.org/10.1016/j.jpowsour.2014.08.090]

[48]   Gao, Y.; Zhou, Y.S.; Qian, M.; He, X.N.; Redepenning, J.; Goodman, P.; Li, H.M.; Jiang, L.; Lu, Y.F. Chemical activation of carbon nano-onions for high-rate supercapacitor electrodes. *Carbon,* **2013**, *51*, 52-58.
       [http://dx.doi.org/10.1016/j.carbon.2012.08.009]

[49]   Plonska-Brzezinska, M.E.; Echegoyen, L. Carbon nano-onions for supercapacitor electrodes: recent developments and applications. *J. Mater. Chem. A Mater. Energy Sustain.,* **2013**, *1*(44), 13703.

[http://dx.doi.org/10.1039/c3ta12628e]

[50] Anjos, D.M.; McDonough, J.K.; Perre, E.; Brown, G.M.; Overbury, S.H.; Gogotsi, Y.; Presser, V. Pseudocapacitance and performance stability of quinone-coated carbon onions. *Nano Energy,* **2013**, *2*(5), 702-712.
[http://dx.doi.org/10.1016/j.nanoen.2013.08.003]

[51] Pech, D.; Brunet, M.; Durou, H.; Huang, P.; Mochalin, V.; Gogotsi, Y.; Taberna, P-L.; Simon, P. Ultrahigh-power micrometre-sized supercapacitors based on onion-like carbon. *Nat. Nanotechnol.,* **2010**, *5*(9), 651-654.
[http://dx.doi.org/10.1038/nnano.2010.162] [PMID: 20711179]

[52] Borgohain, R.; Li, J.; Selegue, J.P.; Cheng, Y.T. Electrochemical study of functionalized carbon nano-onions for high-performance supercapacitor electrodes. *J. Phys. Chem. C,* **2012**, *116*(28), 15068-15075.
[http://dx.doi.org/10.1021/jp301642s]

[53] Muniraj, V.K.A.; Kamaja, C.K.; Shelke, M.V. $RuO_2 \cdot nH_2O$ Nanoparticles Anchored on Carbon Nano-onions: An Efficient Electrode for Solid State Flexible Electrochemical Supercapacitor. *ACS Sustain. Chem.& Eng.,* **2016**, *4*(5), 2528-2534.
[http://dx.doi.org/10.1021/acssuschemeng.5b01627]

[54] Mykhailiv, O.; Imierska, M.; Petelczyc, M.; Echegoyen, L.; Plonska-Brzezinska, M.E. Chemical versus electrochemical synthesis of carbon nano-onion/polypyrrole composites for supercapacitor electrodes. *Chemistry,* **2015**, *21*(15), 5783-5793.
[http://dx.doi.org/10.1002/chem.201406126] [PMID: 25736714]

[55] Plonska-Brzezinska, M.E.; Brus, D.M.; Molina-Ontoria, A.; Echegoyen, L. Synthesis of carbon nano-onion and nickel hydroxide/oxide composites as supercapacitor electrodes. *RSC Advances,* **2013**, *3*(48), 25891.
[http://dx.doi.org/10.1039/c3ra44249g]

[56] Mohapatra, D.; Dhakal, G.; Sayed, M.S.; Subramanya, B.; Shim, J-J.; Parida, S. Sulfur Doping: Unique strategy to improve the supercapacitive performance of carbon nano-onions. *ACS Appl. Mater. Interfaces,* **2019**, *11*(8), 8040-8050.
[http://dx.doi.org/10.1021/acsami.8b21534] [PMID: 30714716]

[57] Mohapatra, D.; Muhammad, O.; Sayed, M.S.; Parida, S.; Shim, J-J. *In situ* nitrogen-doped carbon nano-onions for ultrahigh-rate asymmetric supercapacitor. *Electrochim. Acta,* **2020**, *331*, 135363.
[http://dx.doi.org/10.1016/j.electacta.2019.135363]

[58] Singh, B.K.; Shaikh, A.; Dusane, R.O.; Parida, S. Nanoporous gold–Nitrogen–doped carbon nano-onions all-solid-state micro-supercapacitor. *Nano-Structures & Nano-Objects,* **2019**, *17*, 239-247.
[http://dx.doi.org/10.1016/j.nanoso.2019.01.011]

[59] Wang, Q.; Sun, X.; He, D.; Zhang, J. Preparation and study of carbon nano-onion for lithium storage. *Mater. Chem. Phys.,* **2013**, *139*(1), 333-337.
[http://dx.doi.org/10.1016/j.matchemphys.2013.02.002]

[60] Han, F-D.; Yao, B.; Bai, Y-J. Preparation of Carbon nano-onions and their application as anode materials for rechargeable lithium-ion batteries. *J. Phys. Chem. C,* **2011**, *115*(18), 8923-8927.
[http://dx.doi.org/10.1021/jp2007599]

[61] Meng, Y.; Wang, G.; Xiao, M.; Duan, C.; Wang, C.; Zhu, F.; Zhang, Y. Ionic liquid-derived Co_3O_4 carbon nano-onions composite and its enhanced performance as anode for lithium-ion batteries. *J. Mater. Sci.,* **2017**, *52*(22), 13192-13202.
[http://dx.doi.org/10.1007/s10853-017-1414-x]

[62] Dhand, V.; Yadav, M.; Kim, S.H.; Rhee, K.Y. A comprehensive review on the prospects of multi-functional carbon nano onions as an effective, high- performance energy storage material. *Carbon,* **2021**, *175*, 534-575.
[http://dx.doi.org/10.1016/j.carbon.2020.12.083]

[63]   Zheng, Y.; Zhu, P. Carbon nano-onions: large-scale preparation, functionalization and their application as anode material for rechargeable lithium ion batteries. *RSC Advances,* **2016**, *6*(95), 92285-92298.
[http://dx.doi.org/10.1039/C6RA19060J]

[64]   Prasanna, K.; Subburaj, T.; Jo, Y.N.; Santhoshkumar, P.; Karthikeyan, S.K.S.S.; Vediappan, K.; Gnanamuthu, R.M.; Lee, C.W. Chitosan complements entrapment of silicon inside nitrogen doped carbon to improve and stabilize the capacity of Li-ion batteries. *Sci. Rep.,* **2019**, *9*(1), 3318.
[http://dx.doi.org/10.1038/s41598-019-39988-4] [PMID: 30824812]

[65]   Ahlawat, J.; Masoudi Asil, S.; Guillama Barroso, G.; Nurunnabi, M.; Narayan, M. Application of carbon nano onions in the biomedical field: recent advances and challenges. *Biomater. Sci.,* **2021**, *9*(3), 626-644.
[http://dx.doi.org/10.1039/D0BM01476A] [PMID: 33241797]

[66]   Tripathi, K.M.; Bhati, A.; Singh, A.; Gupta, N.R.; Verma, S.; Sarkar, S.; Sonkar, S.K. From the traditional way of pyrolysis to tunable photoluminescent water soluble carbon nano-onions for cell imaging and selective sensing of glucose. *RSC Advances,* **2016**, *6*(44), 37319-37329.
[http://dx.doi.org/10.1039/C6RA04030F]

[67]   Park, S.; Boo, H.; Chung, T.D. Electrochemical non-enzymatic glucose sensors. *Anal. Chim. Acta,* **2006**, *556*(1), 46-57.
[http://dx.doi.org/10.1016/j.aca.2005.05.080] [PMID: 17723330]

[68]   Bakker, E.; Qin, Y. Electrochemical sensors. *Anal. Chem.,* **2006**, *78*(12), 3965-3984.
[http://dx.doi.org/10.1021/ac060637m] [PMID: 16771535]

[69]   Shao, Y.; Wang, J.; Wu, H.; Liu, J.; Aksay, I.A.; Lin, Y. Graphene Based Electrochemical Sensors and Biosensors: A Review. *Electroanalysis,* **2010**, *22*(10), 1027-1036.
[http://dx.doi.org/10.1002/elan.200900571]

[70]   Shan, C.; Yang, H.; Song, J.; Han, D.; Ivaska, A.; Niu, L. Direct electrochemistry of glucose oxidase and biosensing for glucose based on graphene. *Anal. Chem.,* **2009**, *81*(6), 2378-2382.
[http://dx.doi.org/10.1021/ac802193c] [PMID: 19227979]

[71]   Shahnavaz, Z.; Lorestani, F.; Alias, Y.; Woi, P.M. Polypyrrole–$ZnFe_2O_4$ magnetic nano-composite with core–shell structure for glucose sensing. *Appl. Surf. Sci.,* **2014**, *317*, 622-629.
[http://dx.doi.org/10.1016/j.apsusc.2014.08.194]

[72]   Chang, G.; Shu, H.; Ji, K.; Oyama, M.; Liu, X.; He, Y. Gold nanoparticles directly modified glassy carbon electrode for non-enzymatic detection of glucose. *Appl. Surf. Sci.,* **2014**, *288*, 524-529.
[http://dx.doi.org/10.1016/j.apsusc.2013.10.064]

[73]   Jia, H.; Chang, G.; Lei, M.; He, H.; Liu, X.; Shu, H.; Xia, T.; Su, J.; He, Y. Platinum nanoparticles decorated dendrite-like gold nanostructure on glassy carbon electrodes for enhancing electrocatalysis performance to glucose oxidation. *Appl. Surf. Sci.,* **2016**, *384*, 58-64.
[http://dx.doi.org/10.1016/j.apsusc.2016.05.020]

[74]   Bai, X.; Chen, W.; Song, Y.; Zhang, J.; Ge, R.; Wei, W.; Jiao, Z.; Sun, Y. Nickel-copper oxide nanowires for highly sensitive sensing of glucose. *Appl. Surf. Sci.,* **2017**, *420*, 927-934.
[http://dx.doi.org/10.1016/j.apsusc.2017.05.174]

[75]   Mijowska, E.; Onyszko, M.; Urbas, K.; Aleksandrzak, M.; Shi, X.; Moszyński, D.; Penkala, K.; Podolski, J.; El Fray, M. Palladium nanoparticles deposited on graphene and its electrochemical performance for glucose sensing. *Appl. Surf. Sci.,* **2015**, *355*, 587-592.
[http://dx.doi.org/10.1016/j.apsusc.2015.07.150]

[76]   Chen, A.; Holt-Hindle, P. Platinum-based nanostructured materials: synthesis, properties, and applications. *Chem. Rev.,* **2010**, *110*(6), 3767-3804.
[http://dx.doi.org/10.1021/cr9003902] [PMID: 20170127]

[77]   Subhramannia, M.; Pillai, V.K. Shape-dependent electrocatalytic activity of platinum nanostructures. *J. Mater. Chem.,* **2008**, *18*(48), 5858.

[http://dx.doi.org/10.1039/b811149a]

[78] Yuan, J.H.; Wang, K.; Xia, X.H. Highly ordered platinum-nanotubule arrays for amperometric glucose sensing. *Adv. Funct. Mater.,* **2005**, *15*(5), 803-809.
[http://dx.doi.org/10.1002/adfm.200400321]

[79] Attard, G.S.; Bartlett, P.N.; Coleman, N.R.; Elliott, J.M.; Owen, J.R.; Wang, J.H. Mesoporous platinum films from lyotropic liquid crystalline phases. *Science,* **1997**, *278*(5339), 838-840.
[http://dx.doi.org/10.1126/science.278.5339.838]

[80] Mu, Y.; Liang, H.; Hu, J.; Jiang, L.; Wan, L. Controllable pt nanoparticle deposition on carbon nanotubes as an anode catalyst for direct methanol fuel cells. *J. Phys. Chem. B,* **2005**, *109*(47), 22212-22216.
[http://dx.doi.org/10.1021/jp0555448] [PMID: 16853891]

[81] Sun, S.; Zhang, G.; Gauquelin, N.; Chen, N.; Zhou, J.; Yang, S.; Chen, W.; Meng, X.; Geng, D.; Banis, M.N.; Li, R.; Ye, S.; Knights, S.; Botton, G.A.; Sham, T-K.; Sun, X. Single-atom catalysis using Pt/graphene achieved through atomic layer deposition. *Sci. Rep.,* **2013**, *3*(1), 3.
[http://dx.doi.org/10.1038/srep01775]

[82] Choi, S.M.; Seo, M.H.; Kim, H.J.; Kim, W.B. Synthesis of surface-functionalized graphene nanosheets with high Pt-loadings and their applications to methanol electrooxidation. *Carbon,* **2011**, *49*(3), 904-909.
[http://dx.doi.org/10.1016/j.carbon.2010.10.055]

[83] Yang, S.; Dong, J.; Yao, Z.; Shen, C.; Shi, X.; Tian, Y.; Lin, S.; Zhang, X. One-pot synthesis of graphene-supported monodisperse Pd nanoparticles as catalyst for formic acid electro-oxidation. *Sci. Rep.,* **2014**, *4*(1), 4501.
[http://dx.doi.org/10.1038/srep04501] [PMID: 24675779]

[84] Zhu, C.; Dong, S. Synthesis of graphene-supported noble metal hybrid nanostructures and their applications as advanced electrocatalysts for fuel cells. *Nanoscale,* **2013**, *5*(22), 10765-10775.
[http://dx.doi.org/10.1039/c3nr03280a] [PMID: 24060985]

[85] Guo, S.; Dong, S.; Wang, E. Three-dimensional Pt-on-Pd bimetallic nanodendrites supported on graphene nanosheet: facile synthesis and used as an advanced nanoelectrocatalyst for methanol oxidation. *ACS Nano,* **2010**, *4*(1), 547-555.
[http://dx.doi.org/10.1021/nn9014483] [PMID: 20000845]

[86] Jiang, H.; Lee, P.S.; Li, C. 3D carbon based nanostructures for advanced supercapacitors. *Energy Environ. Sci.,* **2013**, *6*(1), 41-53.
[http://dx.doi.org/10.1039/C2EE23284G]

[87] Wang, C.; Ma, L.; Liao, L.; Bai, S.; Long, R.; Zuo, M.; Xiong, Y. A unique platinum-graphene hybrid structure for high activity and durability in oxygen reduction reaction. *Sci. Rep.,* **2013**, *3*(1), 2580.
[http://dx.doi.org/10.1038/srep02580] [PMID: 23999570]

[88] Onoe, T.; Iwamoto, S.; Inoue, M. Synthesis and activity of the Pt catalyst supported on CNT. *Catal. Commun.,* **2007**, *8*(4), 701-706.
[http://dx.doi.org/10.1016/j.catcom.2006.08.018]

[89] Sano, N.; Wang, H.; Chhowalla, M.; Alexandrou, I.; Amaratunga, G.A. Synthesis of carbon 'onions' in water. *Nature,* **2001**, *414*(6863), 506-507.
[http://dx.doi.org/10.1038/35107141] [PMID: 11734841]

[90] Delgado, J.L.; Herranz, M.A.; Martin, N. The nano-forms of carbon. *J. Mater. Chem.,* **2008**, *18*(13), 1417.
[http://dx.doi.org/10.1039/b717218d]

[91] Bartelmess, J.; Giordani, S. Carbon nano-onions (multi-layer fullerenes): chemistry and applications. *Beilstein J. Nanotechnol.,* **2014**, *5*(1), 1980-1998.
[http://dx.doi.org/10.3762/bjnano.5.207] [PMID: 25383308]

[92] Breczko, J.; Plonska-Brzezinska, M.E.; Echegoyen, L. Electrochemical oxidation and determination of dopamine in the presence of uric and ascorbic acids using a carbon nano-onion and poly(diallyldimethylammonium chloride) composite. *Electrochim. Acta,* **2012**, *72*, 61-67.
[http://dx.doi.org/10.1016/j.electacta.2012.03.177]

[93] Yang, J.; Zhang, Y.; Kim, D.Y. Electrochemical sensing performance of nanodiamond-derived carbon nano-onions: Comparison with multiwalled carbon nanotubes, graphite nanoflakes, and glassy carbon. *Carbon,* **2016**, *98*, 74-82.
[http://dx.doi.org/10.1016/j.carbon.2015.10.089]

[94] Bartolome, J.P.; Echegoyen, L.; Fragoso, A. Reactive carbon nano-onion modified glassy carbon surfaces as dna sensors for human papillomavirus oncogene detection with enhanced sensitivity. *Anal. Chem.,* **2015**, *87*(13), 6744-6751.
[http://dx.doi.org/10.1021/acs.analchem.5b00924] [PMID: 26067834]

[95] Gupta, J.; Mohapatra, J.; Bhargava, P.; Bahadur, D. A pH-responsive folate conjugated magnetic nanoparticle for targeted chemo-thermal therapy and MRI diagnosis. *Dalton Trans.,* **2016**, *45*(6), 2454-2461.
[http://dx.doi.org/10.1039/C5DT04135J] [PMID: 26685824]

[96] Yakoh, A.; Pimpitak, U.; Rengpipat, S.; Hirankarn, N.; Chailapakul, O.; Chaiyo, S. Paper-based electrochemical biosensor for diagnosing COVID-19: Detection of SARS-CoV-2 antibodies and antigen. *Biosens. Bioelectron.,* **2021**, *176*, 112912.
[http://dx.doi.org/10.1016/j.bios.2020.112912] [PMID: 33358057]

CHAPTER 4

Graphene Based Hybrid Nanocomposites for Solar Cells

Sachin Kadian[1,2], Manjinder Singh[1] and Gaurav Manik[1,*]

[1] *Department of Polymer and Process Engineering, Indian Institute of Technology Roorkee, Roorkee, Uttarakhand, 247667, India*

[2] *Department of Electrical and Computer Engineering, University of Alberta, Edmonton, AB T6G 1H9, Canada*

Abstract: Over the last few years, due to its exceptional two-dimensional (2D) structure, graphene has played a key role in developing conductive transparent devices and acquired significant attention from scientists to get placed as a boon material in the energy industry. Graphene-based materials have played several roles, including interfacial buffer layers, electron/hole transport material, and transparent electrodes in photovoltaic devices. Apart from charge extraction and electron transportation, graphene protects the photovoltaic devices from atmospheric degradation through its 2D network and offers long-term air or environmental stability. This chapter focuses on the recent advancements in graphene and its nanocomposites-based solar cell devices, including dye-sensitized solar cells (DSSCs), organic solar cells (OSCs), and perovskite solar cells (PSCs). We further discuss the impact of incorporating graphene-based materials on the power conversion efficiency for each type of solar cell. The last section of this chapter highlights the potential challenges and future research scope of graphene-based nanocomposites for solar cell applications.

Keywords: Bulk heterojunction, Carbon nanotubes, Dye-sensitized solar cells, Electron transport layer, Graphene, Graphene oxide, Graphene quantum dots, Hole transport layer, Organic solar cells, Open circuit voltage, Perovskite solar cells, Reduced graphene oxide.

INTRODUCTION

The rapidly increasing demand for clean energy, led by the escalation of high-speed, wearable, and portable optoelectronic devices, has accelerated the development of advanced techniques to combat unprecedented challenges.

* **Corresponding author Gaurav Manik:** Department of Polymer and Process Engineering, Indian Institute of Technology Roorkee, Roorkee, Uttarakhand, India; E-mail: gaurav.manik@pe.iitr.ac.in

For decades, sunlight has been explored as a safe, clean, and economical source of energy that could generate electricity deprived of creating any environmental issues. In this, photovoltaic materials aid the conversion of solar energy into electricity through the photovoltaic effect, and thus, a solar cell is also known as a photovoltaic cell [1]. Further, the structure and properties of these photovoltaic materials can be tailored with the help of nanotechnology to develop highly efficient devices. Therefore, several new technologies, materials, and synthetic approaches have been discovered and continue to be implemented. Recently, due to graphene's attractive characteristics, including high carrier mobility, good optical transparency, zero-band gap, and desired mechanical strength, graphitic materials have been extensively explored for solar cell applications [2]. More specifically, due to the several functional groups including carbonyl, carboxylic acid, hydroxyl, and epoxy, reduced graphene oxide (rGO) and graphene oxide (GO) have been practically employed as attractive derivatives of graphene to synthesise graphene-based nanocomposites [3, 4]. Furthermore, the existing oxygen-containing groups facilitate more active sites for further functionalization of several other molecules, organic and inorganic moieties on GO, to develop novel nanocomposites with desired properties [5, 6]. Despite these several attractive characteristics of graphene, its commercial applications for the fabrication of next-generation optoelectronic devices rely on its optical response [7, 8]. Therefore, research on the synthesis of zero-dimensional (0D) fluorescent graphene sheets (*i.e.*, graphene quantum dots) has recently gained significant attention to compile the state-of-the-art involvement of graphene-based materials in the development of advanced photovoltaics and other optoelectronic devices [9, 10].

Herein, we discuss the progress of graphene and its nanocomposites in solar cells such as dye-sensitized solar cells (DSSCs), organic solar cells (OSCs), and perovskite solar cells (PSCs) amidst their cell configuration, working principles, and challenges of individual device. Next, the performance of advanced solar cell devices prepared after incorporating the graphene-based nanocomposites as active layers, electrodes, electrons, and hole transport layers is demonstrated and compared with the performance of commercially available conventional solar cell devices. Also, the role of graphene-based nanocomposites in solar cell devices is emphasized by comparing the most compelling outcomes. Decisive insights into the existing problems and future scopes of graphene and its nanocomposites in photovoltaic applications are discussed in the last section.

APPLICATION OF GRAPHENE AND ITS NANOCOMPOSITES IN OSCS

In pursuit of developing novel solar cell devices, this section will systematically discuss the significance of graphene and its nanocomposites as an electrode, electron-transport layer (ETL), and hole-transport layer (HTL) in OSCs devices.

Graphene-based Materials as Electrode in OSCs

The materials used as OSCs' electrodes are needed to be highly conductive and transparent. Thus, Fluorine tin oxide (FTO) and Indium tin oxide (ITO) have widely been used as electrodes in commercially available OSC devices. The transmittance and resistance of ITO are 80% and 10-15 Ω/sq, respectively, on a glass substrate, while on the polyethylene terephthalate substrates, the resistance varies from 60 to 300 Ω/sq [11]. Nevertheless, ITO has several limitations such as high sensitivity towards acidic and basic media, difficulty in patterning, high manufacturing cost due to inadequacy of indium, and its brittle nature. These issues signify that the solar cell market is still looking for a promising material having high conductivity, acceptable transparency, desired flexibility, and good chemical stability. Metal nanowires, metallic grids, oxides, and carbon nanotubes have been studied as a suitable substitute to replace the transnationally used electrodes [12 - 15]. However, because of their poor transparency, the power conversion efficiency of resulting solar cell device align="center"was not enhanced, and thus, all these materials were unable to replace the ITO. Recently, graphene and its nanocomposites have been explored in OSCs as new electrode material [16 - 19]. Park and colleagues reported that the OSCs prepared from anode and cathode electrodes of graphene exhibited 6.1% and 7.1% power conversion efficiencies (PCE), respectively [20]. Wang *et al.* reported that graphene prepared from chemical vapor deposition (CVD) exhibits higher conductance and transparency than Hummer's method [21]. The higher conductance and transparency align="center"are ascribed to the presence of sp^3-carbons and structural defects in the graphene synthesized *via* Hummer's method, which affects the π-bond connectivity and, thus, its electron transfer capability. In contrast, graphene sheets prepared from the CVD method demonstrate high conductivity with large and controlled sheet align="center"size, resulting in comparatively higher performance of OSCs. In general, the performance of OSC devices mainly depends on the transmittance, conductance, and thickness of thin films of graphene. Further, Gómez-Navarro *et al.* reported that the morphological defects present in GO and rGO sheets help in light transmittance. Hence, GO based electrodes display higher transmittance than those of CVD graphene [22]. Later on, Liu *et al.* prepared a four-layer CVD graphene with 90% transmittance and used the same in fabricating a flexible photovoltaic device as a top transparent electrode [23]. Because of the slow air diffusion through layered graphene, the as-prepared device exhibited slower degradation of organic materials with high stability and thereby demonstrated that a few layers of graphene are beneficial over a single-layer graphene.

Further, as we know that resistance and transmittance are associated with the number of layers and optical conductivities, an optimum balance among layers, conductivity, and transparency is needed for an efficient electrode [26]. The performance with a single and a few-layered graphene and its nanocomposites-based electrodes are listed in Table **1**. As shown in the comparison table and discussed earlier, the performance of a photovoltaic device is strongly associated with the balance among resistance and transmittance of the electrode. It has also been reported that the strain existing in multi-layer graphene persuades a competition between interlayer transport and coupling, which helps in creating a mini-bandgap and tuning of electronic properties of graphene [27, 28].

Table 1. Comparison of resistance (R_s), transmittance (T), and solar cell parameters (PCE, fill factor (*FF*), and short-circuit current density (J_{sc}) for OSCs developed using graphene-based electrodes.

Device Configuration	Graphene Electrodes R_s and T		Solar Cell Parameters			Refs.
	R_s (Ω/sq)	T (%)	PCE (%)	*FF* (%)	J_{sc} (mA/cm^2)	
Graphene/ZnO/PTB7:PC71BM/MoO$_3$/Ag	300	92	7.1	69.5	14.1	[20]
Graphene/PEDOT:PSS/P3HT:PC61BM/LiF/Al	80	90	3.0	57.0	9.2	[21]
Graphene/PEDOT:PSS/P3HT:PCBM/Ag	276	97	2.48	41.9	10.26	[23]
(Single-layer) graphene/Au/PEDOT:PSS/P3HT:PCBM/Ag	179	97	2.91	45.7	10.78	[23]
(Double-layer) graphene/Au/PEDOT:PSS/P3HT:PCBM/Ag	92	95	3.17	50.0	10.61	[23]
(Triple-layer) graphene/Au/PEDOT:PSS/P3HT:PCBM/Ag	75	92	3.08	49.4	10.38	[23]
(4-layer) graphene/Au/PEDOT:PSS/P3HT:PCBM/Ag	68	90	2.8	49.7	9.88	[23]
PET/rGO/PEDOT:PSS/P3HT:PCBM/TiO$_2$/Al	3.2k	65	0.78	32	4.39	[24]
Graphene/PEDOT:PSS/P3HT:PCBM/ TiO$_x$/Al	610	87	2.58	0.48	9.03	[25]

Graphene-based Materials as Hole and Electron Transport Layer in OSCs

In the OSCs, a nanoscale composite of electron acceptor-donor known as bulk heterojunction (BHJ) absorbs photons and offers an easy path for electrons towards the electrodes for effective carrier collection [29 - 31]. The interaction among donor-acceptor materials and electrodes causes recombination due to the current leakage. Therefore, to assist electron extraction and collection without current leakage, additional efficient buffer layers are needed at the interaction point of the electrode and photoactive coating. These interfacial/buffer layers are selective to permit only one type of carrier (hole or electron) and restrict another one (electron or hole). Generally, the HTL is a p-type wide bandgap component that creates an Ohmic connection with donor polymer and tunes the work function to chunk the charge carrier. It is essential for ETL and HTL applications to generate n-type and p-type graphene, respectively. For instance, B-, S-, Au- and Pt-doped graphene are p-type, whereas Al-, Cl-, N-, and Cu-doped are n-type [32, 33]. In addition to this, p- and n-type metal oxides immobilized graphene are also known as p- and n-type graphene composite, respectively [34]. Among the various p-type metal oxides (*i.e.* NiO, MoO_3, Cu_2O, V_2O_5), poly(styrene sulfonic acid) (PSS) doped organic poly(3,4-ethylenedioxithiophene) (PEDOT) is the most widely applied HTL in OSCs [35]. Its acidic and hygroscopic nature can *etch* the ITO and permit water absorption that eventually degrades the device's lifetime. Therefore, to address these challenges, recently, p-type graphene-based materials have been widely applied as a competent HTL that exhibits high electron blocking capability and ensures high-efficiency. For example, Stratakis *et al.* prepared and used chlorinated GO (GO-Cl, p-type graphene) films as HTL to develop efficient OSCs [36]. Next, Yun and colleagues have used rGO, GO, and pr-GO (p-TosNHNH$_2$ reduced GO) as electron restricting layers in Al/Ca/P3HT:PCBM/HTL/ITO device. The conversion performance of graphene-based devices was found to be enhanced from PCE= 0.72% (for only ITO) to 3.7% (after incorporating pr-GO as HTL). Additionally, the open-circuit voltage (V_{oc}), J_{sc}, and *FF* outcomes were also comparable with PEDOT:PSS [37]. Since the thickness of GO films can also affect the PCE of the device, Li *et al.* reported that the thinnest film exhibits the best results because of increased serial resistance and lower J_{sc} and *FF* values. Also, the lifetime decay results for thin GO HTL, PEDOT:PSS HTL and ITO-only were estimated to be 11.6, 9.6 and 8.1 μs, respectively (Fig. **1**). This change in lifetimes suggests the effective separation of carriers *via* blocking of electrons and effective extraction of holes to ITO and suppression of leakage current [38].

Next, to enhance the efficiency of cathode electrodes in extracting and collecting the negative charge carriers, n-type graphene has been effectively used as ETL [39]. For instance, Lee and colleagues have demonstrated 36% enhancement in the relative efficiency of inverted devices using rGO and zinc oxide (ZnO) composites as an n-type interface in the Ag/PEDOT:PSS/P3HT:PCBM/ZnO:rGO/ITO device [39]. It was found that devices comprising of ZnO-only revealed an efficiency of 3.06%, whereas the incorporation of rGO displays considerable improvement in *FF* and J_{sc} value. Further, Wang *et al.* developed PCDTBT:PC$_{71}$BM BHJ solar cells containing an ETL of str*etch*able GO with enhanced J_{sc} and *PCE* as compared to the device without a buffer layer [40]. The GO was directly coated onto the BHJ surface through the stamping process. Also, because of the synergistic effect of enhanced optical field amplitude and improved charge transport, the BHJ solar cell with ETL of GO/TiOx composite displayed a 29% increment in the *PCE* as compared to a solar cell with single coated GO or TiOx.

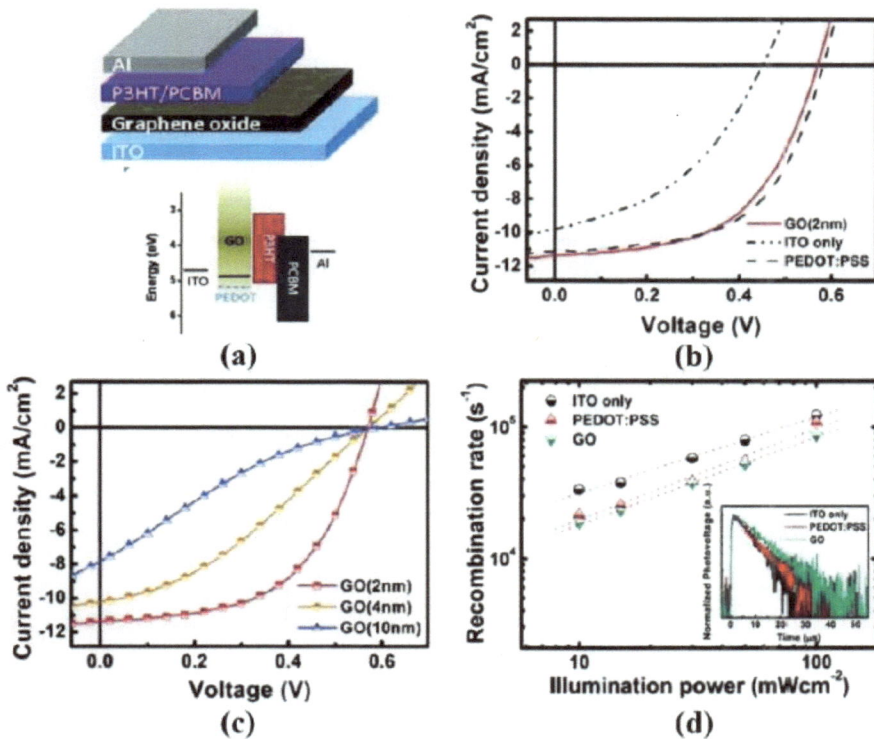

Fig. (1). (a) Schematic diagram of solar cell device structure and energy level diagram (b) I-V characteristics of solar cell device with (PEDOT:PSS and 2 nm GO) and without (ITO Only) HTL (c) I-V characteristics of solar cell device with different thickness of GO (2, 4, 10 nm) (d) light intensities versus charge recombination rate constant (Inset: transient photovoltage decay of solar cell device) [38]. Republished with copyright permission from ACS publication.

APPLICATION OF GRAPHENE AND ITS NANOCOMPOSITES IN DSSCS

Since the emergence of DSSCs in 1991, due to cost-effective production and high efficiency, they have become a potential alternate for Si-based photovoltaic devices [41, 42]. A typical DSSC comprises a translucent FTO anode decorated with a thin layer of TiO_2, which provides an advantageous porous structure with a large surface area. Next, the coated plate is submerged in a solution of photosensitive material. Similarly, another plate is prepared, which comprises a thin layer of iodide electrolyte cast onto a Pt metal sheet. Both the plates are then fixed tightly to avoid the leakage of electrolyte (Fig. **2a**) [43]. DSSCs work on the photosynthesis principle (*i.e.*, photogeneration of electron through dye) instead of the conventional p-n junction principle. In this, the photoexcitation of a photosensitive dye injects an electron into the TiO_2 followed by restoration of the ground state of sensitizer dye through electron donated by electrolyte during reduction of existing oxidants. Owing to the excellent catalytic activity and conductivity to avail efficient redox reaction, graphene and its nanocomposites have been widely used as an alternate to Pt in the counter electrode (Fig. **2b**) [44]. Jang and coworkers have reported that in comparison to pristine graphene, graphene-based materials, including GO, rGO, and 3D graphene scaffolds, have been widely used as a potential candidate for the counter electrode. This may due to a deficiency of active oxygen containing sites to enhance electrocatalytic reactions [45]. Further, Chen and colleagues also reached the same conclusion and reported that PCE of GO based DSSCs is directly proportional to the degree of reduction of GO [46]. Yu *et al*. developed a DSSC using 3D grown graphene as a counter electrode and observed an equivalent efficiency of 7.63% compared to Pt (8.48%) based counter electrode due to high porosity, surface area and conductivity enhanced electrode-electrolyte interface [47].

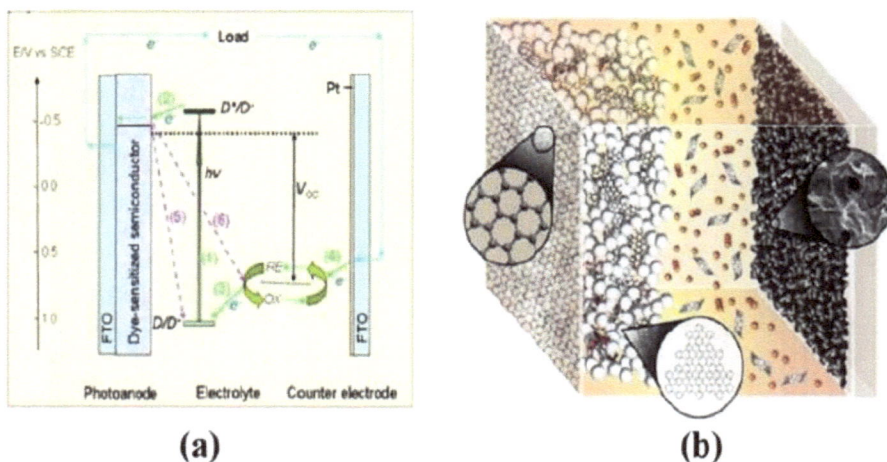

Fig. (2). (a) Working principle of DSSCs [43] and **(b)** Illustration of incorporation of graphene in each part of DSSC device [44]. Republished with copyright permission from ACS publication.

Next, the metal, polymer and metal oxide doped graphene-based materials have also been used as a suitable alternative to Pt based counter electrode. In a report, Dao and coworkers described that Pt nanoparticles decorated GO displayed better performance as compared to simple Pt counter electrode [48]. They observed that graphene-Pt nanohybrid exhibited higher efficiency of 8.56% as compared to Pt alone (8.18%) and GO-only (4.48%). In another study, a counter electrode made of a nanocomposite of single-walled carbon nanotubes and rGO exhibited PCE of 8.37% and V_{oc} of 0.86 V higher than graphene only (0.78 V, 7.19%) and Pt-based

(0.77 V, 7.79%) control DSSC devices [49]. In contrast, to counter electrodes, graphene-based materials have also been used as a potential photoanode. Zhao and his team reported that DSCCs made up of graphene/TiO_2 photoanode exhibited a PCE of 6.97% superior to only TiO_2 (5.01%) solar cells [50]. In this, graphene facilitated a bridge in TiO_2 scaffold, which hindered the recombination through high conductivity and offered high light scattering [50]. Similarly, another nanocomposite (TiO_2/CNT/graphene) was also applied as a photoanode and observed that the PCE of nanocomposite was enhanced by 31% in comparison to only TiO_2 based solar cells [51].

In addition to photoanode and counter electrodes, liquid electrolytes are also an important part of DSSCs for the transport of electrons. These electrolytes exhibit several interesting features, including low viscosity, high conductivity, easy preparation, decent interfacial wetting among electrodes and electrolytes, and hence, exhibit excellent efficiency of DSSCs [43]. Nevertheless, there are some limitations of liquid electrolytes, such as solvent evaporation, electrolyte leakages, temperature instability, *etc*. Therefore, to address these issues, organic-inorganic hybrids and polymer electrolytes based solid-state electrolytes are introduced [43]. In this way, Neo and Ouyang prepared 3-methoxypropionitrile (MPN)-GO organogels composite and used it as a quasi-solid state electrolyte [52]. They observed that such DSSCs having MPN-GO electrolytes displayed a performance efficiency of 6.70%, which was comparable to the efficiency of liquid electrolyte based DSSCs (7.18%). In another study, the same group obtained an improvement in the efficiency of DSSCs (7.12%) due to the incorporation of 1D multi-walled carbon nanotubes into MPN-GO organogels composite with a significant increase in V_{oc} and J_{sc} values [53]. Hence, graphene-based gel electrolyte materials have exhibited excellent device efficiency, and thus, have strong potential to replace the liquid electrolytes.

APPLICATION OF GRAPHENE AND ITS NANOCOMPOSITES IN PSCS

Organometal halide perovskite $CH_3NH_3PbX_3$ (X=Br, I) material was initially employed by Kojima and the team in the year 2009 to develop a photovoltaic device with PCE of 3.8% and 3.1% for bromine and iodine, respectively [54]. Because of their unique set of characteristics, including high carrier mobility, large absorption coefficient, long term stability and long charge diffusion length, perovskite materials-based solar cells have been known to be excellent light-harvesting materials [55 - 57]. Within the five years of its emergence, the PCE of PSCs reached to 19.3% [58]. The architecture of PSCs is animated from DSSCs that need high-temperature processing, thus limiting their industrial application. To address this issue, the configurations of OSCs have also been inspected because of their flexible, simple and low-temperature processable designs [59,

60]. Therefore, motivated by the flexible design of OSCs, the PCBM and PEDOT:PSS components were used as electron and hole transport materials in PSCs [61, 62]. They observed that the as-prepared rGO scaffold enhanced the electron transportation and thus, offered improved device performance with a high PCE of 17.2%, *FF* of 72%, V_{oc} of 1.05 V and J_{sc} of 22.8 mA/cm². The developed PSC device exhibited longer stability and lower hysteresis effects as compared to simple TiO_2 based solar cells (Fig. **3**) [63].

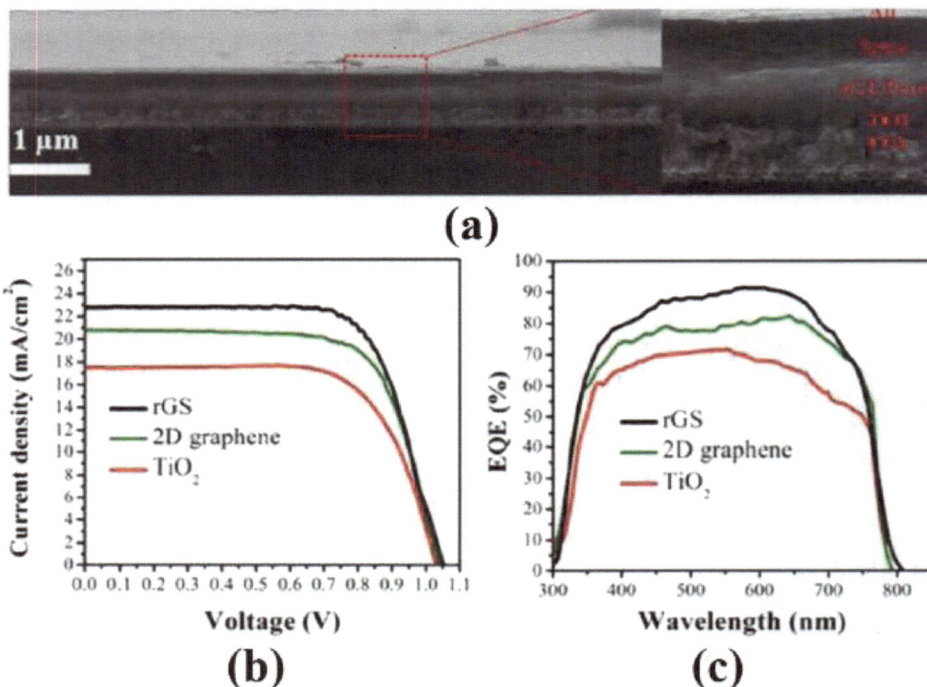

(a)

(b)

(c)

Fig. (3). (a) FE SEM image (b) J−V curves and (c) EQE spectrum of rGS-based device [63]. Republished with copyright permission from ACS publication.

In another attempt, Wang and his team fabricated a low-temperature mesostructured PSC using graphene/TiO_2 nanocomposite as an electron transport material [64]. All experiments were conducted at temperature < 150 and the final device exhibited a high PCE of 15.6% with *FF* of 73%, J_{sc} of 21.9 mA/cm² and V_{oc} of 1.05 V. The solar cells prepared from high-temperature annealed TiO_2, only TiO_2 and graphene exhibited PCEs of 14.1%, 10% and 5.9%, respectively, which suggest that use of graphene with TiO_2 as an electron transport layer in PSCs can avoid the high temperature annealing process to attain the high efficiency. Further, Fan and colleagues used GQDs between $CH_3NH_3PbI_3$ and TiO_2 layers to develop a PSC (Fig. **4**) and observed that the performance of the developed device was enhanced from 8.81% to 10.15% due to the faster electron extraction *via* GQDs [65].

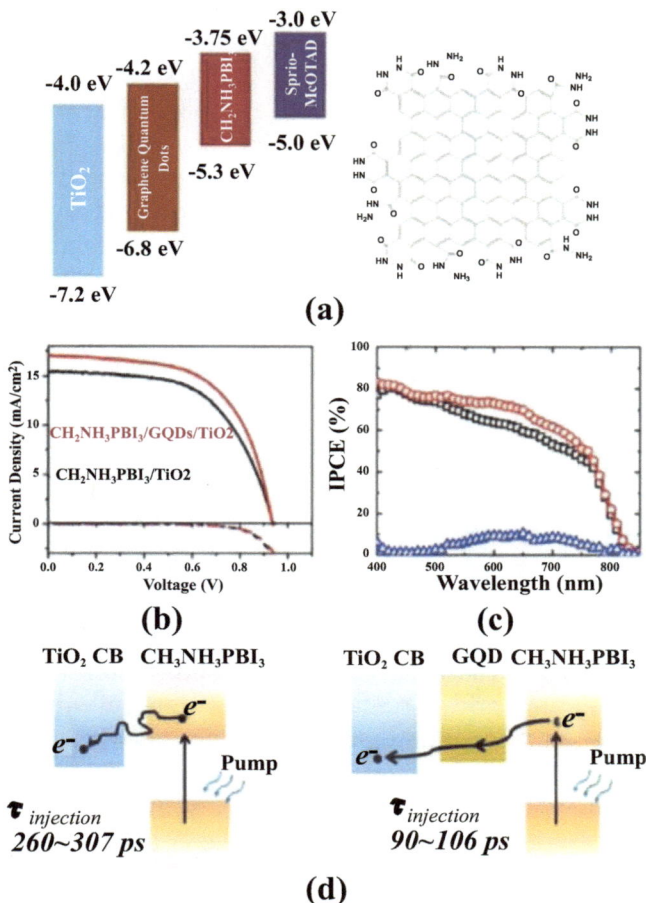

Fig. (4). (a) Schematic illustration of device structure and energy band diagram (b) J–V curves (c) IPCE spectrum of device and (d) illustration of electron generation and extraction at interfaces [65]. Republished with copyright permission from ACS publication.

They measured the electron extraction speed through transient absorption and found a significant change in the charge extraction speed with (90-106 ps) and without (260-307 ps) GQDs, which align="center"suggest that GQDs interlayer positively assisted the $CH_3NH_3PbI_3$ for faster charge transfer to TiO_2. Xie and his team developed a device with lower hysteresis using tin dioxide (SnO_2):GQDs nanocomposite as an electron transport material [66]. In this, GQDs resolved the issue of electron traps in SnO_2 by simply transferring the charge carriers to the conduction band of SnO_2 and reducing the recombination at ETL/perovskite interface. Similar to electron transport material, graphene-based materials have also been used as hole transport materials in PSCs. Palma and coworkers applied rGO as a hole transport layer in PSCs and found a noteworthy improvement in the lifetime (1987 h) as compared to OMeTAD-Spiro-based solar cells [67]. Next, Feng and the team prepared a planer PSC using solution processed GO:NH$_3$-PEDOT:PSS as a hole transport material [68]. They observed that the incorporation of ammonia functionalized GO into PSCs enhanced the PCE of device by 30%, along with the structural stability of perovskite. These enhancements in device performance are ascribed to good crystallization, energy-level arrangement at the interface, and orientation of perovskite.

CONCLUSION AND FUTURE PROSPECTIVE

In summary, graphene and its nanocomposites have been widely used in developing improved photovoltaic devices instead of traditional materials due to their exceptional characteristics, including packed 2D network, low resistivity, high carrier mobility, and transmittance. In this chapter, we discussed the application of graphene-based materials in making state-of-the-art solar cell devices. The exciting features of graphene make it a potential alternative to commercial ITO for a conductive electrode in flexible photovoltaic devices. Graphene and its nanocomposites are also able to serve as an efficient charge transport component in the next-generation photovoltaic devices. Due to its highly packed and stable 2D structure, graphene-based materials can enhance the atmospheric degradation and air stability of solar cell devices. Despite the substantial advancements in graphene based solar cell devices, a few challenges still remain and need to be answered in the future. The advancement in the synthesis process is still needed for the large-scale industrial production of graphene. The integration of purification and separation processes needs more preciseness to control the device fabrication. The incorporation of graphene and its nanocomposites as a buffer layer into the development of solar cell devices with high yield is also an important point that needs to be increasingly focused on in the future. Conclusively, if engineered carefully, graphene-based materials have strong potential to open a new window in the development of solar cell devices.

CONSENT FOR PUBLICATION

Not applicable.

CONFLICT OF INTEREST

The authors declare no conflict of interest, financial or otherwise.

ACKNOWLEDGEMENTS

The first author would like to thank Science and Engineering Research Board

(SERB), Department of Science and Technology, Government of India for facilitating overseas visiting doctoral fellowship (OVDF). First author would also like to thank Dean of Resources & Alumni Affairs (DORA), Indian Institute of Technology Roorkee for financial support. Authors gratefully acknowledge the Director, IIT Roorkee for providing the research infrastructure.

REFERENCES

[1]   Singh, E.; Nalwa, H.S. Graphene-based bulk-heterojunction solar cells: A review. *J. Nanosci. Nanotechnol.*, **2015**, *15*(9), 6237-6278.
       [http://dx.doi.org/10.1166/jnn.2015.11654] [PMID: 26716184]

[2]   Kadian, S.; Sethi, S.K.; Manik, G. Recent advancements in synthesis and property control of graphene quantum dots for biomedical and optoelectronic applications. *Mater. Chem. Front.*, **2021**, *5*(2), 627-658.
       [http://dx.doi.org/10.1039/D0QM00550A]

[3]   Kalkal, A.; Pradhan, R.; Kadian, S.; Manik, G.; Packirisamy, G. Biofunctionalized graphene quantum dots based fluorescent biosensor toward efficient detection of small cell lung cancer. *ACS Appl. Bio Mater.*, **2020**, *3*(8), 4922-4932.
       [http://dx.doi.org/10.1021/acsabm.0c00427] [PMID: 35021736]

[4]   Kumar, S.; Kumar, J.; Sharma, S.N. Investigation of charge transfer properties in MEHPVV and rGO-AA nanocomposites for Green organic photovoltaic application. *Optik (Stuttg.)*, **2020**, *208*, 164540.
       [http://dx.doi.org/10.1016/j.ijleo.2020.164540]

[5]   Kadian, S.; Manik, G.; Das, N.; Nehra, P.; Chauhan, R.P.; Roy, P. Synthesis, characterization and investigation of synergistic antibacterial activity and cell *via*bility of silver-sulfur doped graphene quantum dot (Ag@S-GQDs) nanocomposites. *J. Mater. Chem. B Mater. Biol. Med.*, **2020**, *8*(15), 3028-3037.
       [http://dx.doi.org/10.1039/C9TB02823D] [PMID: 32186305]

[6]   Kadian, S.; Manik, G.; Das, N.; Roy, P. Targeted bioimaging and sensing of folate receptor-positive cancer cells using folic acid-conjugated sulfur-doped graphene quantum dots. *Mikrochim. Acta*, **2020**, *187*(8), 458.
       [http://dx.doi.org/10.1007/s00604-020-04448-8] [PMID: 32683509]

[7]   Singh, E.; Nalwa, H.S. Graphene-based dye-sensitized solar cells: A review. *Sci. Adv. Mater.*, **2015**, *7*(10), 1863-1912.
       [http://dx.doi.org/10.1166/sam.2015.2438]

[8]   Kadian, S.; Manik, G. A highly sensitive and selective detection of picric acid using fluorescent sulfur-doped graphene quantum dots. *Luminescence*, **2020**, *35*(5), 763-772.
       [http://dx.doi.org/10.1002/bio.3782] [PMID: 31984670]

[9]   Kadian, S.; Manik, G.; Ashish, K.; Singh, M.; Chauhan, R.P. Effect of sulfur doping on fluorescence and quantum yield of graphene quantum dots: an experimental and theoretical investigation. *Nanotechnology*, **2019**, *30*(43), 435704.
       [http://dx.doi.org/10.1088/1361-6528/ab3566] [PMID: 31342919]

[10]   Kadian, S.; Manik, G. Sulfur doped graphene quantum dots as a potential sensitive fluorescent probe for the detection of quercetin. *Food Chem.*, **2020**, *317*, 126457.
       [http://dx.doi.org/10.1016/j.foodchem.2020.126457] [PMID: 32106009]

[11]   Stotter, J.; Show, Y.; Wang, S.; Swain, G. Comparison of the electrical, optical, and electrochemical properties of diamond and indium tin oxide thin-film electrodes. *Chem. Mater.*, **2005**, *17*(19), 4880-4888.
       [http://dx.doi.org/10.1021/cm050762z]

[12]   Bonaccorso, F.; Sun, Z.; Hasan, T.; Ferrari, A.C. Graphene photonics and optoelectronics. *Nat. Photonics,* **2010**, *4*(9), 611-622.
[http://dx.doi.org/10.1038/nphoton.2010.186]

[13]   Park, J.H.; Kang, S.J.; Na, S.I.; Lee, H.H.; Kim, S.W.; Hosono, H.; Kim, H-K. Indium-free, acid-resistant anatase Nb-doped TiO$_2$ electrodes activated by rapid-thermal annealing for cost-effective organic photovoltaics. *Sol. Energy Mater. Sol. Cells,* **2011**, *95*(8), 2178-2185.
[http://dx.doi.org/10.1016/j.solmat.2011.03.021]

[14]   Choi, Y.Y.; Choi, K.H.; Lee, H.; Lee, H.; Kang, J.W.; Kim, H.K. Nano-sized Ag-inserted amorphous ZnSnO$_3$ multilayer electrodes for cost-efficient inverted organic solar cells. *Sol. Energy Mater. Sol. Cells,* **2011**, *95*(7), 1615-1623.
[http://dx.doi.org/10.1016/j.solmat.2011.01.013]

[15]   Park, J.H.; Ahn, K.J.; Na, S.I.; Kim, H.K. Effects of deposition temperature on characteristics of Ga-doped ZnO film prepared by highly efficient cylindrical rotating magnetron sputtering for organic solar cells. *Sol. Energy Mater. Sol. Cells,* **2011**, *95*(2), 657-963.
[http://dx.doi.org/10.1016/j.solmat.2010.09.036]

[16]   Yin, Z.; Wu, S.; Zhou, X.; Huang, X.; Zhang, Q.; Boey, F.; Zhang, H. Electrochemical deposition of ZnO nanorods on transparent reduced graphene oxide electrodes for hybrid solar cells. *Small,* **2010**, *6*(2), 307-312.
[http://dx.doi.org/10.1002/smll.200901968] [PMID: 20039255]

[17]   Park, H.; Chang, S.; Zhou, X.; Kong, J.; Palacios, T.; Gradečak, S. Flexible graphene electrode-based organic photovoltaics with record-high efficiency. *Nano Lett.,* **2014**, *14*(9), 5148-5154.
[http://dx.doi.org/10.1021/nl501981f] [PMID: 25141259]

[18]   Zhang, D.; Choy, W.C.H.; Wang, C.C.D.; Li, X.; Fan, L.; Wang, K.; Zhu, H. Polymer solar cells with gold nanoclusters decorated multi-layer graphene as transparent electrode. *Appl. Phys. Lett.,* **2011**, *99*(22), 259.
[http://dx.doi.org/10.1063/1.3664120]

[19]   Lee, B.H.; Lee, J.H.; Kahng, Y.H.; Kim, N.; Kim, Y.J.; Lee, J.; Lee, T.; Lee, K. Graphene-conducting polymer hybrid transparent electrodes for efficient organic optoelectronic devices. *Adv. Funct. Mater.,* **2014**, *24*(13), 1847-1856.
[http://dx.doi.org/10.1002/adfm.201302928]

[20]   Park, H.; Chang, S.; Zhou, X.; Kong, J.; Palacios, T.; Gradecak, S. Flexible Graphene Electrode-Based Organic Photovoltaics with Record-High Efficiency. *ECS Trans.,* **2015**, *69*(14), 10.
[http://dx.doi.org/10.1149/06914.0077ecst]

[21]   Wang, Y.; Tong, S.W.; Xu, X.F.; Özyilmaz, B.; Loh, K.P. Interface engineering of layer-by-layer stacked graphene anodes for high-performance organic solar cells. *Adv. Mater.,* **2011**, *23*(13), 1514-1518.
[http://dx.doi.org/10.1002/adma.201003673] [PMID: 21449053]

[22]   Gómez-Navarro, C.; Meyer, J.C.; Sundaram, R.S.; Chuvilin, A.; Kurasch, S.; Burghard, M.; Kern, K.; Kaiser, U. Atomic structure of reduced graphene oxide. *Nano Lett.,* **2010**, *10*(4), 1144-1148.
[http://dx.doi.org/10.1021/nl9031617] [PMID: 20199057]

[23]   Liu, Z.; Li, J.; Yan, F. Package-free flexible organic solar cells with graphene top electrodes. *Adv. Mater.,* **2013**, *25*(31), 4296-4301.
[http://dx.doi.org/10.1002/adma.201205337] [PMID: 23553739]

[24]   Yin, Z.; Sun, S.; Salim, T.; Wu, S.; Huang, X.; He, Q.; Lam, Y.M.; Zhang, H. Organic photovoltaic devices using highly flexible reduced graphene oxide films as transparent electrodes. *ACS Nano,* **2010**, *4*(9), 5263-5268.
[http://dx.doi.org/10.1021/nn1015874] [PMID: 20738121]

[25]   Choe, M.; Lee, B.H.; Jo, G.; Park, J.; Park, W.; Lee, S.; Hong, W-K.; Seong, M-J.; Kahng, Y.H.; Lee,

K. Efficient bulk-heterojunction photovoltaic cells with transparent multi-layer graphene electrodes. *Org. Electron.*, **2010**, *11*(11), 1864-1869.
[http://dx.doi.org/10.1016/j.orgel.2010.08.018]

[26]   Li, X.; Zhang, G.; Bai, X.; Sun, X.; Wang, X.; Wang, E.; Dai, H. Highly conducting graphene sheets and Langmuir-Blodgett films. *Nat. Nanotechnol.*, **2008**, *3*(9), 538-542.
[http://dx.doi.org/10.1038/nnano.2008.210] [PMID: 18772914]

[27]   Ott, A.; Verzhbitskiy, I.A.; Clough, J.; Eckmann, A.; Georgiou, T.; Casiraghi, C. Tunable D peak in gated graphene. *Nano Res.*, **2014**, *7*(3), 338-344.
[http://dx.doi.org/10.1007/s12274-013-0399-2]

[28]   Xu, K.; Wang, K.; Zhao, W.; Bao, W.; Liu, E.; Ren, Y.; Wang, M.; Fu, Y.; Zeng, J.; Li, Z.; Zhou, W.; Song, F.; Wang, X.; Shi, Y.; Wan, X.; Fuhrer, M.S.; Wang, B.; Qiao, Z.; Miao, F.; Xing, D. The positive piezoconductive effect in graphene. *Nat. Commun.*, **2015**, *6*(1), 8119.
[http://dx.doi.org/10.1038/ncomms9119] [PMID: 26360786]

[29]   Kumar, S.; Sharma, S.N.; Kumar, J. Comparative Charge Transport Study of MEHPPV-TiO$_2$ and P3HT-TiO$_2$ Nanocomposites for Hybrid Bulk Heterojunction Solar Cells. *J. Nanosci. Nanotechnol.*, **2019**, *19*(6), 3408-3419.
[http://dx.doi.org/10.1166/jnn.2019.16130] [PMID: 30744768]

[30]   Kumar, S.; Kumar, J.; Narayan Sharma, S.; Srivastava, S. rGO integrated MEHPPV and P3HT polymer blends for bulk hetero junction solar cells: A comparative insight. *Optik (Stuttg.)*, **2019**, *178*, 411-421.
[http://dx.doi.org/10.1016/j.ijleo.2018.09.148]

[31]   Kadian, S.; Arya, B.D.; Kumar, S.; Sharma, S.N.; Chauhan, R.P.; Srivastava, A.; Chandra, P.; Singh, S.P. Synthesis and Application of PHT-TiO$_2$ nanohybrid for amperometric glucose detection in human saliva sample. *Electroanalysis*, **2018**, *30*(11), 2793-2802.
[http://dx.doi.org/10.1002/elan.201800207]

[32]   Hou, Y.X.; Geng, X.M.; Li, Y.Z.; Dong, B.; Liu, L.W.; Sun, M.T. Electrical and Raman properties of p-type and n-type modified graphene by inorganic quantum dot and organic molecule modification. *Sci. China Phys. Mech. Astron.*, **2011**, *54*(3), 416-419.
[http://dx.doi.org/10.1007/s11433-011-4253-9]

[33]   Mahmoudi, T.; Seo, S.; Yang, H.Y.; Rho, W.Y.; Wang, Y.; Hahn, Y.B. Efficient bulk heterojunction hybrid solar cells with graphene-silver nanoparticles composite synthesized by microwave-assisted reduction. *Nano Energy*, **2016**, *28*, 179-187.
[http://dx.doi.org/10.1016/j.nanoen.2016.08.018]

[34]   Steim, R.; Kogler, F.R.; Brabec, C.J. Interface materials for organic solar cells. *J. Mater. Chem.*, **2010**, *20*(13), 2499-2512.
[http://dx.doi.org/10.1039/b921624c]

[35]   Irwin, M.D.; Buchholz, D.B.; Hains, A.W.; Chang, R.P.H.; Marks, T.J. p-Type semiconducting nickel oxide as an efficiency-enhancing anode interfacial layer in polymer bulk-heterojunction solar cells. *Proc. Natl. Acad. Sci. USA*, **2008**, *105*(8), 2783-2787.
[http://dx.doi.org/10.1073/pnas.0711990105]

[36]   Stratakis, E.; Savva, K.; Konios, D.; Petridis, C.; Kymakis, E. Improving the efficiency of organic photovoltaics by tuning the work function of graphene oxide hole transporting layers. *Nanoscale*, **2014**, *6*(12), 6925-6931.
[http://dx.doi.org/10.1039/C4NR01539H] [PMID: 24839176]

[37]   Yun, J.M.; Yeo, J.S.; Kim, J.; Jeong, H.G.; Kim, D.Y.; Noh, Y.J.; Kim, S.S.; Ku, B.C.; Na, S.I. Solution-processable reduced graphene oxide as a novel alternative to PEDOT:PSS hole transport layers for highly efficient and stable polymer solar cells. *Adv. Mater.*, **2011**, *23*(42), 4923-4928.
[http://dx.doi.org/10.1002/adma.201102207] [PMID: 21954085]

[38]   Li, S.S.; Tu, K.H.; Lin, C.C.; Chen, C.W.; Chhowalla, M. Solution-processable graphene oxide as an

efficient hole transport layer in polymer solar cells. *ACS Nano,* **2010**, *4*(6), 3169-3174.
[http://dx.doi.org/10.1021/nn100551j] [PMID: 20481512]

[39]   Woo Lee, H.; Young Oh, J.; Il Lee, T.; Soon Jang, W.; Bum Yoo, Y.; Sang Chae, S.; Ho Park, J.; Min Myoung, J.; Moon Song, K.; Koo Baik, H. Highly efficient inverted polymer solar cells with reduced graphene-oxide-zinc-oxide nanocomposites buffer layer. *Appl. Phys. Lett.,* **2013**, *102*(19), 193903.
[http://dx.doi.org/10.1063/1.4804645]

[40]   Wang, D.H.; Kim, J.K.; Seo, J.H.; Park, I.; Hong, B.H.; Park, J.H.; Heeger, A.J. Transferable graphene oxide by stamping nanotechnology: electron-transport layer for efficient bulk-heterojunction solar cells. *Angew. Chem. Int. Ed. Engl.,* **2013**, *52*(10), 2874-2880.
[http://dx.doi.org/10.1002/anie.201209999] [PMID: 23341253]

[41]   O'Regan, B.; Grätzel, M. A low-cost, high-efficiency solar cell based on dye-sensitized colloidal TiO$_2$ films. *Nature,* **1991**, *353*(6346), 737-740.
[http://dx.doi.org/10.1038/353737a0]

[42]   Grätzel, M. Photoelectrochemical cells. *Nature,* **2001**, *414*(6861), 338-344.
[http://dx.doi.org/10.1038/35104607] [PMID: 11713540]

[43]   Wu, J.; Lan, Z.; Lin, J.; Huang, M.; Huang, Y.; Fan, L.; Luo, G. Electrolytes in dye-sensitized solar cells. *Chem. Rev.,* **2015**, *115*(5), 2136-2173.
[http://dx.doi.org/10.1021/cr400675m] [PMID: 25629644]

[44]   Roy-Mayhew, J.D.; Aksay, I.A. Graphene materials and their use in dye-sensitized solar cells. *Chem. Rev.,* **2014**, *114*(12), 6323-6348.
[http://dx.doi.org/10.1021/cr400412a] [PMID: 24814731]

[45]   Jang, S.Y.; Kim, Y.G.; Kim, D.Y.; Kim, H.G.; Jo, S.M. Electrodynamically sprayed thin films of aqueous dispersible graphene nanosheets: highly efficient cathodes for dye-sensitized solar cells. *ACS Appl. Mater. Interfaces,* **2012**, *4*(7), 3500-3507.
[http://dx.doi.org/10.1021/am3005913] [PMID: 22724560]

[46]   Chen, C.C.; Chang, W.H.; Yoshimura, K.; Ohya, K.; You, J.; Gao, J.; Hong, Z.; Yang, Y. An efficient triple-junction polymer solar cell having a power conversion efficiency exceeding 11%. *Adv. Mater.,* **2014**, *26*(32), 5670-5677.
[http://dx.doi.org/10.1002/adma.201402072] [PMID: 25043698]

[47]   Nechiyil, D.; Vinayan, B.P.; Ramaprabhu, S. Tri-iodide reduction activity of ultra-small size PtFe nanoparticles supported nitrogen-doped graphene as counter electrode for dye-sensitized solar cell. *J. Colloid Interface Sci.,* **2017**, *488*, 309-316.
[http://dx.doi.org/10.1016/j.jcis.2016.11.011] [PMID: 27838555]

[48]   Dao, V.D.; Hoa, N.T.Q.; Larina, L.L.; Lee, J.K.; Choi, H.S. Graphene-platinum nanohybrid as a robust and low-cost counter electrode for dye-sensitized solar cells. *Nanoscale,* **2013**, *5*(24), 12237-12244.
[http://dx.doi.org/10.1039/c3nr03219a] [PMID: 24146088]

[49]   Zheng, H.; Neo, C.Y.; Ouyang, J. Highly efficient iodide/triiodide dye-sensitized solar cells with gel-coated reduce graphene oxide/single-walled carbon nanotube composites as the counter electrode exhibiting an open-circuit voltage of 0.90 V. *ACS Appl. Mater. Interfaces,* **2013**, *5*(14), 6657-6664.
[http://dx.doi.org/10.1021/am401392k] [PMID: 23786582]

[50]   Zhao, J.; Wu, J.; Zheng, M.; Huo, J.; Tu, Y. Improving the photovoltaic performance of dye-sensitized solar cell by graphene/titania photoanode. *Electrochim. Acta,* **2015**, *156*, 261-266.
[http://dx.doi.org/10.1016/j.electacta.2015.01.045]

[51]   Yen, M.Y.; Hsiao, M.C.; Liao, S.H.; Liu, P.I.; Tsai, H.M.; Ma, C.C.M.; Pu, N-W.; Ger, M-D. Preparation of graphene/multi-walled carbon nanotube hybrid and its use as photoanodes of dye-sensitized solar cells. *Carbon,* **2011**, *49*(11), 3597-3606.
[http://dx.doi.org/10.1016/j.carbon.2011.04.062]

[52]   Neo, C.Y.; Ouyang, J. The production of organogels using graphene oxide as the gelator for use in

high-performance quasi-solid state dye-sensitized solar cells. *Carbon,* **2013**, *54*, 48-57.
[http://dx.doi.org/10.1016/j.carbon.2012.11.002]

[53]   Neo, C.Y.; Gopalan, N.K.; Ouyang, J. Graphene oxide/multi-walled carbon nanotube nanocomposites
       as the gelator of gel electrolytes for quasi-solid state dye-sensitized solar cells. *J. Mater. Chem. A
       Mater. Energy Sustain.,* **2014**, *2*(24), 9226-9235.
       [http://dx.doi.org/10.1039/c4ta00232f]

[54]   Kojima, A.; Teshima, K.; Shirai, Y.; Miyasaka, T. Organometal halide perovskites as visible-light
       sensitizers for photovoltaic cells. *J. Am. Chem. Soc.,* **2009**, *131*(17), 6050-6051.
       [http://dx.doi.org/10.1021/ja809598r] [PMID: 19366264]

[55]   Stranks, SD; Eperon, GE; Grancini, G; Menelaou, C; Alcocer, MJP; Leijtens, T Electron-hole
       diffusion lengths exceeding 1 micrometer in an organometal trihalide perovskite absorber. *Science,*
       2013, *342*, 341-344.
       [http://dx.doi.org/10.1126/science.1243982]

[56]   Kadian, S.; Tailor, N.K.; Chaulagain, N.; Shankar, K.; Satapathi, S.; Manik, G. Effect of sulfur-doped
       graphene quantum dots incorporation on morphological, optical and electron transport properties of
       CH3NH3PbBr3 perovskite thin films. *J. Mater. Sci. Mater. Electron.,* **2021**, *32*(13), 1-12.
       [http://dx.doi.org/10.1007/s10854-021-06272-z]

[57]   Sethi, S.K.; Kadian, S.; Anubhav, ; Goel, ; Chauhan, R.P.; Manik, G. Anubhav, Goel, Chauhan RP,
       Manik G. Fabrication and analysis of zno quantum dots based easy clean coating: A combined
       theoretical and experimental investigation. *ChemistrySelect,* **2020**, *5*(29), 8942-8950.
       [http://dx.doi.org/10.1002/slct.202001092]

[58]   Zhou, H; Chen, Q; Li, G; Luo, S; Song, TB; Duan, HS Interface engineering of highly efficient
       perovskite solar cells. *Science,* **2014**, *345*, 542-6.
       [http://dx.doi.org/10.1126/science.1254050]

[59]   Yeo, J.S.; Kang, R.; Lee, S.; Jeon, Y.J.; Myoung, N.S.; Lee, C.L.; Kim, D-Y.; Yun, J-M.; Seo, Y-H.;
       Kim, S-S.; Na, S-I. Highly efficient and stable planar perovskite solar cells with reduced graphene
       oxide nanosheets as electrode interlayer. *Nano Energy,* **2015**, *12*, 96-104.
       [http://dx.doi.org/10.1016/j.nanoen.2014.12.022]

[60]   Liang, P.W.; Liao, C.Y.; Chueh, C.C.; Zuo, F.; Williams, S.T.; Xin, X.K.; Lin, J.; Jen, A.K. Additive
       enhanced crystallization of solution-processed perovskite for highly efficient planar-heterojunction
       solar cells. *Adv. Mater.,* **2014**, *26*(22), 3748-3754.
       [http://dx.doi.org/10.1002/adma.201400231] [PMID: 24634141]

[61]   Kuang, C.; Tang, G.; Jiu, T.; Yang, H.; Liu, H.; Li, B.; Luo, W.; Li, X.; Zhang, W.; Lu, F.; Fang, J.;
       Li, Y. Highly efficient electron transport obtained by doping PCBM with graphdiyne in planar-
       heterojunction perovskite solar cells. *Nano Lett.,* **2015**, *15*(4), 2756-2762.
       [http://dx.doi.org/10.1021/acs.nanolett.5b00787] [PMID: 25803148]

[62]   Adam, G.; Kaltenbrunner, M.; Głowacki, E.D.; Apaydin, D.H.; White, M.S.; Heilbrunner, H.; Tombe,
       S.; Stadler, P.; Ernecker, B.; Klampfl, C.W.; Sariciftci, N.S.; Scharber, M.C. Solution processed
       perovskite solar cells using highly conductive PEDOT:PSS interfacial layer. *Sol. Energy Mater. Sol.
       Cells,* **2016**, *157*, 318-325.
       [http://dx.doi.org/10.1016/j.solmat.2016.05.011]

[63]   Tavakoli, M.M.; Tavakoli, R.; Hasanzadeh, S.; Mirfasih, M.H. Interface engineering of perovskite
       solar cell using a reduced-graphene scaffold. *J. Phys. Chem. C,* **2016**, *120*(35), 19531-19536.
       [http://dx.doi.org/10.1021/acs.jpcc.6b05667]

[64]   Wang, J.T.W.; Ball, J.M.; Barea, E.M.; Abate, A.; Alexander-Webber, J.A.; Huang, J.; Saliba, M.;
       Mora-Sero, I.; Bisquert, J.; Snaith, H.J.; Nicholas, R.J. Low-temperature processed electron collection
       layers of graphene/TiO$_2$ nanocomposites in thin film perovskite solar cells. *Nano Lett.,* **2014**, *14*(2),
       724-730.
       [http://dx.doi.org/10.1021/nl403997a] [PMID: 24341922]

[65]   Zhu, Z.; Ma, J.; Wang, Z.; Mu, C.; Fan, Z.; Du, L.; Bai, Y.; Fan, L.; Yan, H.; Phillips, D.L.; Yang, S. Efficiency enhancement of perovskite solar cells through fast electron extraction: the role of graphene quantum dots. *J. Am. Chem. Soc.,* **2014**, *136*(10), 3760-3763.
[http://dx.doi.org/10.1021/ja4132246] [PMID: 24558950]

[66]   Xie, J.; Huang, K.; Yu, X.; Yang, Z.; Xiao, K.; Qiang, Y.; Zhu, X.; Xu, L.; Wang, P.; Cui, C.; Yang, D. Enhanced electronic properties of SnO_2 *via* electron transfer from graphene quantum dots for efficient perovskite solar cells. *ACS Nano,* **2017**, *11*(9), 9176-9182.
[http://dx.doi.org/10.1021/acsnano.7b04070] [PMID: 28858471]

[67]   Palma, A.L.; Cinà, L.; Pescetelli, S.; Agresti, A.; Raggio, M.; Paolesse, R.; Bonaccorso, F.; Di Carlo, A. Reduced graphene oxide as efficient and stable hole transporting material in mesoscopic perovskite solar cells. *Nano Energy,* **2016**, *22*, 349-360.
[http://dx.doi.org/10.1016/j.nanoen.2016.02.027]

[68]   Feng, S.; Yang, Y.; Li, M.; Wang, J.; Cheng, Z.; Li, J.; Ji, G.; Yin, G.; Song, F.; Wang, Z.; Li, J.; Gao, X. High-performance perovskite solar cells engineered by an ammonia modified graphene oxide interfacial layer. *ACS Appl. Mater. Interfaces,* **2016**, *8*(23), 14503-14512.
[http://dx.doi.org/10.1021/acsami.6b02064] [PMID: 27229127]

CHAPTER 5

New Frontiers of Graphene Based Nanohybrids for Energy Harvesting Applications

Sriparna De[1,*], **Arpita Adhikari**[2] and **Dipankar Chattopadhyay**[2]

[1] *Department of Allied Health Sciences, Brainware University, Kolkata, West Bengal 700125, India*

[2] *Department of Polymer Science and Technology, University of Calcutta, 92 A.P.C. Road, Kolkata, 700009, India*

Abstract: Graphene has gained recognition within the research community owing to its fascinating properties in the plethora of energy-related applications. The properties include high thermal and electrical conductivity, greater mechanical strength, optical translucency, intrinsic flexibility, massive surface area, and distinctive two-dimensional structure. Graphene is highly competent in enriching the functional performance, endurance, stability of many applications. However, still ample research diversity will be desirable for graphene commercialization in energy sectors. This intuitive scrutinization reconnoitered the talented employment arena of graphene in various energy storage and harvesting fields. The amplification of the versatile applicability of graphene and comprehensive perception regarding pros and cons of graphene based nanohybrids could critically pinpoint current constrictions by upgrading its characteristics performance. The chapter provides an insight into the unique features of graphene and amalgamation with nanomaterials to enlighten its various energy-related applications, including supercapacitors, biosensors, solar cells, batteries. With the breakneck miniaturization in the employment of graphene in various energy-relevant applications, it is crucial to epitomize align="center" and figure out the progressive momentum of graphene and its nanohybrids in several energy-related application territories.

Keywords: Batteries, Conductivity, Energy storage, Energy harvesting, Fuel cell, Graphene, Nanomaterial, Nanohybrid, Surface area, Sensor, Supercapacitor, Solar cell.

INTRODUCTION

According to the Paris settlement signed by 2015, the foremost objective was to cogitate reasonable ways of contending environmental fluctuations by recognizing

---

[*] **Corresponding author Sriparna De:** Department of Allied Health Sciences, Brainware University, Kolkata, West Bengal 700125, India; E-mail: sriparna.de2@gmail.com

Gaurav Manik and Sushanta Kumar Sahoo (Eds.)

the unconventional form of energy generation [1]. In the light of environmental sustainability of tackling energy, the fossil artifacts impacted profoundly for usage in energy harvesting regime. Depletion of fossil fuels is now alarming for the emerging energy harvesting and storage process. Renewable energy is a copious cradle of energy and alternative for minimizing the ozone layer depletion caused by the emission of fossil products [2, 3]. Meanwhile, solar energy and wind energy are remarkable and in dynamic phases, irregular alteration of phases requires the integrative approach with energy storage and converting device in a cost-effective manner [4]. Recent trend align="center"in energy storage and sustainable development is underway to explore more effective, sustainable energy harvesting systems through fuel cell technology [5]. Likewise, dye synthesized solar cells (DSSCs) are also alluring energy conversion tools. Graphene align="center"is a two dimensional (2D) single graphitic sheet-like morphology that align="center"possesses remarkable and fascinating properties *e.g.*, extraordinary conductivity, good electrocatalytic efficiency, mechanical stability, and solar radiation absorptivity [6]. It offers phenomenal concert in different application meadows such as sensors, catalysis, adsorption as well as energy storage and conversion. Integration of nanomaterials within the graphene matrix provides enhanced mechanical strength and electrical conductivity of the nanocomposites. Additionally, these nanocomposites illustrated superior thermal stability and electrochemical activity, as well as gas barrier properties [7].

For instance, the accumulation of graphene in these composites enriches the propagation of the electrolytic substance through its outward facet and caters electrically conductive ionic paths for dynamic nanomaterials decorated on it [8]. By designing graphene based nanohybrids, the distinctive properties of graphene and dynamic nanomaterials and their good synergism can be exploited, thereby significantly showcasing the perspective of graphene-based nanohybrids in various energy-related applications.

UNIQUE FEATURES OF GRAPHENE

Graphene, a crystalline allotrope of carbon, is well-advised as the advanced type of carbon nanomaterials owing to its two-dimension solitary leaf of a carbon atom within a hexagonal networking arrangement [9]. Two dimensionally oriented honeycomb lattice of graphene is a unit cell of all dimensional varieties of carbonaceous materials resembling 0D (fullerene), 1D (carbon nanotube) and 3D (graphite) showing zero band gap semi-conductor with extraordinary charge carrier kinesis (up to $10\,0,0\,0\,0$ cm 2 V $^{-1}$ s $^{-1}$) at relativistic speed $\square 10^6$ m s $^{-1}$ [9, 10] (Fig. **1**). Due to its peculiar structural features *e.g.*, quantum Hall effect (QHE), bipolar electric field effect and electron wave propagation, graphene has acquired tremendous superiority for a wide variety of energy-related applications, *e.g.*, energy harvesting applications [11]. The underlying cause behind the relevancy of graphene in fabricating energy devices lies in its huge surface area, extraordinary electrical mobility and extended electron channeling network. Nonetheless, the architecture of graphene governs the definite surface area of graphene structure. Nevertheless, the major drawback includes agglomerates or restacking due to strong π-π stacking, which somehow confines the pristine usage of graphene materials as electrode material in energy devices [12]. Besides that, in semiconducting bilayer graphene, charge transporters comply with parabolic energy dispersal and unveil tunable zero band gap. The electronic and electrochemical features of graphene are intensely influenced by edge configuration [13, 14]. This can be diversified by the unique architectural framework of graphene molecule, which strongly mutates the threshold energies and forces within its circumference. align="center"Hence, an edge to edges force can be achieved by termination or doping with nitrogen, boron, and functional groups including oxygen, hydrogen, metals using oxygen gas plasma hydrogenation process [14, 15]. Interestinglyalign="center", peculiar characteristic thermal conductivity is identified within a solitary align="center"layer of graphene due to acoustic phonons [16, 17]. These phenomena encourage the significance and utility of graphene in the burgeoning of nanoelectric energy devices futuristically [18].

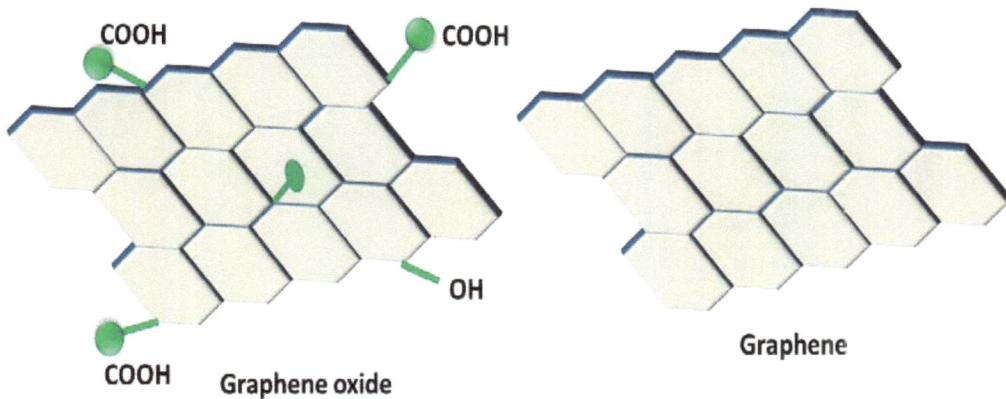

Fig. (1). Structural representation of graphene oxide and graphene.

FABRICATION OF GRAPHENE AND ITS VARIOUS ASSEMBLY

Two strategies were adopted to yield graphene are classified as top-down and bottom-up approaches. Bulk graphite material is the pivotal precursor of single and multilayer graphene. In the top-down approach, exfoliation of graphitic oxide from bulk graphite material can be achieved by mechanical and chemical ways [19]. Fig. (**2**) displays a probable synthesis spectrum of graphene and its nanohybrid. Adopting mechanical exfoliation, graphene of different thicknesses align="center"is produced by removing layers of highly ordered pyrolytic graphite (HOPG) [20]. Whereas in chemical exfoliation, graphene intercalated compounds (GICs) are synthesized by enhancing interlayer spacing *via* minimizing the van der Waals forces. Simultaneously, liquid phase exfoliation [21] and electrochemical exfoliation [22] also belong to top-down approach. On the contrary, bottom-up route ensures the epitaxial and chemical vapour deposition (CVD) graphene growth from organic precursors elements *i.e.* hydrocarbons [23, 24]. CVD-grown graphene offers efficient featured and large surface area promising electrode surface fabrication. This can be done by a thin film deposition process. Graphene growth on nickel surfaces was first explored and scrutinized. In 1975, single layered graphitic materials were stacked onto the platinum surface. Additionally, graphene matured on Iridium (Ir) surface was also investigated in 1984 [25]. Simultaneously, the probability of graphene synthesis on Ni foil camphor was adopted.

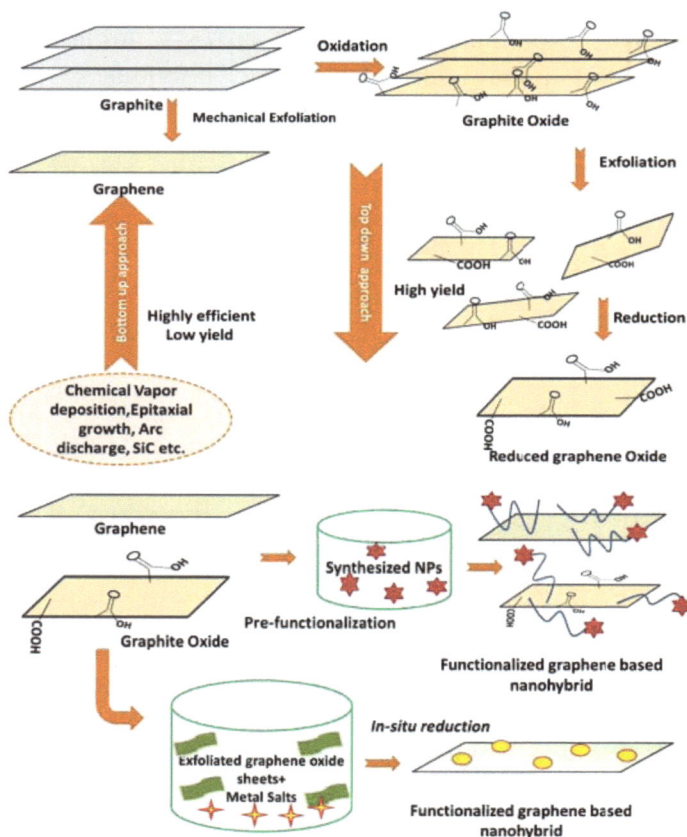

Fig. (2). Various synthesis schemes of graphene and its nanohybrid.

To tune the exposed surface area of graphene, extensive surveys have been undertaken to cultivate various features of nanomaterials *e.g.*, carbon nanoparticles [26, 27], metal nanoparticles (NPs) [28], polymers [29] and clays between graphene sheets to separate them. To address the restacking issues, newly fabricated nanocomposites could be an acceptable alternative to deploy these materials with enriched performances and properties in various energy related applications. Owing to their distinctive structural arrangement and alluring properties, a fanatical research effort has been conducted to assemble graphene based composites with favorable morphological architecture. Numerous synthesis strategies have been adopted to prepare three-dimensional (3D) graphene structures, including *e.g.*, self-assembly scheme, aerosolization, template guided synthesis, direct deposition technique, *etc.*

Self-assembly Methods

To obtain three-dimensional graphene or graphene nanohybrids, this technique is one of the recurrently utilized methodologies [30]. In this method, gelation process is utilized wherein van der Waals attraction forces work to promote steady GO suspension and distinctly well dispersed GO sheets and also an electrostatic repulsive force originates from its neighboring functional groups [31]. Afterward, when the force balance is fragmented, GO sheets get interconnected, resulting in the formation of a highly porous hydrogel network. The prime align="center"characteristic topography of this fabricated structure includes a huge accessible surface area favorable for active material garnishing purposes [32]. Due align="center"to the robust interlayer affinity, this graphene hydrogel imparts great physicochemical properties, including mechanical and thermal stability, which conveys remarkable prospects and open wide applications avenue in various energy sectors.

Template Methods

This template method is also assisted by Chemical Vapour Deposition (CVD) technique wherein direct cultivation of graphene is monitored on 3D oriented template sheet [20]. Analogous to the self-assembly scheme, template guided graphene nanohybrids are customarily originated by various techniques *e.g.*, physicochemical installation of nanoparticles/nanomaterials on the three-dimensional graphene surface. For instance, porous Ni foam can be utilized as both the template and the catalyst resulting in the evolution of graphene network successfully grown on the Ni surface [33]. Another experiment was carried out on polystyrene film wherein three dimensional RGO film was produced upon subsequent pyrolysis procedure [34]align="center". Further versatile anatomical and functional modification are much needed, especially doping, which could be predominantly adopted to upgrade three-dimensional graphene properties for various applica tions.

Aerosolization Process

To obtain crumpled graphene (CG) balls, aerosolization process is well-documented wherein physical deformation was noticed by successive solvent evaporation. In this process, structural alteration from a two-dimensional (2D) sheet-like arrangement to a three-dimensional (3D) hollow type morphology of GO was noticed [35]. To magnify the prospective applications of CG, various functional nanomaterials have been amalgamated, which offered terrific opportunities for energy applications. For example, the aerosolization process has been adopted to develop numerous CG-based composites wherein pre-synthesized GO and NPs including Pt, Au, Ag, carbon black, TiO_2, Fe_3O_4, *etc.*, were mixed together [36, 37]. This fabricated graphene-based composites allow the direct miniaturization of electrode surface for various energy storage and conversion applications.

Direct Deposition Technique

Vertically-aligned graphene (VG) sheets are typically synthesized by adopting a plasma enriched chemical vapor deposition (PECVD) method [38]. Plasma treatment facilitates controlled maturation of VG, which can be successfully accomplished in a lesser temperature and catalyst-free fashion with governable structural properties through refinement of the growing characteristics [38, 39]. By amalgamating other nanomaterials within the VG matrix, VG-based composites were intensively explored and showed their great potential in various energy and sustainable applications. Apart from PECVD, electrochemical reduction technique could also be a worthy alternative for direct deposition of reduced graphene oxide on the electrode periphery [39].

SIGNIFICANCE OF GRAPHENE-BASED NANOHYBRIDS

Major structural deformities, including dislocation, vacancy, grain boundary, topological, and other structured protracted deformities on the basal plane of graphene nanosheets, are monitored. This necessitates the evolution of doping modification theory, and functionalization with various nanostructured materials. The dislocation defects and vacancy defects are due to non-hexagonal structure of C atom and atomic dislocations, respectively [40, 41]. In addition, the grain boundary defects are mainly instigated by the crystal grains derived from pentagonal and heptagonal structures. Besides this, octagonal and pentagonal carbon rings are responsible for the evolution of topological defects. Also, the chemical instability of graphene is mainly due to protracted defects of dangling bonds bestowing on the edge boundaries of the graphene nanoribbons. Agglomeration is also an enduring demur for graphene and its derivatives. To address this issue graphene and graphene oxide based nanohybrid/nanomaterial could be a promising agent. Bestowing to the doping modification theory, further functionalizing of G and GO through van der Waals forces, hydrogen bonds, ionic bonds, coordination bonds, and π-π stacking interactions shows various tuned characteristic properties favourable for energy applications [42]. Likewise, N-doping of graphene material leads to further enhancement in electrocatalytic performance by decorating various transition metals to N^{th} position. Numerous metal-based nanoparticles (*e.g.* Au, Ag, Pt, Pd, TiO_2 *etc.*) has been devoted and revealed superior catalytic activity due to their quantum confinement which ultimately results in catalytic stability and dispersibility of graphene. Besides that, to modulate the charge transfer kinetics and electroactive channeling across the graphene surface, semiconducting material will be incorporated, aiding the fabrication of unique electrode set-up utilizing 2D van der Waals nanohybrid for promising applications in various plethora of electrochemical array [43]. Keeping in view the exploration and state-of-the-art avant-grade graphene based nanohybrid could be an encouraging candidate for next generation energy harvesting applications.

SUPERCAPACITOR APPLICATIONS

Supercapacitors are cutting-edge electrochemical tools highly favorable for energy restoration that endure recurrent charge/discharge cycles at huge power and within the short time interval. Also, it shows superior capacitance activity as compared to regular capacitors. Double-layer capacitors (EDLCs) and pseudocapacitors are the two types of supercapacitor wherein EDLCs percolate with the electric double-layer mechanism and surface adsorption of ions with opposite charges. Simultaneously reversible faradaic-type charge transfer mechanism is embodied in pseudocapacitors. To tune supercapacitors performance such as mass to volume ratio, specific capacitance, rate ability, and energy densities, nanocarbon-based material, graphene employed owing to its enhanced electrical conductivity, vast surface area, and cost-effective nature [44 - 46].

Various morphologies of three-dimensional graphene such as graphene hydrogels [44] template-based 3DGs [47], hollow CGs [48, 49] and VGs [50, 51] were remarkably used as electrode surfaces in supercapacitor applications. Shi and his group amalgamated graphene hydrogel with nickel foam (G-Gel-NF) and modified electrode by depositing GO on the Ni framework foam followed by free-drying. The capacitance obtained for this fabricated network is 45.6 mF cm^{-2} at a discharge current of 0.67 mA cm^{-2} [52]. To tune the functional properties of graphene based composite electrodes, graphene plays a pivotal role by improving the specific loading of energy enriched pseudo-species, increasing the conveyance of charged particles within pseudospecies and electrodes for a higher power. The rate kinetics is also tuned by the adhesion of pseudo-species, thereby improving the cycle stability and reproducibilityalign="center". Therefore, various frequently used pseudo-species include Mn [53, 54align="center"], Ni [55align="center"], Co [56align="center"] and V [57, 58align="center"] based transition metal oxides with versatile morphological arrangements. Additionally, Ni derivatives such as sulphides and hydroxides were developed on three-dimensional graphene grids and potentially used in supercapacitor devices [59]. Various transition metal nitrides possess good electrochemical properties as well as good chemical stability highly encouraging for fabrication of supercapacitor electrode [60 - 62].

Interestingly, binary nanohybrid was formulated with Mn-anchored Co_3O_4 nanoparticles ($MnCo_3O_4$ NPs) embedded on r-GO nanosheets *via* a one-step hydrothermal process (Fig. **3**). The synthesized nanohybrid shows outstanding conductivity (1.84×104 Sm^{-1}) with a greater BET surface area (147 m^2 g−1). With the increase in current density from 1 to 5 Ag−1, the nanohybrid electrode exhibits an excellent capacitance rate capability of 84.4%, along with exceptional cycling stability [63].

Lithium-ion Battery Field

Lithium-ion batteries (LIBs) have drawn substantial awareness owing to its characteristics off-grid strategy, durable life-cycle ability and profound energy density. LIB shows a wide applications arena from renewable energy, opportune electronic devices, to electric vehicles [64]. Enormous effort has been provided to explore unique intercalation material for both anode and cathode for minimizing the high cost and scanty performance of stationary power systems. Advancement has been underway *via* hybrids of NPs reinforced on the uniform structure of RGO for LIBs [65 - 67]. The promising candidature for anode material of LIB possesses interesting features such as great Li storage capability and distinctive structural architecture for repetitive charge-discharge cycles [68 - 70]. Many innovative composites comprising three-dimensional graphene with several active nanoparticles (*e.g.*, Sn, Si, SnO_2, Si, Mn, Fe_3O_4) have been validated for Li-battery devices. Incorporation of low-cost, high-capacity materials as worthy component. This includes Fe_2O_3 [71, 72] MnO [73], MnO_2 [74, 75], TiO_2 [75, 76], NiO [77], MoO_2 [78], V_2O_5 [79], Co_3O_4 [80], CoO [81], MoS [82], Ni_3S_2 [83], and so on. Among these, Sn NP anchored graphene shells align="center"have been utilized as a LIB anode material. Owing to align="center"their tuned electrical conductivity, extensive surface area, and good mechanical flexibility, the 3D porous graphene framework fostered the flexibility of core–shell Sn/G structure imparting subsequent switches in the electrical conductivity and structural veracity of electrode material [83] (Fig. **4**).

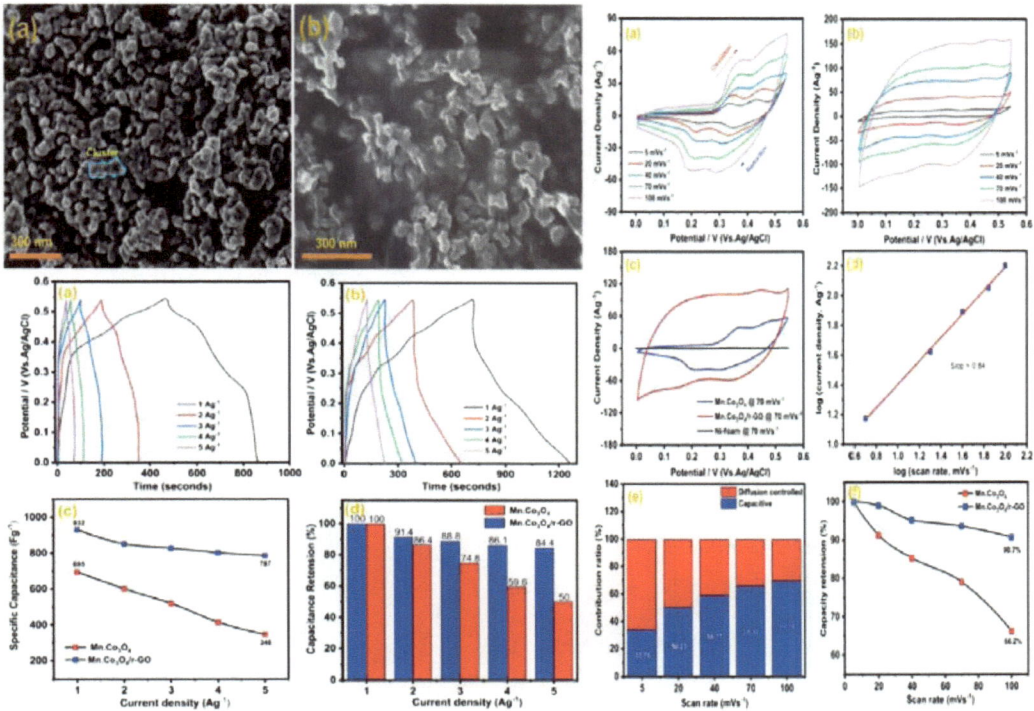

Fig. (3). **(A)** Surface SEM images of **(a)** Mn.CO$_3$O$_4$ and **(b)** Mn.Co$_3$O$_4$/r-GO nanohybrid **(B)** CV voltagrammes of **(a)** Mn.Co$_3$O$_4$, **(b)** Mn.Co$_3$O$_4$/r-GO and **(c)** Comparative study of CV sketch of NF, **(d)** plot of scan rate *vs* current density, **(e)** various parameters study at scan rate variations surface contribution ratio (%) and diffusion controlled capacitance and **(f)** capacity preservation (%) *vs.* scan rates; **(C)** Galvanostatic charge/discharge (GCD) outlines of **(a)** Mn.Co$_3$O$_4$/r-GO and **(b)** Mn.Co$_3$O$_4$, **(c)** Specific capacitance variations, and **(d)** % of capacitance retention *vs.* current density. Reprinted (adapted) with permission from ref 63.

align="center"S. Mao *et al.* developed SnO$_2$ anchored CG nanohybrids and accumulated them on a copper current accumulator foil, utilized as an anode material in LIB [84]. Various research studies reported synthesized 3DG-SnO$_2$ composites using two methodologies, *e.g.* hydrothermal processes [85, 86] ortemplate-assisted method [88] for LIB applications. Huang *et al.* showcased a dual-step fabrication technique wherein SnO$_2$NPs are uniformly decorated on the graphene surface. The resulting 3D SnO$_2$-GFs exhibited huge surface area, plentiful porous morphology and a low molecular density [85]. These fabricated nanocomposites could effectually configure the aspect ratio of SnO$_2$ NPs throughout the cycling process and offer multi-faceted canals for electron passage and electrolytic access. This ultimately facilitates the rapid tunneling of lithium ions between the electrolyte and electrode system. SnO_2-GFs unveiled a high capacity of 830 mA h g−1. Even after 70 charge–discharge cycles, 100 mA.g$^{\text{align="center"}-1}$ aptitude of current was achieved significantly. At a high current density of 500 mA.g$^{\text{align="center"}-1}$, SnO_2-GFs sustained a higher reversible capacitance of 621 mA.h.g$^{\text{align="center"}-1}$ with good charge/discharge constancy. Various researchers have utilized SnO_2 nanoparticles to form a 3D aerogel network with good rate kinetics ability and prolonged cyclic stability [85, 86].

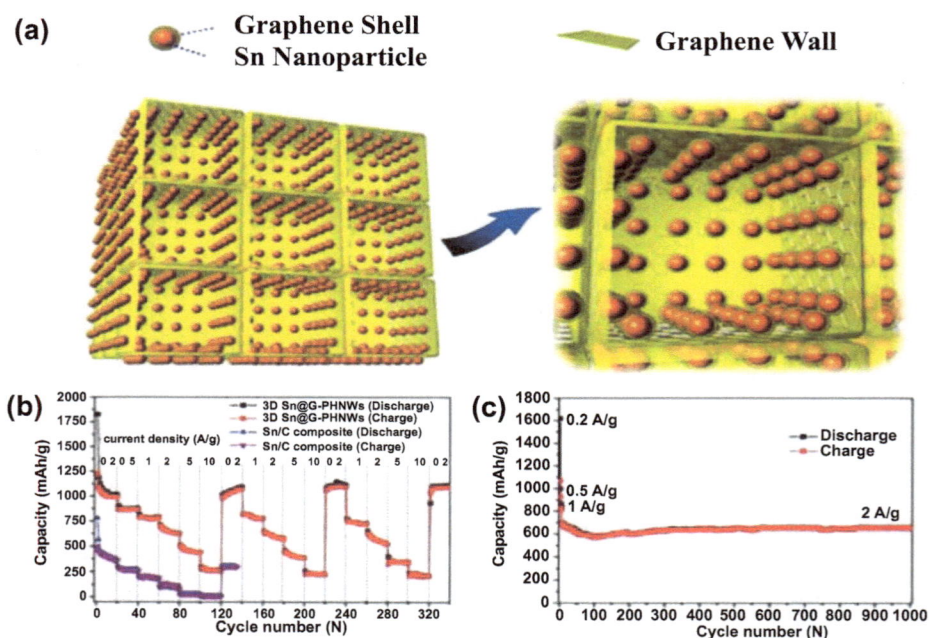

Fig. (4). (a) Schematic diagram of the Sn NP decorated 3D-graphene composite. **(b)** Rate cycle capability study of 3D Sn@G-PGNWs and Sn/C composite electrode at rates ranging from 0.2 to 10 C for 340 cycles duration. **(c)** Capacity evaluation with respect to cycle number of 3D Sn@G-PGNW composite electrode at various current densities for the initial six cycles. Reprinted (adapted) with permission from ref.81.

Zhou *et al.* fabricated ternary composites comprised of silicon, carbon and graphene molecule (abridged as Si–C/G), wherein carbon NPs decorated graphene sheets induced the electron channeling and facilitating conducting matrix [87]. The Si–C/G composite revealed an alterable capacity of 1521 mA h.g^{-1} at 0.2 °C after 200 operating cycles and a skillful magnitude of the current response. In another study Xin and his coworkers adopted a spray-drying method and designed a 3D porous composite of Si-graphene, which displayed a capacity of 1000 $mA.h.g^{-1}$ even after 30 cycles. Several studies have reported that Si imprinted graphene composite showed higher lithium-storage proficiencies hinge on various parameters specific capacity, long life cycle ability, high-rate efficiency and lack of memory effect [88]. Fe_3O_4 is also utilized as a budding candidate for anode material in LIB owing to its unique features such as elevated specific capacity, plentiful nature, cost-effective and nontoxicity [89].

Dye-sensitized Solar Cell

Solar energy is considered the utmost bounteous renewable energy precursor that validated a high prospective to supplant fossil fuel and thus shield the atmosphere from immense imitations that ensued from the conventional practice of fossil fuel usage. Dye-sensitized solar cells (DSSCs) have been deliberated as a methodological and economical substitute for uninterrupted harvesting the solar energy. The basic designing of DSSC includes three counterparts viz.; dye tagged TiO_2 photo anode, a counter electrode and iodide electrolytic system. Amalgamation of novel materials and advanced methodological practices has been vigorously trailed to tune the characteristics performance and lower the cost of DSSCs. CVD grown graphene and reduced graphene oxide have been explored as translucent conducting support [90] in the semiconducting stratum [91] and also act as a sensitizer for the photoanodic material of a DSSC. Owing to excellent electrical conductivity and the porous structural architecture, three-dimensional graphene could be a talented nominee for fabricating DSSC with enhanced photovoltaic properties. To tune the photocurrent density by augmenting electron passaging, light scattering efficiency as well as dye adsorption ability, graphene has been fused within the semiconducting layer. Typically, the uninterrupted networking system between graphene template and photoanode material offers good favorable electron switching for the photo-induced electrons from TiO_2 NPs (photoanode material) to the conductive surface. Various functional nanoparticles have been impregnated onto graphene hydrogel system providing a superior matrix for electron channeling. Han *et al.* adopted a modified hydrothermal method wherein CdS and TiO_2 NPs are anchored onto graphene aerogel [92].

Encouraged by the distinctive features of graphene (*e.g.*, remarkable conductivity, high surface area, and tuned flexibility), various graphene materials and their nanocomposites have been predominantly utilized in the fabrication of DSSC counter electrodes [93 - 97]. Interestingly honeycomb-like structured graphene exhibited energy conversion efficiency of 7.8% [97] whereas CVD assisted the growth of p-doped graphene nano-networks also offered admirable proficiency towards electrode fabrication, showing tuned photo-conversion efficiency around 8.46% [94].

Graphene-based nanocomposites have been widely acknowledged and explored as electrode modifiers for DSSCs. Several metal sulphides such as Ni, Cu, Co, Mo, W *etc.* are considerably effective for counter-electrode catalysts for DSSCs [97]. Amongst them, sulphides derivatives of Ni and Co decorated graphene revealed good electrochemical activity and photovoltaic conversion efficacy as high as 5.04% and 5.25%, respectively [94 - 96]. Moreover, graphene induced honeycomb framework enriched the light-harvesting efficacy and upgraded the photoelectric transfiguration practice. However, at photoanode in DSSC, endurance of the liquid electrolytic medium is extremely questionable due to several factors *e.g.* percolation, corrosion and vaporization. To address this developmental drawback Prabakaran K. *et al.* have adopted polymer electrolyte membrane comprised of graphene with amalgams of polyvinylidene fluoride (PVDF), polyethylene oxide (PEO, hexa fluoro propylene HEP) [95]. Incorporation of graphene leads to an increase in optimum performance percentage of 0.8% and stability of 60% after 1000h cell operation. The synergism of composite films and graphene could offer good electrochemical performance, wide electrode/electrolyte electron exchange region, and steady cycling during the photo-conversion process which could further holding pledge in the plethora of solar energy conversion.

Fuel Cell Applications

Fuel cells are recognized as a sequence of energy conversion tools that typically generate electric power by utilizing the chemical energy of supplied fuels along with low venting of pollutants. The technology behind it has exposed the foremost encouraging sustainable substitute for future energy cradle. Adaptation of appropriate material for fabrication of fuel cell counterparts is highly crucial and imposes multiple constrictions related to electrochemical efficiency, endurance, reproducibility, *etc.* Graphene has captivated incredible devotion owing to its fascinating physicochemical properties. Specifically, graphene also exhibits superior electronic quality as it spectacles half integral quantum Hall effect at ambient temperature [97, 98]. However, voluminous research is still a prerequisite for the further advancement of miniaturization of key components such as proton exchange membranes with high proton conductivity, anodic and cathodic substance, water impermeability as well as biopolar stability *etc.* (Fig. **5**). Graphene oxide (GO) is derived from graphene that comprises functional groups offering large surface area, which assists in the proton exchange through channels and also maintains water permeability favorable for fuel cell performance.

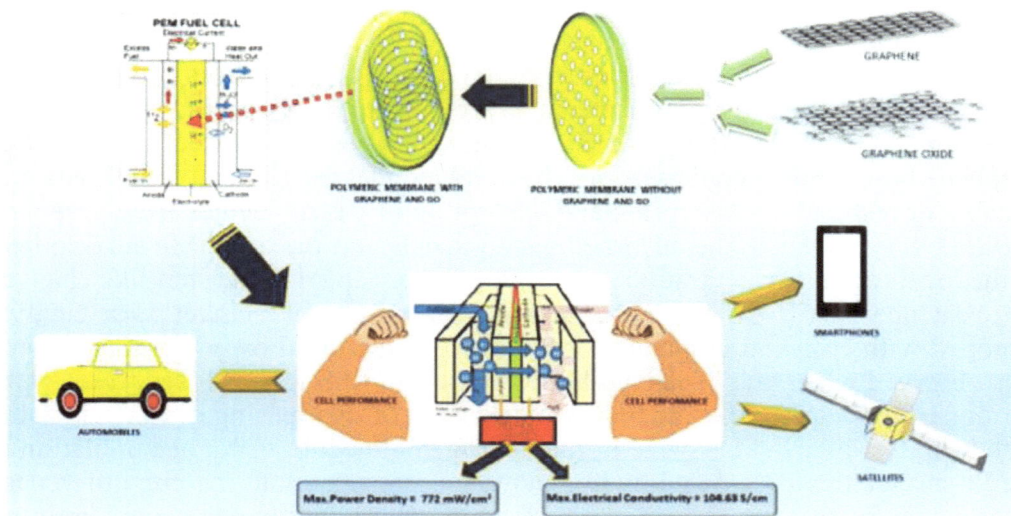

Fig. (5). Schematic illustration of graphene based fuel cell Technology. Reprinted (adapted) with permission from ref. 98.

Several transition metals such as platinum and ruthenium (Pt &Ru) are extensively cast-off for fuel cell's purposes. Pt and its alloys, *e.g.*, Pt–Ni, Pt–Co and Pt–Cu, are the most appreciable for oxidation/ reduction reaction catalysts [99 - 102]. Besides that, graphene possesses good catalytic activities and its preferable catalytic efficiency due to greater conductivity and extensive surface area promote as a potential candidature for fuel cell applications. The non-agglomerated morphological architecture and highly porous assembly of 3DGs enable the packing and diffusion of catalyst through the graphene webs endorsing potent electronic switching between the reaction medium and the electrode surface. PdCu nanocubes anchored onto 3DG framework showed better electrocatalytic activity and stability than Pt-based catalysts [103]. Moreover, this fabricated Pt-PdCu/3DGF modified electrode has shown outstanding peak current response. Nonetheless, a gradual decrease of the current dwindling pattern of Pt-PdCu/3DGF electrode as compared to other electrodes, signifying good stability response and reproducibility of this synthesized catalyst base for long-term maneuver [103]. At the cathode site, various NPs doped graphene nanohybrids have been predominantly used as ORR catalysts. Fe_3O_4 and CoO NPs encapsulated N-doped graphene hydrogel exhibits good synergism wherein the exclusive 3D and hollow structured graphene furnish wide surface zone and uniform matrix providing tuned catalytic efficiency and lowest Tafel slope [104].

Sensor and Sensing Devices

Graphene and its derivatives have hit the skyline as a very glimmering candidate for fabricating trialblazing sensor devices. Graphene based sensors and biosensors have drawn paramount importance in diverse spheres of sensing research like environmental, clinical, food science, *etc.* Schedin *et al.*, the pioneer researcher of graphene based microscopic sensor. They sighted that their miniaturized sensor is capable of responding adsorption and desorption of gas molecules from graphene surfaces very quickly. The adsorbed gas molecules on the graphene surface lead to the local carrier concentration change in graphene, paving the step-like changes in resistance [105]. Chu *et al.* employed fine platinum blanket as a catalyst wrapped with epitaxial graphene for hydrogen detection. Lowering of resistance was observed when graphene engulfed with Pt-thin film was exposed to 1% H_2 at various temperatures. The underlying mechanism of this sensing device lies in the metal catalyst-assisted cleavage of the hydrogen molecule. After accumulation at the Pt-surface the dissociated hydrogen atoms propagate into the circumferences of graphene/pt surface. Herein, the formation of covalent bonds occurs between hydrogen and graphene, which results in an increased work function. Fermi-level shift becomes larger due to the separation of increased distance between graphene and Pt [106]. By using partially reduced GO sheets, Lu *et al.* fabricated a highly efficient gas detecting device. They developed this device by casting well-dispersed GO onto integrated gold (Au) electrodes. During the gas detection study, the magnitude of current variation was measured by administering a covalent DC bias to the fabricated device. This miniaturized device showed a response to the low concentration of NO_2 and p-type semiconducting nature was observed. The sensing study is ascribed to the fact that transferring of an electron occurs from rGO to adsorbed NO_2, which influenced increased hole concentration and increased electrical tunneling within RGO layers [107].

Graphene having magnificent electrochemical and physical properties like high conductivity, large surface area, huge defect sites, fast electron transfer rate, superior electrocatalytic activity, is very adept at evincing gripping performance in the realm of enzymatic biosensors as well as nonenzymatic biosensors [108] Among the vast examples of enzymatic biosensors, graphene-based glucose sensors have drawn unrivaled importance [109]. For glucose detection, the electrode deposited with graphene-based materials is generally modified by the glucose oxidase enzyme in the last stage of electrode preparation. Kang *et al.* developed CS-GR/Gox [Chitosan-Graphene/glucose oxidase enzyme] wrapped sensing platform for electrochemical response study of glucose molecules. This sensor manifests excellent sensitivity of 37.93 $\mu AmM^{-1}Cm^{-2}$ and a wide linear range from 0.08 mM to 12 mM with a detection limit of 0.02 mM. The outstanding performance of the sensor is ascribed to tuned conductivity and wide aspect ratio of graphene as well as good biocompatibility of chitosan molecule. Those coupled factors enhance the absorption of enzyme and fosters the direct transfers of electrons between redox couple enzymes and the electrode surface [110]. Besides that, cerium oxide (CeO_2) nanoparticles anchored RGO will provide an advanced potential electrocatalytic surface for glucose detection [111]. A green biosensing platform comprised of nano-CeO_2/RGO nanohybrid anchored chitosan nanocomposites exhibited a very low detection limit (LOD) of 0.05mM with a varied detection range of 0.05–6.5mM. The synergistic effect of CeO_2 NPs and RGO nanosheet may endorse the electrochemical switching within the electrode surface and the electrolyte imparting high sensitivity of 7.198 µA mM^{-1} cm^{-2}. Xu *et al.* fabricated a CS-GR (chitosan-graphene) coated electrode for electrochemical sensing of Hb (Hemoglobin). They investigated the direct transfer of electron and bioelectrocatalytic activity of hemoglobin after embodiment of composite film [112]. Dey and Raj fabricated high performance amperometric H_2O_2 biosensor wherein graphene and PtNP composite were utilized for detection purpose. Successful entrapment of cholesterol oxidase and cholesterol esterase was done on the GR/PtNP modified electrode surface to develop a cholesterol biosensor. The sensitivity towards cholesterol ester was found 2.07 ± 0.1 µAµM^{-1} cm^{-2} with detection limit 0.2 µM [113]. F. Li. *et al.* explored a novel sensing podium utilizing synthesized Pt/ionic liquid/graphene nanohybrid for the detection of ascorbic acid and dopamine simultaneously by using cyclic voltammetry profiling. The difference between the peak potentials is above 200 mV, which is very helpful for discriminating DA from AA. This sensor also exhibits good reproducibility and life time [114].

In the agriculture section fabrication of plant biosensors is also a paramount important achievement wherein detection of auxin hormone could be essential for the plant growth process. Tryptophan (Trp) is a key regulator for auxin biosynthesis. Electrochemically deposited polydopamine-reduced graphene oxide (RGO)/MnO_2 electrode offered high electrocatalytic activity. The synergism within PDA and RGO/MnO_2 induced the electrocatalytic switching, thereby flourishing the electrochemical performance [115]. In plant biosensor the sensitivity showed 0.39-1.66 mA mM^{-1} within Trp detection range of 1-300mM. A precise low detection range of 0.22-0.39 µM was noticed in the fabricated sensor. In addition to validating the practical applicability, the feasibility of the sensor was also monitored through analysing Trp content in fresh fruit juices. Therefore, the recommendation of graphene-based materials considerably ameliorates the bioanalytical performance of various biosensors, thereby exploring their utilizations in the biosensing plethora for the future realm.

CONCLUDING REMARKS AND FUTURE OUTLOOK

The benefits and prime characteristics of 2D/3D graphene based nanohybrids confer versatile virtues and application avenues in numerous energy storage/conversion applications. Research interests meticulously delve into the modern aspect on synthesis strategies, unique characterizations as well as day-to-day applications. In the recent chapter, we have concentrated on morphologically unique graphene and its nanohybrid/composites and explored their performance and versatility in energy related applications. Also, intense research on graphene sponges and align="center"their scalability significantly urge adaptation of a novel synthesis scheme. Special emphasis and different unique functional aspects can be explored to further advancement of graphene-based composites so that they can withstand and claim the unlike petition in energy related applications. Interestingly in plane holes, namely holey graphene has endowed numerous potential magnitudes for energy storage applications [116]. These resulting in-plane nanoholes can not only hinder restacking but also align="center"enlarging their surface areas by furnishing copious ion/mass transport channeling network. In comparison to pristine graphene, holey graphene offers a distinctive design and architecture with functional, active edges, significant hole-dependent semiconducting band-gap and optical properties. This further explores the application of graphene-based materials for tuned energy storage, catalysis, semiconductor devices, gas separation, bio separation, water purification, *etc.* Simultaneously graphene foam based 2D $CuMnO_2$-Mn_xO_y hybrid nanomaterial is fabricated by G. Saeed *et al.* wherein these electrodes unveil a tremendous areal capacitance of 7.82 F cm^{-2} at a current density value of 3 mA.cm^{-2} [117]. The CuMnO2-MnxOy@GF modified electrode employs a high capacitance of 92.15% after maximum 10,000 cycles. Impressively 3D advanced well-designed coronal hybrid materials with porous graphene-(PG-) based nanowebs were miniaturized, which illustrated three-fold inflation in energy storage capacity. These fascinating coronal hybrids can convey 776 Cg^{-1}, 984 Cg^{-1}, and 1056 Cg^{-1} for Co-Mn LDH@3D-PG, Ni-Mn LDH@3D-PG, and Ni-Co LDH@3D-PG, respectively [118]. Another paramount technical advancement for energy devices was scrutinized by utilizing porous Nb_4N_5/rGO nanocomposite wherein enhanced specific energy and specific power obtained was 295.1 W.h.kg^{-1} and 112.3 W.kg^{-1}, respectively, even after 4000 charge-discharge cycles [119]. In a nutshell, the use of several transition metal oxides (*e.g* CuO, MnO_2, TiO_2, ZnO, *etc.*) based graphene nanohybrid may have showcased promising application in the area of advanced pseudo/hybrid supercapacitors, Li-ion rechargeable batteries, fuel cells and sensor. The production scalability and cost-effectiveness of the graphene-based composites is a pivotal bottleneck for their practical applications because of the high operative materials requirement for energy devices.

Comprehensive understanding of anatomical and functional features of graphene and its nanohybrid could anticipate that versatile energy and environmental application plethora will emerge in future endeavor.

CONSENT FOR PUBLICATION

Not applicable.

CONFLICT OF INTEREST

The authors declare no conflict of interest, financial or otherwise.

ACKNOWLEDGEMENTS

The authors acknowledge Prof. Dipankar Chattopadhyay for their esteemed edits, suggestions and contributions to this book chapter.

REFERENCES

[1]   Sudhakar, K.; Winderl, M.; Priya, S.S. Net-zero building designs in hot and humid climates: A state-of-art. *Case Stud. Therm. Eng.,* **2019**, *13*, 100400.
[http://dx.doi.org/10.1016/j.csite.2019.100400]

[2]   Abbas, N.; Awan, M. B.; Amer, M.; Ammar, S. M.; Sajjad, U.; Ali, H. M.; Zahra, N.; Hussain, M.; Badshah, M. A.; Jafry, A. T. Applications of Nanofluids in Photovoltaic Thermal Systems: A Review of Recent Advances. *Phys. A Stat. Mech. its Appl,* **2019**536, 122513.
[http://dx.doi.org/10.1016/j.physa.2019.122513]

[3]   Hussain, A.; Mehdi, S.M.; Abbas, N.; Hussain, M.; Naqvi, R.A. Synthesis of graphene from solid carbon sources: A focused review. *Mater. Chem. Phys.,* **2020**, *248*, 122924.
[http://dx.doi.org/10.1016/j.matchemphys.2020.122924]

[4]   Roslan, N.; Ya'acob, M.E.; Radzi, M.A.M.; Hashimoto, Y.; Jamaludin, D.; Chen, G. Dye sensitized solar cell (dssc) greenhouse shading: New insights for solar radiation manipulation. *Renew. Sustain. Energy Rev.,* **2018**, *92*, 171-186.
[http://dx.doi.org/10.1016/j.rser.2018.04.095]

[5]   Kim, S.K.; Cho, K.H.; Kim, J.Y.; Byeon, G. Field Study on Operational Performance and Economics of Lithium-Polymer and Lead-Acid Battery Systems for Consumer Load Management. *Renew. Sustain. Energy Rev.,* **2019**, *113*, 109234.
[http://dx.doi.org/10.1016/j.rser.2019.06.041]

[6]   Novoselov, K. S. Electric Field Effect in Atomically Thin Carbon Films. *Science (80-.),* **2004**, *306*(5696), 666-669.
[http://dx.doi.org/10.1126/science.1102896]

[7]   Ramakrishnan, S.; Pradeep, K.R.; Raghul, A.; Senthilkumar, R.; Rangarajan, M.; Kothurkar, N.K. One-step synthesis of pt-decorated graphene–carbon nanotubes for the electrochemical sensing of dopamine, uric acid and ascorbic acid. *Anal. Methods,* **2015**, *7*(2), 779-786.
[http://dx.doi.org/10.1039/C4AY02487G]

[8]   Yu, W.; Sisi, L.; Haiyan, Y. Jie, **L.;** Progress in the functional modification of graphene/graphene oxide: a review. *RSC Advances,* **2020**, *10*, 15328-15345.
[http://dx.doi.org/10.1039/D0RA01068E]

[9]   Geim, A. K. Graphene: Status and Prospects. *Science (80-.),* **2009**, *324*(5934), 1530-1534.

[http://dx.doi.org/10.1126/science.1158877]

[10]   Green, A.A.; Hersam, M.C. Emerging methods for producing monodisperse graphene dispersions. *J. Phys. Chem. Lett.,* **2010**, *1*(2), 544-549.
        [http://dx.doi.org/10.1021/jz900235f] [PMID: 20657758]

[11]   Geim, A.K.; Novoselov, K.S. The rise of graphene. *Nat. Mater.,* **2007**, *6*(3), 183-191.
        [http://dx.doi.org/10.1038/nmat1849] [PMID: 17330084]

[12]   Zhang, Y.; Tang, T-T.; Girit, C.; Hao, Z.; Martin, M.C.; Zettl, A.; Crommie, M.F.; Shen, Y.R.; Wang, F. Direct observation of a widely tunable bandgap in bilayer graphene. *Nature,* **2009**, *459*(7248), 820-823.
        [http://dx.doi.org/10.1038/nature08105] [PMID: 19516337]

[13]   Craciun, M.F.; Russo, S.; Yamamoto, M.; Oostinga, J.B.; Morpurgo, A.F.; Tarucha, S. Trilayer graphene is a semimetal with a gate-tunable band overlap. *Nat. Nanotechnol.,* **2009**, *4*(6), 383-388.
        [http://dx.doi.org/10.1038/nnano.2009.89] [PMID: 19498401]

[14]   Lu, Y.H.; Feng, Y.P. Band-Gap Engineering with Hybrid Graphane–Graphene Nanoribbons. *J. Phys. Chem. C,* **2009**, *113*(49), 20841-20844.
        [http://dx.doi.org/10.1021/jp9067284]

[15]   Stampfer, C.; Güttinger, J.; Hellmüller, S.; Molitor, F.; Ensslin, K.; Ihn, T. Energy gaps in etched graphene nanoribbons. *Phys. Rev. Lett.,* **2009**, *102*(5), 056403.
        [http://dx.doi.org/10.1103/PhysRevLett.102.056403] [PMID: 19257529]

[16]   Saito, K.; Nakamura, J.; Natori, A. Ballistic thermal conductance of a graphene sheet. *Phys. Rev. B Condens. Matter Mater. Phys.,* **2007**, *76*(11), 115409.
        [http://dx.doi.org/10.1103/PhysRevB.76.115409]

[17]   Ghosh, S.; Bao, W.; Nika, D.L.; Subrina, S.; Pokatilov, E.P.; Lau, C.N.; Balandin, A.A. Dimensional crossover of thermal transport in few-layer graphene. *Nat. Mater.,* **2010**, *9*(7), 555-558.
        [http://dx.doi.org/10.1038/nmat2753] [PMID: 20453845]

[18]   Ghosh, S.; Calizo, I.; Teweldebrhan, D.; Pokatilov, E.P.; Nika, D.L.; Balandin, A.A.; Bao, W.; Miao, F.; Lau, C.N. Extremely high thermal conductivity of graphene: prospects for thermal management applications in nanoelectronic circuits. *Appl. Phys. Lett.,* **2008**, *92*(15), 151911.
        [http://dx.doi.org/10.1063/1.2907977]

[19]   Allen, M.J.; Tung, V.C.; Kaner, R.B. Honeycomb carbon: a review of graphene. *Chem. Rev.,* **2010**, *110*(1), 132-145.
        [http://dx.doi.org/10.1021/cr900070d] [PMID: 19610631]

[20]   Reina, A.; Jia, X.; Ho, J.; Nezich, D.; Son, H.; Bulovic, V.; Dresselhaus, M.S.; Kong, J. Large area, few-layer graphene films on arbitrary substrates by chemical vapor deposition. *Nano Lett.,* **2009**, *9*(1), 30-35.
        [http://dx.doi.org/10.1021/nl801827v] [PMID: 19046078]

[21]   Hernandez, Y.; Nicolosi, V.; Lotya, M.; Blighe, F.M.; Sun, Z.; De, S.; McGovern, I.T.; Holland, B.; Byrne, M.; Gun'Ko, Y.K.; Boland, J.J.; Niraj, P.; Duesberg, G.; Krishnamurthy, S.; Goodhue, R.; Hutchison, J.; Scardaci, V.; Ferrari, A.C.; Coleman, J.N. High-yield production of graphene by liquid-phase exfoliation of graphite. *Nat. Nanotechnol.,* **2008**, *3*(9), 563-568.
        [http://dx.doi.org/10.1038/nnano.2008.215] [PMID: 18772919]

[22]   Park, S.; Ruoff, R.S. Chemical methods for the production of graphenes. *Nat. Nanotechnol.,* **2009**, *4*(4), 217-224.
        [http://dx.doi.org/10.1038/nnano.2009.58] [PMID: 19350030]

[23]   Choi, W.; Lahiri, I.; Seelaboyina, R.; Kang, Y.S. Synthesis of Graphene and Its Applications: A Review. *Crit. Rev. Solid State Mater. Sci.,* **2010**, *35*(1), 52-71.
        [http://dx.doi.org/10.1080/10408430903505036]

[24]   Berger, C. Electronic Confinement and Coherence in Patterned Epitaxial Graphene. *Science,* **2006**,

312(5777), 1191-1196.
[http://dx.doi.org/10.1126/science.1125925]

[25]   Gall, N. Rut'kov, E.; Tontegode, A. Y. Intercalation of Nickel Atoms under Two-Dimensional Graphene Film on (111)Ir. *Carbon N. Y.,* **2000**, *38*(5), 663-667.
[http://dx.doi.org/10.1016/S0008-6223(99)00135-9]

[26]   Fan, Z.; Yan, J.; Zhi, L.; Zhang, Q.; Wei, T.; Feng, J.; Zhang, M.; Qian, W.; Wei, F. A three-dimensional carbon nanotube/graphene sandwich and its application as electrode in supercapacitors. *Adv. Mater.,* **2010**, *22*(33), 3723-3728.
[http://dx.doi.org/10.1002/adma.201001029] [PMID: 20652901]

[27]   Zhao, M-Q.; Liu, X-F.; Zhang, Q.; Tian, G-L.; Huang, J-Q.; Zhu, W.; Wei, F. Graphene/single-walled carbon nanotube hybrids: one-step catalytic growth and applications for high-rate Li-S batteries. *ACS Nano,* **2012**, *6*(12), 10759-10769.
[http://dx.doi.org/10.1021/nn304037d] [PMID: 23153374]

[28]   Chang, J.; Huang, X.; Zhou, G.; Cui, S.; Hallac, P.B.; Jiang, J.; Hurley, P.T.; Chen, J. Multilayered Si nanoparticle/reduced graphene oxide hybrid as a high-performance lithium-ion battery anode. *Adv. Mater.,* **2014**, *26*(5), 758-764.
[http://dx.doi.org/10.1002/adma.201302757] [PMID: 24115353]

[29]   Kuilla, T.; Bhadra, S.; Yao, D.; Kim, N.H.; Bose, S.; Lee, J.H. Recent advances in graphene based polymer composites. *Prog. Polym. Sci.,* **2010**, *35*(11), 1350-1375.
[http://dx.doi.org/10.1016/j.progpolymsci.2010.07.005]

[30]   Bagri, A.; Mattevi, C.; Acik, M.; Chabal, Y.J.; Chhowalla, M.; Shenoy, V.B. Structural evolution during the reduction of chemically derived graphene oxide. *Nat. Chem.,* **2010**, *2*(7), 581-587.
[http://dx.doi.org/10.1038/nchem.686] [PMID: 20571578]

[31]   Xu, Y.; Sheng, K.; Li, C.; Shi, G. Self-assembled graphene hydrogel *via* a one-step hydrothermal process. *ACS Nano,* **2010**, *4*(7), 4324-4330.
[http://dx.doi.org/10.1021/nn101187z] [PMID: 20590149]

[32]   Wilson, P.M.; Mbah, G.N.; Smith, T.G.; Schmidt, D.; Lai, R.Y.; Hofmann, T.; Sinitskii, A. Three-Dimensional Periodic Graphene Nanostructures. *J. Mater. Chem. C Mater. Opt. Electron. Devices,* **2014**, *2*(10), 1879.
[http://dx.doi.org/10.1039/c3tc32277g]

[33]   Wu, L.; Li, W.; Li, P.; Liao, S.; Qiu, S.; Chen, M.; Guo, Y.; Li, Q.; Zhu, C.; Liu, L. Powder, paper and foam of few-layer graphene prepared in high yield by electrochemical intercalation exfoliation of expanded graphite. *Small,* **2014**, *10*(7), 1421-1429.
[http://dx.doi.org/10.1002/smll.201302730] [PMID: 24323826]

[34]   Jiang, Y.; Wang, W-N.; Biswas, P.; Fortner, J.D. Facile aerosol synthesis and characterization of ternary crumpled graphene-TiO$_2$-magnetite nanocomposites for advanced water treatment. *ACS Appl. Mater. Interfaces,* **2014**, *6*(14), 11766-11774.
[http://dx.doi.org/10.1021/am5025275] [PMID: 24983817]

[35]   Luo, J.; Zhao, X.; Wu, J.; Jang, H.D.; Kung, H.H.; Huang, J. Crumpled Graphene-Encapsulated Si Nanoparticles for Lithium Ion Battery Anodes. *J. Phys. Chem. Lett.,* **2012**, *3*(13), 1824-1829.
[http://dx.doi.org/10.1021/jz3006892] [PMID: 26291867]

[36]   Sohn, K.; Joo Na, Y.; Chang, H.; Roh, K-M.; Dong Jang, H.; Huang, J. Oil absorbing graphene capsules by capillary molding. *Chem. Commun. (Camb.),* **2012**, *48*(48), 5968-5970.
[http://dx.doi.org/10.1039/c2cc32049e] [PMID: 22569878]

[37]   Yu, K.; Wen, Z.; Pu, H.; Lu, G.; Bo, Z.; Kim, H.; Qian, Y.; Andrew, E.; Mao, S.; Chen, J. Hierarchical vertically oriented graphene as a catalytic counter electrode in dye-sensitized solar cells. *J. Mater. Chem. A Mater. Energy Sustain.,* **2013**, *1*(2), 188-193.
[http://dx.doi.org/10.1039/C2TA00380E]

[38]   Yang, C.; Bi, H.; Wan, D.; Huang, F.; Xie, X.; Jiang, M. Direct PECVD growth of vertically erected graphene walls on dielectric substrates as excellent multifunctional electrodes. *J. Mater. Chem. A Mater. Energy Sustain.,* **2013**, *1*(3), 770-775.
[http://dx.doi.org/10.1039/C2TA00234E]

[39]   Davami, K.; Shaygan, M.; Kheirabi, N.; Zhao, J.; Kovalenko, D.A.; Rümmeli, M.H.; Opitz, J.; Cuniberti, G.; Lee, J-S.; Meyyappan, M. Synthesis and characterization of carbon nanowalls on different substrates by radio frequency plasma enhanced chemical vapor deposition. *Carbon N. Y.,* **2014**, *72*, 372-380.
[http://dx.doi.org/10.1016/j.carbon.2014.02.025]

[40]   Ma, J.; Alfè, D.; Michaelides, A.; Wang, E. Stone-wales defects in graphene and other planar Sp^2 -bonded materials. *Phys. Rev. B Condens. Matter Mater. Phys.,* **2009**, *80*(3), 033407.
[http://dx.doi.org/10.1103/PhysRevB.80.033407]

[41]   Susi, T.; Kaukonen, M.; Havu, P.; Ljungberg, M.P.; Ayala, P.; Kauppinen, E.I. Core level binding energies of functionalized and defective graphene. *Beilstein J. Nanotechnol.,* **2014**, *5*, 121-132.
[http://dx.doi.org/10.3762/bjnano.5.12] [PMID: 24605278]

[42]   Mohan, V.B.; Lau, K.; Hui, D.; Bhattacharyya, D. Graphene-Based Materials and Their Composites: A Review on Production, Applications and Product Limitations. *Compos., Part B Eng.,* **2018**, *142*, 200-220.
[http://dx.doi.org/10.1016/j.compositesb.2018.01.013]

[43]   Du, A. In Silico Engineering of Graphene-Based van Der Waals Heterostructured Nanohybrids for Electronics and Energy Applications. *Wiley Interdiscip. Rev. Comput. Mol. Sci.,* **2016**, *6*(5), 551-570.
[http://dx.doi.org/10.1002/wcms.1266]

[44]   Zhang, L.L.; Zhao, X.S. Carbon-based materials as supercapacitor electrodes. *Chem. Soc. Rev.,* **2009**, *38*(9), 2520-2531.
[http://dx.doi.org/10.1039/b813846j] [PMID: 19690733]

[45]   Futaba, D.N.; Hata, K.; Yamada, T.; Hiraoka, T.; Hayamizu, Y.; Kakudate, Y.; Tanaike, O.; Hatori, H.; Yumura, M.; Iijima, S. Shape-engineerable and highly densely packed single-walled carbon nanotubes and their application as super-capacitor electrodes. *Nat. Mater.,* **2006**, *5*(12), 987-994.
[http://dx.doi.org/10.1038/nmat1782] [PMID: 17128258]

[46]   Wen, Z.; Wang, X.; Mao, S.; Bo, Z.; Kim, H.; Cui, S.; Lu, G.; Feng, X.; Chen, J. Crumpled nitrogen-doped graphene nanosheets with ultrahigh pore volume for high-performance supercapacitor. *Adv. Mater.,* **2012**, *24*(41), 5610-5616.
[http://dx.doi.org/10.1002/adma.201201920] [PMID: 22890786]

[47]   Talebian, S.; Mehrali, M. Raad,R.; Safaei,F.; Xi,J.; Liu,Z and Foroughi,J. Electrically conducting hydrogel graphene nanocomposite biofibers for biomedical applications. *Front Chem.,* **2020**, *27*
[http://dx.doi.org/10.3389/fchem.2020.00088] [PMID: 32175306]

[48]   Wu, Z-S.; Winter, A.; Chen, L.; Sun, Y.; Turchanin, A.; Feng, X.; Müllen, K. Three-dimensional nitrogen and boron co-doped graphene for high-performance all-solid-state supercapacitors. *Adv. Mater.,* **2012**, *24*(37), 5130-5135.
[http://dx.doi.org/10.1002/adma.201201948] [PMID: 22807002]

[49]   Xu, Z.; Li, Z.; Holt, C.M.B.; Tan, X.; Wang, H.; Amirkhiz, B.S.; Stephenson, T.; Mitlin, D. Electrochemical Supercapacitor Electrodes from Sponge-like Graphene Nanoarchitectures with Ultrahigh Power Density. *J. Phys. Chem. Lett.,* **2012**, *3*(20), 2928-2933.
[http://dx.doi.org/10.1021/jz301207g] [PMID: 26292228]

[50]   Zhi, J.; Zhao, W.; Liu, X.; Chen, A.; Liu, Z.; Huang, F. Highly Conductive Ordered Mesoporous Carbon Based Electrodes Decorated by 3D Graphene and 1D Silver Nanowire for Flexible Supercapacitor. *Adv. Funct. Mater.,* **2014**, *24*(14), 2013-2019.
[http://dx.doi.org/10.1002/adfm.201303082]

[51]   Luo, J.; Jang, H.D.; Huang, J. Effect of sheet morphology on the scalability of graphene-based ultracapacitors. *ACS Nano*, **2013**, *7*(2), 1464-1471.
[http://dx.doi.org/10.1021/nn3052378] [PMID: 23350607]

[52]   Chen, J.; Sheng, K.; Luo, P.; Li, C.; Shi, G. Graphene hydrogels deposited in nickel foams for high-rate electrochemical capacitors. *Adv. Mater.*, **2012**, *24*(33), 4569-4573.
[http://dx.doi.org/10.1002/adma.201201978] [PMID: 22786775]

[53]   Li, L.; Hu, Z.; Yang, Y.; Liang, P.; Lu, A.; Xu, H.; Hu, Y.; Wu, H. Hydrothermal self-assembly synthesis of Mn_3O_4 /reduced graphene oxide hydrogel and its high electrochemical performance for supercapacitors. *Chin. J. Chem.*, **2013**, *31*(10), 1290-1298.
[http://dx.doi.org/10.1002/cjoc.201300324]

[54]   Yuan, J.; Zhu, J.; Bi, H.; Zhang, Z.; Chen, S.; Liang, S.; Wang, X. Self-assembled hydrothermal synthesis for producing a $MnCO_3$/graphene hydrogel composite and its electrochemical properties. *RSC Advances*, **2013**, *3*(13), 4400.
[http://dx.doi.org/10.1039/c3ra23477k]

[55]   Cao, X.; Shi, Y.; Shi, W.; Lu, G.; Huang, X.; Yan, Q.; Zhang, Q.; Zhang, H. Preparation of novel 3D graphene networks for supercapacitor applications. *Small*, **2011**, *7*(22), 3163-3168.
[http://dx.doi.org/10.1002/smll.201100990] [PMID: 21932252]

[56]   Yuan, J.; Zhu, J.; Bi, H.; Meng, X.; Liang, S.; Zhang, L.; Wang, X. Graphene-based 3D composite hydrogel by anchoring Co_3O_4 nanoparticles with enhanced electrochemical properties. *Phys. Chem. Chem. Phys.*, **2013**, *15*(31), 12940-12945.
[http://dx.doi.org/10.1039/c3cp51710a] [PMID: 23812434]

[57]   Wang, H.; Yi, H.; Chen, X.; Wang, X. One-Step strategy to three-dimensional graphene/VO_2 nanobelt composite hydrogels for high performance supercapacitors. *J. Mater. Chem. A Mater. Energy Sustain.*, **2014**, *2*(4), 1165-1173.
[http://dx.doi.org/10.1039/C3TA13932H]

[58]   Zhang, H.; Xie, A.; Wang, C.; Wang, H.; Shen, Y.; Tian, X. Bifunctional reduced graphene oxide/V_2O_5 composite hydrogel: fabrication, high performance as electromagnetic wave absorbent and supercapacitor. *ChemPhysChem*, **2014**, *15*(2), 366-373.
[http://dx.doi.org/10.1002/cphc.201300822] [PMID: 24318771]

[59]   Zhang, Z.; Xiao, F.; Guo, Y.; Wang, S.; Liu, Y. One-pot self-assembled three-dimensional TiO_2-graphene hydrogel with improved adsorption capacities and photocatalytic and electrochemical activities. *ACS Appl. Mater. Interfaces*, **2013**, *5*(6), 2227-2233.
[http://dx.doi.org/10.1021/am303299r] [PMID: 23429833]

[60]   Qi, H.; Yick, S.; Francis, O.; Murdock, A.; van der Laan, T.; Ostrikov, K. (Ken); Bo, Z.; Han, Z.; Bendavid, A. Nanohybrid TiN/Vertical Graphene for High-Performance Supercapacitor Applications. *Energy Storage Mater.*, **2020**, *26*, 138-146.
[http://dx.doi.org/10.1016/j.ensm.2019.12.040]

[61]   Amani-Ghadim, A.R.; Khodam, F.; Aber, S.; Seyed Dorraji, M.S. Mesoporous CuZnAl-Layered Double Hydroxide/Graphene Oxide Nanohybrid as an Energy Storage Electrode for Supercapacitor Application. *Bull. Mater. Sci.*, **2021**, *44*(1), 61.
[http://dx.doi.org/10.1007/s12034-020-02345-7]

[62]   Chakrabarty, N.; Dey, A.; Krishnamurthy, S.; Chakraborty, A.K. CeO_2/Ce_2O_3 Quantum Dot Decorated Reduced Graphene Oxide Nanohybrid as Electrode for Supercapacitor. *Appl. Surf. Sci.*, **2021**, *536*, 147960.
[http://dx.doi.org/10.1016/j.apsusc.2020.147960]

[63]   Aadil, M.; Zulfiqar, S.; Agboola, P.O.; Aly Aboud, M.F.; Shakir, I.; Warsi, M.F. Fabrication of graphene supported binary nanohybrid with multiple approaches for electrochemical energy storage applications. *Synth. Met.*, **2021**, *272*, 116645.
[http://dx.doi.org/10.1016/j.synthmet.2020.116645]

[64]   Tarascon, J-M.; Armand, M. Issues and challenges facing rechargeable lithium batteries. *Nature,* **2001**, *414*(6861), 359-367.
        [http://dx.doi.org/10.1038/35104644] [PMID: 11713543]

[65]   Bruce, P.G. Energy storage beyond the horizon: rechargeable lithium batteries. *Solid State Ion.,* **2008**, *179*(21–26), 752-760.
        [http://dx.doi.org/10.1016/j.ssi.2008.01.095]

[66]   Su, F-Y.; You, C.; He, Y-B.; Lv, W.; Cui, W.; Jin, F.; Li, B.; Yang, Q-H.; Kang, F. Flexible and planar graphene conductive additives for lithium-ion batteries. *J. Mater. Chem.,* **2010**, *20*(43), 9644.
        [http://dx.doi.org/10.1039/c0jm01633k]

[67]   Zhou, G.; Wang, D-W.; Li, F.; Zhang, L.; Li, N.; Wu, Z-S.; Wen, L.; Lu, G.Q. (Max); Cheng, H.-M. Graphene-Wrapped Fe_3O_4 anode material with improved reversible capacity and cyclic stability for lithium ion batteries. *Chem. Mater.,* **2010**, *22*(18), 5306-5313.
        [http://dx.doi.org/10.1021/cm101532x]

[68]   Cao, X.; Zheng, B.; Rui, X.; Shi, W.; Yan, Q.; Zhang, H. Metal oxide-coated three-dimensional graphene prepared by the use of metal-organic frameworks as precursors. *Angew. Chem. Int. Ed. Engl.,* **2014**, *53*(5), 1404-1409.
        [http://dx.doi.org/10.1002/anie.201308013] [PMID: 24459058]

[69]   Xiao, L.; Wu, D.; Han, S.; Huang, Y.; Li, S.; He, M.; Zhang, F.; Feng, X. Self-assembled Fe_2O_3/graphene aerogel with high lithium storage performance. *ACS Appl. Mater. Interfaces,* **2013**, *5*(9), 3764-3769.
        [http://dx.doi.org/10.1021/am400387t] [PMID: 23551107]

[70]   Jiang, X.; Yang, X.; Zhu, Y.; Jiang, H.; Yao, Y.; Zhao, P.; Li, C. 3D Nitrogen-doped graphene foams embedded with ultrafine TiO_2 nanoparticles for high-performance lithium-ion batteries. *J. Mater. Chem. A Mater. Energy Sustain.,* **2014**, *2*(29), 11124.
        [http://dx.doi.org/10.1039/c4ta01348d]

[71]   Shen, L.; Zhang, X.; Li, H.; Yuan, C.; Cao, G. Design and tailoring of a three-dimensional TiO_2–Graphene–carbon nanotube nanocomposite for fast lithium storage. *J. Phys. Chem. Lett.,* **2011**, *2*(24), 3096-3101.
        [http://dx.doi.org/10.1021/jz201456p]

[72]   Choi, S.H.; Ko, Y.N.; Lee, J-K.; Kang, Y.C. Rapid continuous synthesis of spherical reduced graphene ball-nickel oxide composite for lithium ion batteries. *Sci. Rep.,* **2014**, *4*(1), 5786.
        [http://dx.doi.org/10.1038/srep05786] [PMID: 25167932]

[73]   Choi, S.H.; Kang, Y.C. Crumpled graphene-molybdenum oxide composite powders: preparation and application in lithium-ion batteries. *ChemSusChem,* **2014**, *7*(2), 523-528.
        [http://dx.doi.org/10.1002/cssc.201300838] [PMID: 24243867]

[74]   Choi, S.H.; Kang, Y.C. Uniform decoration of vanadium oxide nanocrystals on reduced graphene-oxide balls by an aerosol process for lithium-ion battery cathode material. *Chemistry,* **2014**, *20*(21), 6294-6299.
        [http://dx.doi.org/10.1002/chem.201400134] [PMID: 24715540]

[75]   Choi, B.G.; Chang, S-J.; Lee, Y.B.; Bae, J.S.; Kim, H.J.; Huh, Y.S. 3D heterostructured architectures of Co3O4 nanoparticles deposited on porous graphene surfaces for high performance of lithium ion batteries. *Nanoscale,* **2012**, *4*(19), 5924-5930.
        [http://dx.doi.org/10.1039/c2nr31438j] [PMID: 22899185]

[76]   Ma, J.; Wang, J.; He, Y-S.; Liao, X-Z.; Chen, J.; Wang, J-Z.; Yuan, T.; Ma, Z-F. A Solvothermal Strategy: One-Step in Situ Synthesis of Self-Assembled 3D Graphene-Based Composites with Enhanced Lithium Storage Capacity. *J. Mater. Chem. A Mater. Energy Sustain.,* **2014**, *2*(24), 9200-9207.
        [http://dx.doi.org/10.1039/C4TA01006J]

[77]   Cao, X.; Shi, Y.; Shi, W.; Rui, X.; Yan, Q.; Kong, J.; Zhang, H. Preparation of MoS2-coated three-dimensional graphene networks for high-performance anode material in lithium-ion batteries. *Small,* **2013,** *9*(20), 3433-3438.
[http://dx.doi.org/10.1002/smll.201202697] [PMID: 23637090]

[78]   Wang, J.; Liu, J.; Chao, D.; Yan, J.; Lin, J.; Shen, Z.X. Self-assembly of honeycomb-like MoS$_2$ nanoarchitectures anchored into graphene foam for enhanced lithium-ion storage. *Adv. Mater.,* **2014,** *26*(42), 7162-7169.
[http://dx.doi.org/10.1002/adma.201402728] [PMID: 25250514]

[79]   Zhao, Y.; Feng, J.; Liu, X.; Wang, F.; Wang, L.; Shi, C.; Huang, L.; Feng, X.; Chen, X.; Xu, L.; Yan, M.; Zhang, Q.; Bai, X.; Wu, H.; Mai, L. Self-adaptive strain-relaxation optimization for high-energy lithium storage material through crumpling of graphene. *Nat. Commun.,* **2014,** *5*(1). 4565.
[http://dx.doi.org/10.1038/ncomms5565] [PMID: 25081187]

[80]   Wang, C.; Su, K.; Wan, W.; Guo, H.; Zhou, H.; Chen, J.; Zhang, X.; Huang, Y. High sulfur loading composite wrapped by 3d nitrogen-doped graphene as a cathode material for lithium–sulfur batteries. *J. Mater. Chem. A Mater. Energy Sustain.,* **2014,** *2*(14), 5018-5023.
[http://dx.doi.org/10.1039/C3TA14921H]

[81]   Qin, J.; He, C.; Zhao, N.; Wang, Z.; Shi, C.; Liu, E-Z.; Li, J. Graphene networks anchored with sn@graphene as lithium ion battery anode. *ACS Nano,* **2014,** *8*(2), 1728-1738.
[http://dx.doi.org/10.1021/nn406105n] [PMID: 24400945]

[82]   Mao, S.; Wen, Z.; Kim, H.; Lu, G.; Hurley, P.; Chen, J. A general approach to one-pot fabrication of crumpled graphene-based nanohybrids for energy applications. *ACS Nano,* **2012,** *6*(8), 7505-7513.
[http://dx.doi.org/10.1021/nn302818j] [PMID: 22838735]

[83]   Huang, Y.; Wu, D.; Han, S.; Li, S.; Xiao, L.; Zhang, F.; Feng, X. Assembly of tin oxide/graphene nanosheets into 3D hierarchical frameworks for high-performance lithium storage. *ChemSusChem,* **2013,** *6*(8), 1510-1515.
[http://dx.doi.org/10.1002/cssc.201300109] [PMID: 23784753]

[84]   Wang, R.; Xu, C.; Sun, J.; Gao, L.; Yao, H. Solvothermal-induced 3D macroscopic SnO$_2$/nitrogen-doped graphene aerogels for high capacity and long-life lithium storage. *ACS Appl. Mater. Interfaces,* **2014,** *6*(5), 3427-3436.
[http://dx.doi.org/10.1021/am405557c] [PMID: 24555873]

[85]   Liu, X.; Cheng, J.; Li, W.; Zhong, X.; Yang, Z.; Gu, L.; Yu, Y. Superior lithium storage in a 3D macroporous graphene framework/SnO$_2$ nanocomposite. *Nanoscale,* **2014,** *6*(14), 7817-7822.
[http://dx.doi.org/10.1039/c4nr01493f] [PMID: 24910323]

[86]   Xin, X.; Zhou, X.; Wang, F.; Yao, X.; Xu, X.; Zhu, Y.; Liu, Z. A 3D Porous Architecture of Si/Graphene Nanocomposite as High-Performance Anode Materials for Li-Ion Batteries. *J. Mater. Chem.,* **2012,** *22*(16), 7724.
[http://dx.doi.org/10.1039/c2jm00120a]

[87]   Zhou, X.; Yin, Y-X.; Cao, A-M.; Wan, L-J.; Guo, Y-G. Efficient 3D conducting networks built by graphene sheets and carbon nanoparticles for high-performance silicon anode. *ACS Appl. Mater. Interfaces,* **2012,** *4*(5), 2824-2828.
[http://dx.doi.org/10.1021/am3005576] [PMID: 22563769]

[88]   Wang, X.; Zhi, L.; Müllen, K. Transparent, conductive graphene electrodes for dye-sensitized solar cells. *Nano Lett.,* **2008,** *8*(1), 323-327.
[http://dx.doi.org/10.1021/nl072838r] [PMID: 18069877]

[89]   Yang, N.; Zhai, J.; Wang, D.; Chen, Y.; Jiang, L. Two-dimensional graphene bridges enhanced photoinduced charge transport in dye-sensitized solar cells. *ACS Nano,* **2010,** *4*(2), 887-894.
[http://dx.doi.org/10.1021/nn901660v] [PMID: 20088539]

[90]   Han, W.; Ren, L.; Gong, L.; Qi, X.; Liu, Y.; Yang, L.; Wei, X.; Zhong, J. Self-Assembled three-

dimensional graphene-based aerogel with embedded multifarious functional nanoparticles and its excellent photoelectrochemical activities. *ACS Sustain. Chem.& Eng.,* **2014**, *2*(4), 741-748.
[http://dx.doi.org/10.1021/sc400417u]

[91]   H.; Song,M.; Ameen.S; Akhtar,S.; ShikShin, H.; New counter electrode of hot filament chemical vapor deposited graphene thin film for dye sensitized solar cell. *Chem. Eng. J.,* **2013**, *2013*(15), 464-471.

[92]   Ahn, H-J.; Kim, I-H.; Yoon, J-C.; Kim, S-I.; Jang, J-H. p-Doped three-dimensional graphene nano-networks superior to platinum as a counter electrode for dye-sensitized solar cells. *Chem. Commun. (Camb.),* **2014**, *50*(19), 2412-2415.
[http://dx.doi.org/10.1039/C3CC48920E] [PMID: 24427775]

[93]   Hong, W.; Xu, Y.; Lu, G.; Li, C.; Shi, G. Transparent Graphene/PEDOT–PSS Composite Films as Counter Electrodes of Dye-Sensitized Solar Cells. *Electrochem. Commun.,* **2008**, *10*(10), 1555-1558.
[http://dx.doi.org/10.1016/j.elecom.2008.08.007]

[94]   Kavan, L.; Yum, J-H.; Nazeeruddin, M.K.; Grätzel, M. Graphene nanoplatelet cathode for Co(III)/(II) mediated dye-sensitized solar cells. *ACS Nano,* **2011**, *5*(11), 9171-9178.
[http://dx.doi.org/10.1021/nn203416d] [PMID: 21995546]

[95]   Wang, H.; Sun, K.; Tao, F.; Stacchiola, D.J.; Hu, Y.H. 3D honeycomb-like structured graphene and its high efficiency as a counter-electrode catalyst for dye-sensitized solar cells. *Angew. Chem. Int. Ed. Engl.,* **2013**, *52*(35), 9210-9214.
[http://dx.doi.org/10.1002/anie.201303497] [PMID: 23897636]

[96]   Mulmudi, H.K.; Batabyal, S.K.; Rao, M.; Prabhakar, R.R.; Mathews, N.; Lam, Y.M.; Mhaisalkar, S.G. Solution processed transition metal sulfides: application as counter electrodes in dye sensitized solar cells (DSCs). *Phys. Chem. Chem. Phys.,* **2011**, *13*(43), 19307-19309.
[http://dx.doi.org/10.1039/c1cp22817j] [PMID: 21964615]

[97]   Prabakaran, K.; Jandas, P.J.; Mohanty, S.; Nayak, S.K. Synthesis, characterization of reduced graphene oxide nanosheets and its reinforcement effect on polymer electrolyte for dye sensitized solar cell applications. *Sol. Energy,* **2018**, *170*, 442-453.
[http://dx.doi.org/10.1016/j.solener.2018.05.008]

[98]   Yadav, R.; Subhash, A.; Chemmenchery, N.; Kandasubramanian, B. Graphene and graphene oxide for fuel cell technology. *Ind. Eng. Chem. Res.,* **2018**, *57*(29), 9333-9350.
[http://dx.doi.org/10.1021/acs.iecr.8b02326]

[99]   Ahn, H.S.; Kim, J.M.; Park, C.; Jang, J-W.; Lee, J.S.; Kim, H.; Kaviany, M.; Kim, M.H. A novel role of three dimensional graphene foam to prevent heater failure during boiling. *Sci. Rep.,* **2013**, *3*(1), 1960.
[http://dx.doi.org/10.1038/srep01960] [PMID: 23743619]

[100] Stamenkovic, V. R.; Fowler, B.; Mun, B. S.; Wang, G.; Ross, P. N.; Lucas, C. A.; Markovic, N. M. Improved Oxygen Reduction Activity on Pt3Ni(111) *via* Increased Surface Site Availability. *Science,* **2007**, *315*(5811), 493-497.
[http://dx.doi.org/10.1126/science.1135941]

[101] Wang, D.; Xin, H.L.; Hovden, R.; Wang, H.; Yu, Y.; Muller, D.A.; DiSalvo, F.J.; Abruña, H.D. Structurally ordered intermetallic platinum-cobalt core-shell nanoparticles with enhanced activity and stability as oxygen reduction electrocatalysts. *Nat. Mater.,* **2013**, *12*(1), 81-87.
[http://dx.doi.org/10.1038/nmat3458] [PMID: 23104154]

[102] Wang, D.; Yu, Y.; Xin, H.L.; Hovden, R.; Ercius, P.; Mundy, J.A.; Chen, H.; Richard, J.H.; Muller, D.A.; DiSalvo, F.J.; Abruña, H.D. Tuning oxygen reduction reaction activity *via* controllable dealloying: a model study of ordered Cu3Pt/C intermetallic nanocatalysts. *Nano Lett.,* **2012**, *12*(10), 5230-5238.
[http://dx.doi.org/10.1021/nl302404g] [PMID: 22954373]

[103] Hu, C.; Cheng, H.; Zhao, Y.; Hu, Y.; Liu, Y.; Dai, L.; Qu, L. Newly-designed complex ternary

Pt/PdCu nanoboxes anchored on three-dimensional graphene framework for highly efficient ethanol oxidation. *Adv. Mater.,* **2012**, *24*(40), 5493-5498.
[http://dx.doi.org/10.1002/adma.201200498] [PMID: 22886893]

[104]   Mao, S.; Wen, Z.; Huang, T.; Hou, Y.; Chen, J. High-Performance Bi-Functional Electrocatalysts of 3D Crumpled Graphene–Cobalt Oxide Nanohybrids for Oxygen Reduction and Evolution Reactions. *Energy Environ. Sci.,* **2014**, *7*(2), 609-616.
[http://dx.doi.org/10.1039/C3EE42696C]

[105]   Schedin, F.; Geim, A.K.; Morozov, S.V.; Hill, E.W.; Blake, P.; Katsnelson, M.I.; Novoselov, K.S. Detection of individual gas molecules adsorbed on graphene. *Nat. Mater.,* **2007**, *6*(9), 652-655.
[http://dx.doi.org/10.1038/nmat1967] [PMID: 17660825]

[106]   Chu, B.H.; Lo, C.F.; Nicolosi, J.; Chang, C.Y.; Chen, V.; Strupinski, W.; Pearton, S.J.; Ren, F. Hydrogen Detection Using Platinum Coated Graphene Grown on SiC. *Sens. Actuators B Chem.,* **2011**, *157*(2), 500-503.
[http://dx.doi.org/10.1016/j.snb.2011.05.007]

[107]   Lu, G.; Ocola, L.E.; Chen, J. Gas detection using low-temperature reduced graphene oxide sheets. *Appl. Phys. Lett.,* **2009**, *94*(8), 083111.
[http://dx.doi.org/10.1063/1.3086896]

[108]   Wu, S.; He, Q.; Tan, C.; Wang, Y.; Zhang, H. Graphene-based electrochemical sensors. *Small,* **2013**, *9*(8), 1160-1172.
[http://dx.doi.org/10.1002/smll.201202896] [PMID: 23494883]

[109]   Jang, H.D.; Kim, S.K.; Chang, H.; Roh, K-M.; Choi, J-W.; Huang, J. A glucose biosensor based on TiO₂-Graphene composite. *Biosens. Bioelectron.,* **2012**, *38*(1), 184-188.
[http://dx.doi.org/10.1016/j.bios.2012.05.033] [PMID: 22705409]

[110]   Kang, X.; Wang, J.; Wu, H.; Aksay, I.A.; Liu, J.; Lin, Y. Glucose oxidase-graphene-chitosan modified electrode for direct electrochemistry and glucose sensing. *Biosens. Bioelectron.,* **2009**, *25*(4), 901-905.
[http://dx.doi.org/10.1016/j.bios.2009.09.004] [PMID: 19800781]

[111]   De, S.; Mohanty, S.; Nayak, S.K. Nano-CeO2 decorated graphene based chitosan nanocomposites as enzymatic biosensing platform: fabrication and cellular biocompatibility assessment. *Bioprocess Biosyst. Eng.,* **2015**, *38*(9), 1671-1683.
[http://dx.doi.org/10.1007/s00449-015-1408-5] [PMID: 25980384]

[112]   Xu, H.; Dai, H.; Chen, G. Direct electrochemistry and electrocatalysis of hemoglobin protein entrapped in graphene and chitosan composite film. *Talanta,* **2010**, *81*(1-2), 334-338.
[http://dx.doi.org/10.1016/j.talanta.2009.12.006] [PMID: 20188928]

[113]   Dey, R.S.; Raj, C.R. Development of an amperometric cholesterol biosensor based on graphene–Pt nanoparticle hybrid material. *J. Phys. Chem. C,* **2010**, *114*(49), 21427-21433.
[http://dx.doi.org/10.1021/jp105895a]

[114]   Li, F.; Chai, J.; Yang, H.; Han, D.; Niu, L. Synthesis of Pt/ionic liquid/graphene nanocomposite and its simultaneous determination of ascorbic acid and dopamine. *Talanta,* **2010**, *81*(3), 1063-1068.
[http://dx.doi.org/10.1016/j.talanta.2010.01.061] [PMID: 20298894]

[115]   Gao, J.; Li, H.; Li, M.; Wang, G.; Long, Y.; Li, P.; Li, C.; Yang, B. Polydopamine/graphene/MnO₂ composite-based electrochemical sensor for in situ determination of free tryptophan in plants. *Anal. Chim. Acta,* **2021**, *1145*, 103-113.
[http://dx.doi.org/10.1016/j.aca.2020.11.008] [PMID: 33453871]

[116]   Chen, Z.; An, X.; Dai, L.; Xu, Y. Holey Graphene-Based Nanocomposites for Efficient Electrochemical Energy Storage. *Nano Energy,* **2020**, *73*, 104762.
[http://dx.doi.org/10.1016/j.nanoen.2020.104762]

[117]   Saeed, G.; Alam, A.; Bandyopadhyay, P.; Lim, S.; Kim, N.H.; Lee, J.H. Development of hierarchically structured nanosheet arrays of CuMnO2-MnxOy@graphene foam as a nanohybrid electrode material

for high-performance asymmetric supercapacitor. *J. Alloys Compd.,* **2021**, *858*, 158343.
[http://dx.doi.org/10.1016/j.jallcom.2020.158343]

[118] Sarigamala, K.K.; Shukla, S.; Struck, A.; Saxena, S. Graphene-based coronal hybrids for enhanced energy storage. *Energy Mater. Adv.,* **2021**, *2021*, 1-15.
[http://dx.doi.org/10.34133/2021/7273851]

[119] Li, S.; Wang, T.; Huang, Y.; Wei, Z.; Li, G.; Ng, D.H.L.; Lian, J.; Qiu, J.; Zhao, Y.; Zhang, X.; Ma, J.; Li, H. Porous Nb$_4$N$_5$/rGO nanocomposite for ultrahigh-energy-density lithium-ion hybrid capacitor. *ACS Appl. Mater. Interfaces,* **2019**, *11*(27), 24114-24121.
[http://dx.doi.org/10.1021/acsami.9b06351] [PMID: 31245983]

CHAPTER 6

Metal Oxide Based Nanocomposites for Solar Energy Harvesting

K. Prabakaran[1,2,*], **P.J. Jandas**[1] and **Jingting Luo**[1]

[1] *Shenzhen Key Laboratory of Advanced Thin Films and Applications, College of Physics and Optoelectronic Engineering, Shenzhen University, 518060, Shenzhen, PR China*

[2] *Department of Physics, KPR Institute of Engineering and Technology, Coimbatore, 641407, Tamil Nadu, India*

Abstract: In recent years, the development of industrialization and the increasing population has increased energy consumption across the globe. So, there is a need for green and sustainable energy generation from solar cells with greater efficiency. Photovoltaic (PV) technology with improved performance is going to be a gamechanger in resolving the energy crisis in an eco-friendly and more sustainable manner. Widely used silicon (Si) based PVs are relatively expensive due to strong requirements for the high purity of crystalline semiconductors. The Si wafer cost covers 50% of the total cost of the align="center"module. In this regard, metal oxide-based semiconductors are stable and environment-friendly materials that are used in photovoltaics as photoelectrodes in dye solar cells (DSCs), quantum dot sensitized solar cells, and build metal oxide p–n junctions. This chapter comprehensively discusses the most recent progress in metal oxide semiconductors in alternative type solar cells, in particular dye-sensitized solar cells (DSSC).

Keywords: Bandgap, Dye-sensitized solar cells, Efficiency, Electron mobility, Metal oxides, Nanorod and nanotube, Photoanode, Semiconductor.

INTRODUCTION

An increase in energy consumption and global warming as a result of the depletion of fossil fuels has prompted a search for alternative energy sources, particularly renewable energy resources, to address the energy crisis. Therefore, the research and development of clean energy resources like solar radiation, tide, wind, geothermal, and biomass energy as an alternative to fossil fuels have become a major task for modern science and technology. Among this wide variety of renewable energy projects developments, utilization of solar energy is the most

---

* **Corresponding author K. Prabakaran:** College of Physics and Optoelectronic Engineering, Shenzhen University, 518060, Shenzhen, PR China; E-mail: praba736@gmail.com

Gaurav Manik and Sushanta Kumar Sahoo (Eds.)
All rights reserved-© 2022 Bentham Science Publishers

promising one as future energy technology because of its ease of availability and easy conversion to other forms of energy [1]. The solar energy is readily available source align="center"across the globe, but efficient utilization still remains a challenge. The conventional p-n junction type solar cell based on single crystal silicon has been widely used to date. The manufacturing cost of a silicon wafer is still relatively expensive due to the strong requirements of the high purity of crystalline semiconductors. The Si wafer cost covers 50% of the total cost of module cost, so improvements in costs, applicability and sustainability are to be addressed [2]. One way of reducing this major cost component is by replacing Si wafer with semiconductor thin films deposited onto the supporting substrate like glass, plastic, *etc.* [3 - 8]. Such films may have purely inorganic materials such as polycrystalline silicon, amorphous silicon, cadmium telluride and copper-indiu--diselenide, or contain organic materials as an important part of the device [9 - 12]. Many efforts have been made for the development of potentially cheaper thin film-solar cells. But these thin film solar cells are still in the pre-commercial phase because of their higher capital cost per unit output for thin film manufacturing facilities.

Metal oxides based nanomaterials are unique materials because of their electronic structure, physical, and electromagnetic properties [13]. The metals are useful materials, especially where they can be used as adsorbents and photocatalysts as well as for the align="center"manufacturing of environmental monitoring devices and dye sensitized solar cells. In 1991, Gratzel extended the concept of PEC solar cell to a novel solar cell based on a dye sensitized porous nanocrystalline TiO_2 as a photo anode [14, 15]. Generally in a dye sensitized solar cell, a photon absorbed by a dye molecule gives rise to electron injection into the conduction band of nanocrystalline semiconducting metal oxide materials such as TiO_2, ZnO, *etc.* used because of their high surface areas. Voltage generated under illumination corresponds to the difference between quasi-fermi level in the oxide and the redox potential of the electrolyte or work function of the hole conductor [16]. The main advantage of photoelectrochemical solar cells is that it is insensitive to defects in the structure. The scheme of DSSC align="center"is shown in Fig. (**1**).

PHOTOANODE MATERIALS

Titanium Dioxide (TiO₂)

In DSSC, the right selection of photo anode materials is important to achieve better solar conversion efficiency. Generally, TiO_2 nanoparticles based photo anode is being used in DSSC by most researchers as its conduction band edge lies slightly below the excited energy level of many dyes. It also provides better electrostatic shielding to the injected electrons from the oxidized dye molecule due to its higher dielectric constant ($\varepsilon = 80$) and can prevent the recombination before reduction of the dye. More than 60% of the research reports of DSSCs are based on TiO_2, because of their easy availability, electrical, optical, chemical stability and biocompatibility. The efficiency of TiO_2 photo electrode based DSSCs has been reported to be 8% more than that align="center"of liquid electrolytes. There are various methods that have been used for the synthesis of TiO_2 in different morphologies. F. Sauvage *et al.*. prepared crystalline TiO_2 beads with mesoporous structure (size 800 nm) using sol-gel followed by solvothermal method. The solar conversion efficiency obtained was about 7.20%, which was higher than standard Degussa P25 TiO_2 (5.66%) [17]. Another similar work also reports the preparation of mesoporous beads with more than one dye and obtained efficiency of 10.5% [18].

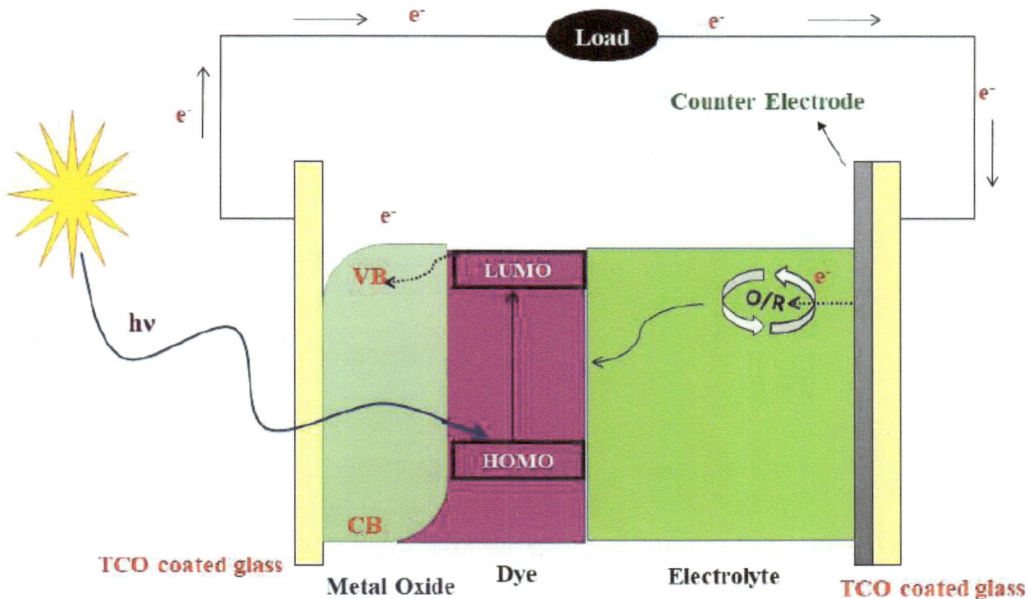

Fig. (1). Schematic diagram of Dye sensitized solar cells.

The mechanism of light absorption in nanoparticle and mesoporous structure is depicted in Fig. (2). However, the mesoporous nanostructure still shows lower electrical conductivity and their pores facilitate recombination reactions at the electrode. In this connection, a new approach of nanowires has been introduced that achieved promising results as compared to mesostructured materials [19]. J. Fan *et al.*. synthesized TiO_2 nanowire and reported a solar conversion efficiency align="center"of about 9.40% [20]. The improved photo conversion efficiency of nanowires arises due to their higher surface area. For further enhancement in

TiO_2, a nanotube like morphology with a hollow cavity structure has been introduced that comparatively provides high active surface area than nanowires. Many researchers have attempted to prepare TiO_2 nanotube to serve as photoanode for DSSC applications and reported photo-current conversion efficiency align="center"of about 4-6% [21, 22]. Another similar work involved the synthesis of TiO_2 nanotube arrays of 20.8 μm length with a reported overall efficiency of 8.1% [23]. S. Dadgostar *et al.*. synthesized TiO_2 hollow spheres and deposited align="center"them on TiO_2 nanoparticles coated FTO glass that was used as photoanode with an efficiency of 8.3% [24].

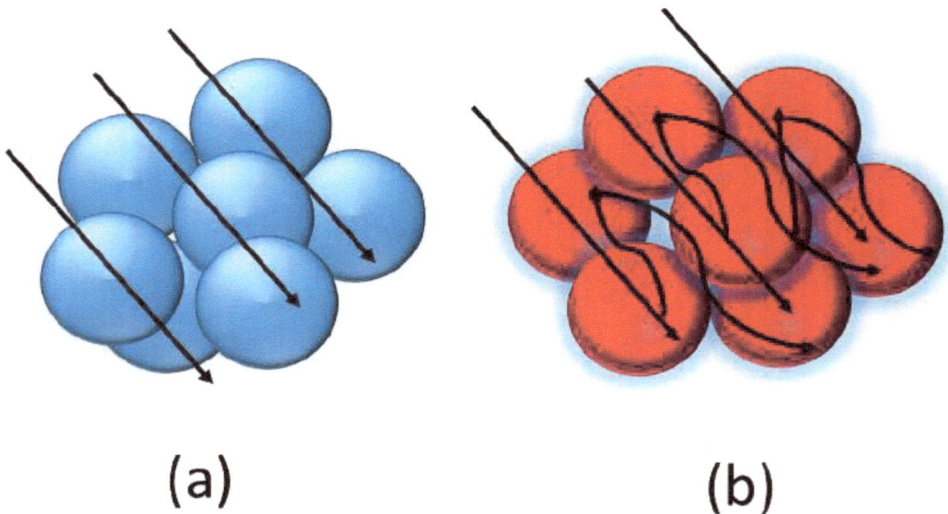

(a)   **(b)**

Fig. (2). Light absorption in normal **(a)** nanoparticles and **(b)** mesoporous structures.

Many research groups showed interest to improve visible light absorption and electron mobility. Suitable cations/anions have been doped to alter the band gap of TiO_2 alongwith enhancement in electro-optical behaviour. Federico proposed a concept of size quantization in late 1980s that can narrow down the band gap of metal oxide semiconductors [25]. Different cations such as nickel [26], tungsten [27], aluminium [28], Lanthanum [29], Niobium [30], *etc.*, have been studied to improve the electro-optical properties. This has not only remained limited to cations but non-metals such as nitrogen [31, 32], carbon [33], fluorine [34] and sulphur [35] have also been introduced in TiO_2.

Tin Oxide (SnO_2)

Besides TiO_2, SnO_2 exhibits various advantages such as high electron mobility and positive conduction band edge position as compared with TiO_2. The electron mobility of 100 $cm^2/(Vs)$ of SnO_2 is over two orders of magnitude higher than that of 0.1 $cm^2/(Vs)$) for TiO_2 [36], suggesting a faster electron migration possibilities in SnO_2. As a result, a reduction in interfacial charge recombination improves the device performance by losing to oxidized redox species in the electrolyte or hole transporter. Further, the bandgap of SnO_2 is about 3.6 eV, which is also higher than TiO_2. This creates fewer oxidative holes in the valence band of SnO_2, thereby causing reduction in the dye-degradation rate and facilitating long-term stability of DSSCs [37]. Various types of SnO_2 nanostructures have been used for photoanode applications in DSSC. Birkel A. *et al.*. [38] prepared different morphologies of tin oxide nanocrystals with different precursors. The different morphologies were achieved using NaOH, KOH, tetramethylammonium hydroxide and NH_4OH. The normalized photoconversion of SnO_2 based cells is shown in Fig. (3). The highest efficiency was 3.16% with FF 64 for optimized SnO_2 nanostructure. The results indicated that open circuit voltage is strongly dependent on the morphology of SnO_2. The significant efficiency of SnO_2 photoanodes arises due to the efficient electron diffusion through the network of interlinked nanorods, flat-band potential (V_{fb}) as well as less recombination at the interfaces. This is expressed as unusual long term stability, which is higher than TiO_2 based photoanode.

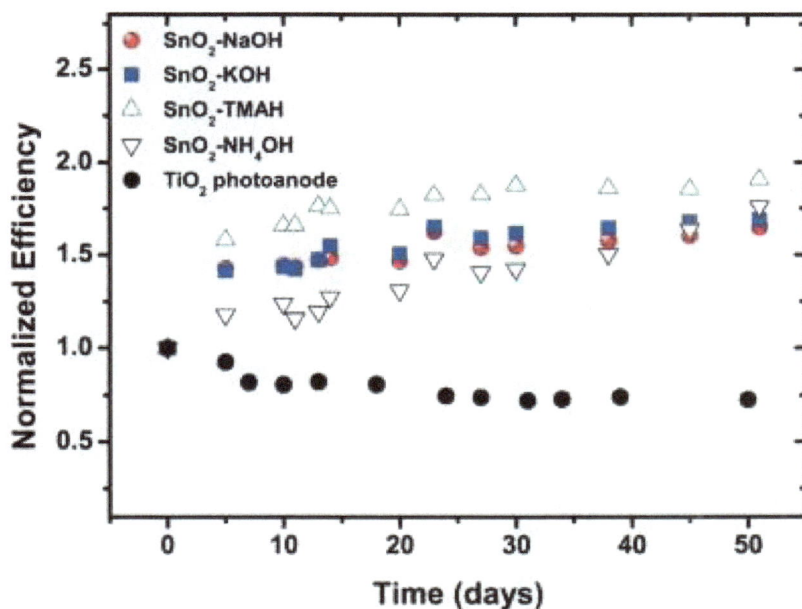

Fig. (3). Normalized efficiency *versus* Time for DSSCs with SnO_2 photoanodes Reprinted from Ref [38]. with permission from Royal Society of chemistry, copyright 2012.

In another work, DSSCs with SnO_2 nanowires as photoanode showed high V_{oc} (0.52–0.56 V), but the short circuit current was lower than SnO_2 nanoparticles based device. Similarly, Krishnamoorthy *et al.*. [39, 40] prepared well-aligned nanofibers and vertically oriented SnO_2 nanowires on an FTO substrate. The aligned SnO_2 nanowires with the nanoparticulate adhesion layer showed a V_{oc} of 0.525 V and J_{sc} of 9.9 mA/cm^2. The aligned one-dimensional fibre exhibited solar conversion efficiency of about 2.53%, *versus* 0.5% of nanoparticle. Similarly, the coral-like porous SnO_2 hollow architecture was also prepared *via* a facile wet-chemical treatment followed by calcination [41]. The DSSCs based on such SnO_2, showed efficiency of 1.04% with V_{oc} of 0.52 V and J_{sc} of 3.60 mA/cm^2. Fig. (**4**) shows SEM image and photovoltaic characteristics curve of a bilayer film with the ZnO nanorod overlayer and the SnO_2 nanoparticle underlayer. Other than morphology, various approaches have been introduced to increase the efficiency of SnO_2 based DSSCs. The modification of SnO_2 surface by other metal oxides/metals is expected to act as a barrier between conduction band and the dye that can reduce the dark current arising from the reduction reaction of electrolyte at the semiconductor surface,and hence, improving the open circuit voltage. Various attempts have been made on the surface modification of SnO_2 as photoanode using TiO_2, ZnO, Al_2O_3, MgO, Y_2O_3, CdO, NiO, CuO, $CaCO_3$ or Zn_2SnO_4 for further increment in the efficiency of cells.

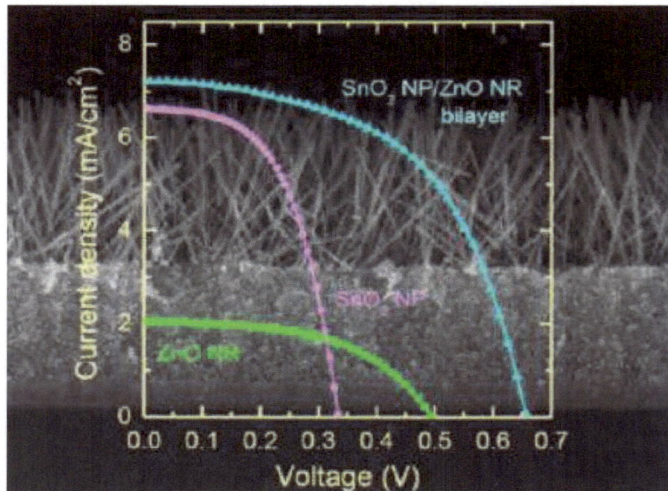

Fig. (4). SEM image and Photovoltaic characteristics curve of a bilayer film with the ZnO nanorods on SnO_2 nanoparticle. Reprinted from Ref [42]. with permission from the American Chemical Society, copyright 2013.

Zinc Oxide (ZnO)

Zinc oxide is also used as an alternative to TiO_2 as the photoanode DSSC, because of similar electron affinity and band gap. However, compared to TiO_2, ZnO exhibits higher electron diffusivity, electron mobility, binding energy, photo-corrosion resistance and less recombination reaction. The irreversible electron injection from organic molecules into the conduction band of a wide band gap semiconductor is first seen in ZnO [43, 44]. The highest efficiency (η) of pure ZnO has been reported as 7.5% [45] with liquid electrolyte, which is much lower than that of pure TiO_2 (12.3%) [46]. Tingli Ma *et al.*. have achieved an efficiency of 6.46% for ZnO nanostructure using solid electrolyte [47]. Zhao *et al.* [48] prepared iodine-doped ZnO (I-ZnO) based DSSC with indoline D205 and N719 as the sensitizers. They achieved an efficiency of 4.01 and 4.44% for N719 and indoline D205 respectively. K. Mahmood *et al.* synthesized Boron-doped ZnO (BZO) porous structured nanosheets with an efficiency of 6.75%. The efficiency obtained for BZO nanosheets is higher than that achieved for BZO films (3.0%) and un-doped ZnO nanosheets (2.62%). A comparison of photovoltaic behaviours of un-doped ZnO and Boron doped ZnO prepared at different temperatures is shown in Fig. (5).

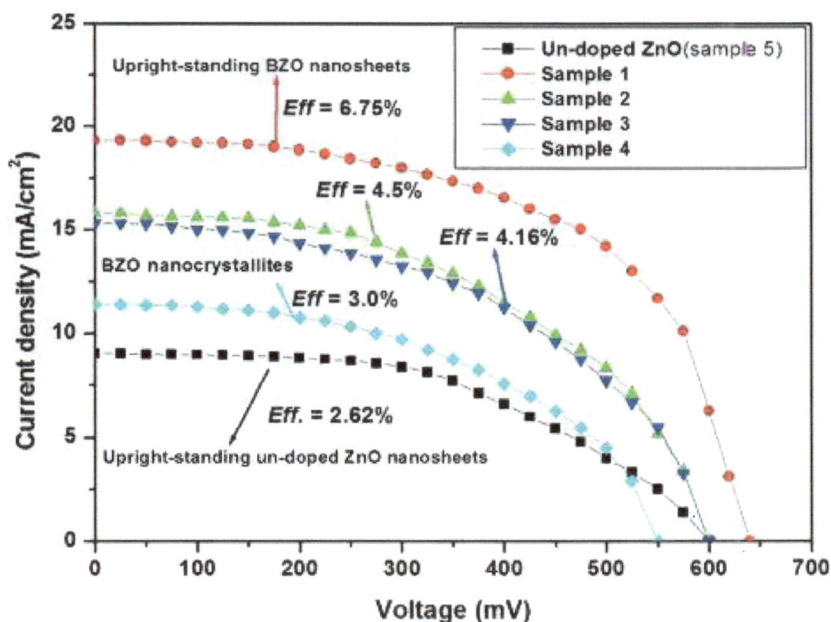

Fig. (5). Photovoltaic characteristics curve of Boron doped ZnO sheets in different temperature sample 1-170 °C, sample 2-180 °C, sample 3-190 °C, sample 1-200 °C, Undoped ZnO sheet at 170 °C. Reprinted from Ref [49]. with permission from the Royal society of chemistry, copyright 2013.

Tao *et al.* prepared aluminium doped ZnO nanorod arrays as photoanode and found that V_{oc} of ZnO nanorod increased significantly from 0.636 V to 0.732 V with the addition of aluminium [50]. The DSSC with the fluorine doped zinc oxide (F-ZnO) as photoanode showed a conversion efficiency (η) of 3.43%, while pure ZnO nanorods showed efficiency of 1.04% [51]. A.S. Gonalves *et al.* prepared a dye-sensitized solar cells based on ZnO:Ga nanostructured photoelectrodes, and studied by using EIS technique with different applied bias voltages at 0.6 V, 0.4 V and 0.2 V [52]. It is seen that the charge transfer resistance reduced with the addition of Ga^{3+} as impurity within ZnO. Q. Zhang *et al.* [53] prepared lithium doped ZnO film as photoanode for DSSC and obtained overall cell efficiency of about 6.1%. A photoanode with N-doped ZnO showed an efficiency of 2.64%, whereas pure ZnO gave η of only 0.67% owing to the decrease in band gap and prolongation in the absorption edge caused by N-doping [54]. Magnesium doped ZnO with the thickness of 4 μm exhibited an efficiency of 4.11% [55]. The energy gaps of ZnO increased by doping with yttrium, aluminium, and tin from 3.0 eV (the energy gap of undoped ZnO) to 3.05, 3.08 and 3.15 eV, respectively.

Other Metal Oxides

Many researchers have made attempts with alternative metal oxides for DSSC photoanode applications. The Nb_2O_5 is a wide bandgap semiconductor with band gap energy 3.49eV, about 0.29 eV larger than that anatase TiO_2. In order to prevent the electron back scattering, different morphologies of Nb_2O_5 such as nanoparticles, nanowires, nanobelts, bilayers, and blocking layers have been explored. Mohammadi Memari *et al.* [56] reported that calcination of Nb_2O_5 significantly increases the short circuit current and solar conversion efficiency. Further, the calcined Nb_2O_5 has been used with TiO_2 as bilayer and found that the efficiency increased to 1.5%. The energy level diagram of TiO_2 and Nb_2O_5 is shown in Fig. (**6**). The photocatalytic decomposition of dye and organic solvents may put an end to the sealed cell, because of gas release [57]. Suresh *et al.* [58, 59] reported that when Nb_2O_5 was combined with metal nanoparticles as blocking layer, photoconversion efficiency increased from 7.6 to 9.24%.

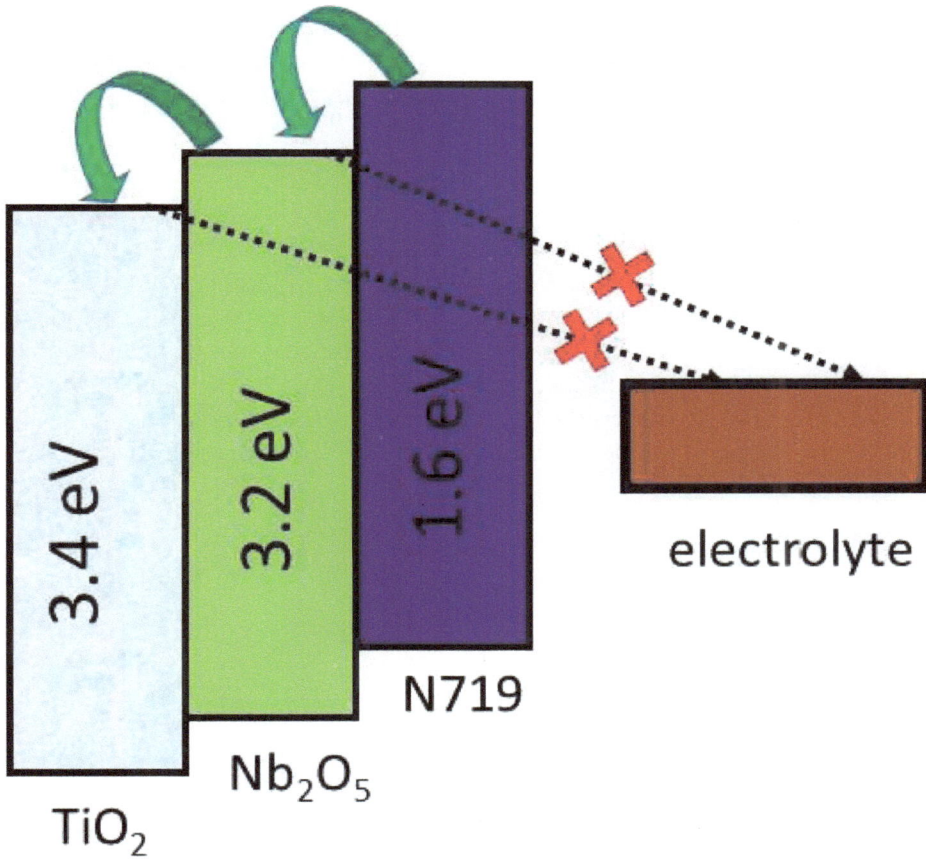

Fig. (6). Energy diagram of the fabricated DSSC with bilayer configuration and proposed enhance discharge transport mechanism.

CdO is an n-type semiconducting material with a band gap (Eg) ranging from 2.2-2.9 eV with high electro-optical properties. In this regard, Min-Hye Kim *et al.* [60] reported cadmium oxide as a blocking layer, which was coated on SnO_2 nanostructure in which the energy conversion efficiency of SnO_2 increased from 2 to 3.23% after coating with CdO layer. Sule Erten-Ela *et al.* prepared CdO composite with ZnO nanorods and achieved higher efficiencies than pure zinc oxide nanorod. The SnO_2/TPD dye showed a short-circuit current of 3.10 mA/cm^2 and V_{oc} of 0.48 V with an efficiency of 0.80%. Similarly, zinc oxide nanorod based DSSC exhibited a J_{sc} of 1.61 mA/cm^2 and a V_{oc} of 600 mV, with the corresponding efficiency of 0.58% [61]. Crystalline cadmium oxide and zinc oxide hexagonal nanocones nanocomposite align="center"were prepared using hydrothermal method for DSSC with overall conversion efficiency of 2.55%. A high open circuit voltage of about 0.810 V was found for CdO/ZnO due to the fast recombination rate and high R_{ct} value.

CONCLUSION

Till date, TiO_2 nanostructure has been explored as the most practical metal oxide semiconducting material for solar cell applications in general and dye sensitized solar cells (DSSC)in particular; however, TiO_2 suffers due to its spectral range of absorption and electron mobility, which limits their application in photovoltaic technology. To address this issue, many efforts have been taken in developing alternative photoanode materials for dye sensitized solar cells. In this chapter, various types of metal oxide alternatives in DSSC and their efficiency under suitable operating conditions are discussed.

CONSENT FOR PUBLICATION

Not applicable.

CONFLICT OF INTEREST

The authors declare no conflict of interest, financial or otherwise.

ACKNOWLEDGEMENTS

Declared none.

REFERENCES

[1]   Green, M.A. Thin-film solar cells: review of materials, technologies and commercial status. *J. Mater. Sci. Mater. Electron.,* **2007**, *18*(S1), 15-19.
        [http://dx.doi.org/10.1007/s10854-007-9177-9]

[2]   Dai, S.; Weng, J.; Sui, Y.; Shi, C.; Huang, Y.; Chen, S.; Pan, X.; Fang, X.; Hu, L.; Kong, F.; Wang, K. Dye-sensitized solar cells, from cell to module. *Sol. Energy Mater. Sol. Cells,* **2004**, *84*(1–4), 125-133.
        [http://dx.doi.org/10.1016/j.solmat.2004.03.002]

[3]   Sankir, N.D.; Dogan, B. Growth of CdS thin films and nanowires for flexible photoelectrochemical cells. *J. Mater. Process. Technol.,* **2011**, *211*(3), 382-387.
        [http://dx.doi.org/10.1016/j.jmatprotec.2010.10.010]

[4]   Sandoval-Paz, M.G.; Ramírez-Bon, R. Optical and Structural Properties of Chemically Deposited CdS Thin Films on Polyethylene Naphthalate Substrates. *Thin Solid Films,* **2011**, *520*(3), 999-1004.
        [http://dx.doi.org/10.1016/j.tsf.2011.08.006]

[5]   Goudarzi, A.; Aval, G.M.; Sahraei, R.; Ahmadpoor, H. Ammonia-Free Chemical Bath Deposition of Nanocrystalline ZnS Thin Film Buffer Layer for Solar Cells. *Thin Solid Films,* **2008**, *516*(15), 4953-4957.
        [http://dx.doi.org/10.1016/j.tsf.2007.09.051]

[6]   Aboul-Enein, S.; Badawi, M.H.; Ghali, M.; Hassan, G. Preparation and Properties of CdS Thin Films Prepared on Cold Substrate as a Window Layer for Solar Cells. *Renew. Energy,* **1998**, *14*(1–4), 113-118.
[http://dx.doi.org/10.1016/S0960-1481(98)00056-1]

[7]   Moses Ezhil Raj, A.; Mary Delphine, S.; Sanjeeviraja, C.; Jayachandran, M. Growth of ZnSe thin layers on different substrates and their structural consequences with bath temperature. *Physica B,* **2010**, *405*(10), 2485-2491.
[http://dx.doi.org/10.1016/j.physb.2010.03.019]

[8]   Tiwari, A.N.; Romeo, A.; Baetzner, D.; Zogg, H. Flexible CdTe solar cells on polymer films. *Prog. Photovolt. Res. Appl.,* **2001**, *9*(3), 211-215.
[http://dx.doi.org/10.1002/pip.374]

[9]   Ouyang, Z.; Pillai, S.; Beck, F.; Kunz, O.; Varlamov, S.; Catchpole, K.R.; Campbell, P.; Green, M.A. Effective light trapping in polycrystalline silicon thin-film solar cells by means of rear localized surface plasmons. *Appl. Phys. Lett.,* **2010**, *96*(26), 261109.
[http://dx.doi.org/10.1063/1.3460288]

[10]   Palanchoke, U.; Jovanov, V.; Kurz, H.; Obermeyer, P.; Stiebig, H.; Knipp, D. Plasmonic effects in amorphous silicon thin film solar cells with metal back contacts. *Opt. Express,* **2012**, *20*(6), 6340-6347.
[http://dx.doi.org/10.1364/OE.20.006340] [PMID: 22418515]

[11]   Mathew, X. Development of CdTe Thin Films on Flexible Substrates—a Review. *Sol. Energy Mater. Sol. Cells,* **2003**, *76*(3), 293-303.
[http://dx.doi.org/10.1016/S0927-0248(02)00281-7]

[12]   Zhang, Z.; Guo, H.; Li, J.; Zhu, C. Preparation and characterization of electrodeposited-annealed CuInSe₂ Thin films for solar cells. *Chin. J. Chem. Phys.,* **2011**, *24*(2), 225-230.
[http://dx.doi.org/10.1088/1674-0068/24/02/225-230]

[13]   Jose, R.; Thavasi, V.; Ramakrishna, S. Metal Oxides for Dye-Sensitized Solar Cells. *J. Am. Ceram. Soc.,* **2009**, *92*(2), 289-301.
[http://dx.doi.org/10.1111/j.1551-2916.2008.02870.x]

[14]   Grätzel, M. Applied physics: Solar cells to dye for. *Nature,* **2003**, *421*(6923), 586-587.
[http://dx.doi.org/10.1038/421586a] [PMID: 12571576]

[15]   Baglio, V.; Girolamo, M.; Antonucci, V.; Aricò, A.S. Influence of TiO₂ film thickness on the electrochemical behaviour of dye-sensitized solar cells. *Int. J. Electrochem. Sci.,* **2011**, *6*(8), 3375-3384.

[16]   Wang, Y.-F.; Li, J-W.; Hou, Y-F.; Yu, X-Y.; Su, C-Y.; Kuang, D-B. Hierarchical tin oxide octahedra for highly efficient dye-sensitized solar cells. *Chemistry,* **2010**, *16*(29), 8620-8625.
[http://dx.doi.org/10.1002/chem.201001333] [PMID: 20593448]

[17]   Sauvage, F.; Chen, D.; Comte, P.; Huang, F.; Heiniger, L-P.; Cheng, Y-B.; Caruso, R.A.; Graetzel, M. Dye-sensitized solar cells employing a single film of mesoporous TiO₂ beads achieve power conversion efficiencies over 10%. *ACS Nano,* **2010**, *4*(8), 4420-4425.
[http://dx.doi.org/10.1021/nn1010396] [PMID: 20731428]

[18]   Cao, K.; Lu, J.; Li, H.; Shen, Y.; Wang, M. Efficient dye-sensitized solar cells using mesoporous submicrometer TiO₂ Beads. *RSC Advances,* **2015**, *5*(77), 62630-62637.
[http://dx.doi.org/10.1039/C5RA11281H]

[19]   Feng, X.; Zhu, K.; Frank, A.J.; Grimes, C.A.; Mallouk, T.E. Rapid charge transport in dye-sensitized solar cells made from vertically aligned single-crystal rutile TiO₂ nanowires. *Angew. Chem.,* **2012**, *124*(11), 2781-2784.
[http://dx.doi.org/10.1002/ange.201108076]

[20]   Fan, J.; Fàbrega, C.; Zamani, R.R.; Hao, Y.; Parra, A.; Andreu, T.; Arbiol, J.; Boschloo, G.; Hagfeldt,

A.; Morante, J.R.; Cabot, A. Enhanced photovoltaic performance of nanowire dye-sensitized solar cells based on coaxial TiO_2@TiO heterostructures with a cobalt(II/III) redox electrolyte. *ACS Appl. Mater. Interfaces,* **2013**, *5*(20), 9872-9877.
[http://dx.doi.org/10.1021/am402344d] [PMID: 24025444]

[21]   Flores, I.C.; de Freitas, J.N.; Longo, C.; De Paoli, M-A.; Winnischofer, H.; Nogueira, A.F. Dye-Sensitized Solar Cells Based on TiO_2 Nanotubes and a Solid State Electrolyte. *J. Photochem. Photobiol. Chem.,* **2007**, *189*(2–3), 153-160.
[http://dx.doi.org/10.1016/j.jphotochem.2007.01.023]

[22]   Park, H.; Kim, W-R.; Jeong, H-T.; Lee, J-J.; Kim, H-G.; Choi, W-Y. Fabrication of Dye-Sensitized Solar Cells by Transplanting Highly Ordered TiO_2 Nanotube Arrays. *Sol. Energy Mater. Sol. Cells,* **2011**, *95*(1), 184-189.
[http://dx.doi.org/10.1016/j.solmat.2010.02.017]

[23]   Liu, Z.; Misra, M. Dye-sensitized photovoltaic wires using highly ordered TiO_2 nanotube arrays. *ACS Nano,* **2010**, *4*(4), 2196-2200.
[http://dx.doi.org/10.1021/nn9015696] [PMID: 20235517]

[24]   Dadgostar, S.; Tajabadi, F.; Taghavinia, N. Mesoporous submicrometer $TiO_{(2)}$ hollow spheres as scatterers in dye-sensitized solar cells. *ACS Appl. Mater. Interfaces,* **2012**, *4*(6), 2964-2968.
[http://dx.doi.org/10.1021/am300329p] [PMID: 22606936]

[25]   Serpone, N.; Lawless, D.; Khairutdinov, R. Size Effects on the Photophysical Properties of Colloidal Anatase TiO_2 Particles: Size Quantization versus Direct Transitions in This Indirect Semiconductor? *J. Phys. Chem.,* **1995**, *99*(45), 16646-16654.
[http://dx.doi.org/10.1021/j100045a026]

[26]   Wang, Y.; Hao, Y.; Cheng, H.; Ma, J.; Xu, B.; Li, W.; Cai, S. The photoelectrochemistry of transition metal-ion-doped TiO_2 nanocrystalline electrodes and higher solar cell conversion efficiency based on Zn2+-doped TiO_2 electrode. *J. Mater. Sci.,* **1999**, *34*(12), 2773-2779.
[http://dx.doi.org/10.1023/A:1004658629133]

[27]   Zhang, X.; Liu, F.; Huang, Q-L.; Zhou, G.; Wang, Z-S. Dye-sensitized w-doped TiO_2 solar cells with a tunable conduction band and suppressed charge recombination. *J. Phys. Chem. C,* **2011**, *115*(25), 12665-12671.
[http://dx.doi.org/10.1021/jp201853c]

[28]   Ko, K.H.; Lee, Y.C.; Jung, Y.J. Enhanced efficiency of dye-sensitized TiO_2 solar cells (DSSC) by doping of metal ions. *J. Colloid Interface Sci.,* **2005**, *283*(2), 482-487.
[http://dx.doi.org/10.1016/j.jcis.2004.09.009] [PMID: 15721923]

[29]   Zhang, J.; Zhao, Z.; Wang, X.; Yu, T.; Guan, J.; Yu, Z.; Li, Z.; Zou, Z. Increasing the Oxygen Vacancy Density on the TiO_2 Surface by La-Doping for Dye-Sensitized Solar Cells. *J. Phys. Chem. C,* **2010**, *114*(43), 18396-18400.
[http://dx.doi.org/10.1021/jp106648c]

[30]   Lü, X.; Mou, X.; Wu, J.; Zhang, D.; Zhang, L.; Huang, F.; Xu, F.; Huang, S. Improved-Performance Dye-Sensitized Solar Cells Using Nb-Doped TiO_2 Electrodes: Efficient Electron Injection and Transfer. *Adv. Funct. Mater.,* **2010**, *20*(3), 509-515.
[http://dx.doi.org/10.1002/adfm.200901292]

[31]   Ma, T.; Akiyama, M.; Abe, E.; Imai, I. High-efficiency dye-sensitized solar cell based on a nitrogen-doped nanostructured titania electrode. *Nano Lett.,* **2005**, *5*(12), 2543-2547.
[http://dx.doi.org/10.1021/nl051885l] [PMID: 16351212]

[32]   Kang, S.H.; Kim, H.S.; Kim, J-Y.; Sung, Y-E. Enhanced Photocurrent of Nitrogen-Doped TiO_2 Film for Dye-Sensitized Solar Cells. *Mater. Chem. Phys.,* **2010**, *124*(1), 422-426.
[http://dx.doi.org/10.1016/j.matchemphys.2010.06.059]

[33]   Chu, D.; Yuan, X.; Qin, G.; Xu, M.; Zheng, P.; Lu, J.; Zha, L. Efficient carbon-doped nanostructured TiO_2 (Anatase) film for photoelectrochemical solar cells. *J. Nanopart. Res.,* **2008**, *10*(2), 357-363.

[http://dx.doi.org/10.1007/s11051-007-9241-7]

[34]   Song, J.; Yang, H.B.; Wang, X.; Khoo, S.Y.; Wong, C.C.; Liu, X.W.; Li, C.M. Improved utilization of photogenerated charge using fluorine-doped TiO(2) hollow spheres scattering layer in dye-sensitized solar cells. *ACS Appl. Mater. Interfaces,* **2012**, *4*(7), 3712-3717.
         [http://dx.doi.org/10.1021/am300801f] [PMID: 22731936]

[35]   Sun, Q.; Zhang, J.; Wang, P.; Zheng, J.; Zhang, X.; Cui, Y.; Zhu, Y. Sulfur-doped TiO$_2$ nanocrystalline photoanodes for dye-sensitized solar cells. *J. Renew. Sustain. Energy,* **2012**, *4*(2), 023104.
         [http://dx.doi.org/10.1063/1.3694121]

[36]   Fonstad, C.G.; Rediker, R.H. Electrical properties of high-quality stannic oxide crystals. *J. Appl. Phys.,* **1971**, *42*(7), 2911-2918.
         [http://dx.doi.org/10.1063/1.1660648]

[37]   Senevirathna, M.K.I.; Pitigala, K.D.D.P.; Premalal, E.V.A.; Tennakone, K.; Kumara, G.R.A.; Konno, A. Stability of the SnO$_2$/MgO dye-sensitized photoelectrochemical solar cell. *Sol. Energy Mater. Sol. Cells,* **2007**, *91*(6), 544-547.
         [http://dx.doi.org/10.1016/j.solmat.2006.11.008]

[38]   Birkel, A.; Lee, Y-G.; Koll, D.; Van Meerbeek, X.; Frank, S.; Choi, M.J.; Kang, Y.S.; Char, K.; Tremel, W. Highly Efficient and Stable Dye-Sensitized Solar Cells Based on SnO$_2$ Nanocrystals Prepared by Microwave-Assisted Synthesis. *Energy Environ. Sci.,* **2012**, *5*(1), 5392-5400.
         [http://dx.doi.org/10.1039/C1EE02115J]

[39]   Gubbala, S.; Russell, H.B.; Shah, H.; Deb, B.; Jasinski, J.; Rypkema, H.; Sunkara, M.K. Surface Properties of SnO$_2$ Nanowires for Enhanced Performance with Dye-Sensitized Solar Cells. *Energy Environ. Sci.,* **2009**, *2*(12), 1302.
         [http://dx.doi.org/10.1039/b910174h]

[40]   Krishnamoorthy, T.; Tang, M.Z.; Verma, A.; Nair, A.S.; Pliszka, D.; Mhaisalkar, S.G.; Ramakrishna, S. A Facile Route to Vertically Aligned Electrospun SnO$_2$ Nanowires on a Transparent Conducting Oxide Substrate for Dye-Sensitized Solar Cells. *J. Mater. Chem.,* **2012**, *22*(5), 2166-2172.
         [http://dx.doi.org/10.1039/C1JM15047B]

[41]   Liu, J.; Luo, T.; Mouli T, S.; Meng, F.; Sun, B.; Li, M.; Liu, J. A novel coral-like porous SnO$_2$ hollow architecture: biomimetic swallowing growth mechanism and enhanced photovoltaic property for dye-sensitized solar cell application. *Chem. Commun. (Camb.),* **2010**, *46*(3), 472-474.
         [http://dx.doi.org/10.1039/B915650J] [PMID: 20066329]

[42]   Huu, N.K.; Son, D-Y.; Jang, I-H.; Lee, C-R.; Park, N-G. Hierarchical SnO$_2$ nanoparticle-ZnO nanorod photoanode for improving transport and life time of photoinjected electrons in dye-sensitized solar cell. *ACS Appl. Mater. Interfaces,* **2013**, *5*(3), 1038-1043.
         [http://dx.doi.org/10.1021/am302729v] [PMID: 23331623]

[43]   Anta, J.A.; Guillén, E.; Tena-Zaera, R. ZnO-Based Dye-Sensitized Solar Cells. *J. Phys. Chem. C,* **2012**, *116*(21), 11413-11425.
         [http://dx.doi.org/10.1021/jp3010025]

[44]   Tributrsch, H.; Calvin, M. Electrochemistry of excited molecules: Photo-electrochemical reactions of Chlorophylls. *Photochem. Photobiol.,* **1971**, *14*(2), 95-112.
         [http://dx.doi.org/10.1111/j.1751-1097.1971.tb06156.x]

[45]   Memarian, N.; Concina, I.; Braga, A.; Rozati, S.M.; Vomiero, A.; Sberveglieri, G. Hierarchically assembled ZnO nanocrystallites for high-efficiency dye-sensitized solar cells. *Angew. Chem. Int. Ed. Engl.,* **2011**, *50*(51), 12321-12325.
         [http://dx.doi.org/10.1002/anie.201104605] [PMID: 21953715]

[46]   Yella, A.; Lee, H.-W.; Tsao, H. N.; Yi, C.; Chandiran, A. K.; Nazeeruddin, M. K.; Diau, E. W.-G.; Yeh, C.-Y.; Zakeeruddin, S. M.; Gratzel, M. Porphyrin-Sensitized Solar Cells with Cobalt (II/III)-Based Redox Electrolyte Exceed 12 Percent Efficiency. *Science,* **2011**, *334*(6056), 629-634.

[http://dx.doi.org/10.1126/science.1209688]

[47]   Shi, Y.; Wang, K.; Du, Y.; Zhang, H.; Gu, J.; Zhu, C.; Wang, L.; Guo, W.; Hagfeldt, A.; Wang, N.; Ma, T. Solid-state synthesis of ZnO nanostructures for quasi-solid dye-sensitized solar cells with high efficiencies up to 6.46%. *Adv. Mater.,* **2013**, *25*(32), 4413-4419.
[http://dx.doi.org/10.1002/adma.201301852] [PMID: 23787557]

[48]   Zhao, J-X.; Zheng, Y-Z.; Lu, X-H.; Chen, J-F.; Tao, X.; Zhou, W. Characterizing the role of iodine doping in improving photovoltaic performance of dye-sensitized hierarchically structured ZnO solar cells. *ChemPhysChem,* **2013**, *14*(9), 1977-1984.
[http://dx.doi.org/10.1002/cphc.201300066] [PMID: 23606406]

[49]   Mahmood, K.; Park, S. Bin. Highly Efficient Dye-Sensitized Solar Cell with an Electrostatic Spray Deposited Upright-Standing Boron-Doped ZnO (BZO) Nanoporous Nanosheet-Based Photoanode. *J. Mater. Chem. A Mater. Energy Sustain.,* **2013**, *1*(15), 4826.
[http://dx.doi.org/10.1039/c3ta10587c]

[50]   Tao, R.; Tomita, T.; Wong, R.A.; Waki, K. Electrochemical and Structural Analysis of Al-Doped ZnO Nanorod Arrays in Dye-Sensitized Solar Cells. *J. Power Sources,* **2012**, *214*, 159-165.
[http://dx.doi.org/10.1016/j.jpowsour.2012.04.071]

[51]   Luo, L.; Tao, W.; Hu, X.; Xiao, T.; Heng, B.; Huang, W.; Wang, H.; Han, H.; Jiang, Q.; Wang, J.; Tang, Y. Mesoporous F-Doped ZnO Prism Arrays with Significantly Enhanced Photovoltaic Performance for Dye-Sensitized Solar Cells. *J. Power Sources,* **2011**, *196*(23), 10518-10525.
[http://dx.doi.org/10.1016/j.jpowsour.2011.08.011]

[52]   Gonalves, A.S.; Góes, M.S.; Fabregat-Santiago, F.; Moehl, T.; Davolos, M.R.; Bisquert, J.; Yanagida, S.; Nogueira, A.F.; Bueno, P.R. Doping saturation in dye-sensitized solar cells based on ZnO:Ga nanostructured photoanodes/ *Electrochim. Acta,* **2011**, *56*, 6503-6509.
[http://dx.doi.org/10.1016/j.electacta.2011.05.003]

[53]   Zhang, Q.; Dandeneau, C.S.; Candelaria, S.; Liu, D.; Garcia, B.B.; Zhou, X.; Jeong, Y-H.; Cao, G. Effects of Lithium Ions on Dye-Sensitized ZnO Aggregate Solar Cells. *Chem. Mater.,* **2010**, *22*(8), 2427-2433.
[http://dx.doi.org/10.1021/cm9009942]

[54]   Zhang, L.; Yang, Y.; Fan, R.; Chen, H.; Jia, R.; Wang, Y.; Ma, L.; Wang, Y. The charge-transfer property and the performance of dye-sensitized solar cells of nitrogen doped zinc oxide. *Mater. Sci. Eng. B,* **2012**, *177*(12), 956-961.
[http://dx.doi.org/10.1016/j.mseb.2012.04.026]

[55]   Raj, C.J.; Prabakar, K.; Karthick, S.N.; Hemalatha, K.V.; Son, M-K.; Kim, H-J. Banyan Root Structured Mg-Doped ZnO Photoanode Dye-Sensitized Solar Cells. *J. Phys. Chem. C,* **2013**, *117*(6), 2600-2607.
[http://dx.doi.org/10.1021/jp308847g]

[56]   Memari, M.; Memarian, N. Designed structure of bilayer TiO_2–Nb_2O_5 Photoanode for increasing the performance of dye-sensitized solar cells. *J. Mater. Sci. Mater. Electron.,* **2020**, *31*(3), 2298-2307.
[http://dx.doi.org/10.1007/s10854-019-02762-3]

[57]   Sayama, K.; Sugihara, H.; Arakawa, H. Photoelectrochemical Properties of a Porous Nb_2O_5 Electrode Sensitized by a Ruthenium Dye. *Chem. Mater.,* **1998**, *10*(12), 3825-3832.
[http://dx.doi.org/10.1021/cm980111l]

[58]   Ghosh, R.; Brennaman, M.K.; Uher, T.; Ok, M-R.; Samulski, E.T.; McNeil, L.E.; Meyer, T.J.; Lopez, R. Nanoforest Nb_2O_5 photoanodes for dye-sensitized solar cells by pulsed laser deposition. *ACS Appl. Mater. Interfaces,* **2011**, *3*(10), 3929-3935.
[http://dx.doi.org/10.1021/am200805x] [PMID: 21919494]

[59]   Suresh, S.; Unni, G.E.; Satyanarayana, M.; Nair, A.S.; Pillai, V.P.M. Ag@Nb_2O_5 plasmonic blocking layer for higher efficiency dye-sensitized solar cells. *Dalton Trans.,* **2018**, *47*(13), 4685-4700.
[http://dx.doi.org/10.1039/C7DT04825D] [PMID: 29537003]

[60] Kim, M-H.; Kwon, Y-U. Semiconductor CdO as a Blocking Layer Material on DSSC Electrode: Mechanism and Application. *J. Phys. Chem. C,* **2009**, *113*(39), 17176-17182. [http://dx.doi.org/10.1021/jp904206a]

[61] Erten-Ela, S. Photovoltaic Performance of ZnO Nanorod and ZnO : CdO Nanocomposite Layers in Dye-Sensitized Solar Cells (DSSCs). *Int. J. Photoenergy,* **2013**, *2013*, 1-6. [http://dx.doi.org/10.1155/2013/436831]

CHAPTER 7

Two-dimensional Functionalized Hexagonal Boron Nitride (2D h-BN) Nanomaterials for Energy Storage Applications

Shamsiya Shams[1] and **B. Bindhu**[1,*]

[1] *Department of Physics, Noorul Islam Centre for Higher Education, Kumaracoil, Tamil Nadu, 629180, India*

Abstract: The conservation of energy and the materials utilized for its storage have gathered a wide range of interest nowadays. Two-dimensional hexagonal boron nitride (2D h-BN), often termed as 'white graphene', exhibits various interesting properties and hence, acts as a promising future candidate for energy sustainment and storage. This material assures exquisite thermal and chemical stability, high chemical inertness, exotic mechanical strength, and good optoelectrical properties. 2D h-BN undergoes physical and chemical modulations, and their properties could be tuned, making them more appropriate for energy storage applications. They could also be incorporated with other 2D materials like graphene, molybdenum disulphide (MoS_2), *etc.*, to improve their properties. It is thus thoroughly and systematically studied for its further usage in field effect transistors (FETs), UV detecting devices and emitters, photoelectric and microelectronic devices, tunnelling devices, *etc.* The comprehensive overview provides an insight into 2D h-BN and its synthesis routes developed within the past years. The different major properties exhibited by 2D h-BN are also reviewed. Hybridization and doping processes are also discussed. Functionalised h-BN and its utilisation in different energy storage applications are elaborated and reviewed. This review chapter will give a quick glance and perspectives on 2D h-BN and its extraordinary characteristic features that could enhance their usage in energy conversion, storage, and utilisation applications.

Keywords: Doping, Electrolytes, Energy storage, Functionalised h-BN, Hexagonal boron nitride (h-BN), Honey-comb arrangement, Hydrogen storage, Secondary batteries, Solar energy, Supercapacitors, Technical advancements, Thermal stability.

---

[*] **Corresponding author B. Bindhu:** Department of Physics, Noorul Islam Centre for Higher Education, Kumaracoil, Tamil Nadu, 629180, India; E-mail: bindhu.krishna80@gmail.com

Gaurav Manik and Sushanta Kumar Sahoo (Eds.)

INTRODUCTION

Recent incessant utilization and consumption of global energy have led to the origin and development of new advanced sources of energy such as solar, hydrothermal, biomass, geothermal, wind, and nuclear energy. Energy storage is the process of capturing and storing energy produced once and maintaining it for future energy demands, usage, and production. The two challenges that are commonly faced are energy storage and conversion of the energy produced. Traditional energy storage devices are vulnerable to natural conditions; hence conventional methods and equipment need to be developed to obtain high efficiency and standard performance. Its application includes rechargeable and renewable batteries, supercapacitors, biofuels, flywheels, *etc*. The major drawbacks experienced in energy storage devices are thermal explosion (thermal runaway) that limits the operational working and safety, working efficiencies, scale and lifespan, cost and availability of raw materials for designing, and environmental hazards caused. Substantial research is carried out by researchers and scientists for the prominent utilisation of nanomaterials in energy storage applications to construct a proficient power absorption and storage material with high thermal conductivity, specific capacitance, and corrosion resistance.

The extraordinary behavior and properties of two-dimensional(2D) nanomaterials offer a wide and attractive platform for them to be used in energy storage applications. Few thick atom crystals with ultra-thin thickness, usually of nanoscale are referred to as 2D materials [1]. 2D materials offer high specific area, thermal stability and innumerable other properties, and hence, can be implemented in energy reserving applications. Different 2D nanocomposites have been identified and evaluation studies were carried out for this application which includes graphene [2], hexagonal boron nitride, transition metal oxides [1 - 3], transition metal dichalcogenides [2, 4] and transition metal carbides and nitrides (MXenes) [5 - 9]. 2D nanomaterials offer high capacitance rates and exorbitant specific surface area. 2D inorganic nanomaterials depict pseudocapacitance and could be modified by doping and chemical functionalisation. These materials also show fast transport of ions, thereby enhancing expansion and contraction of layers. A fast ion adsorption mechanism is seen between the nanosheets of 2D channels that fabricates the energy storage application of 2D nanomaterials accredited by the edge in water molecules followed by an accelerated enlargement and confinement of multifaceted and bendable layers of 2D nanomaterials.

The resilience of 2D nanomaterials, favourable mechanical and packing density properties, high volumetric capacitance make them appropriate for the designing and manufacturing energy storage devices. Zero-bandgap graphene shows suitable electrical conductivity and hence, could be applied to fabricate electrically conducting supplements that are free from electrodes. Electrical conducting properties are not shown by most of the 2D materials, thus hampering their further usage in energy storage fields [4]. Atomic layer numbers and the crystal structure is directly persuaded by the electronic structure of 2D nanomaterials. Different methods are implemented to enhance the electrical conductivity by varying 2D nanomaterials and their electronic structures. One of the major problem faced is to impose the full potential of 2D materials in storing energy by preventing restacking of nanosheets which averts full utilization of surface area domains and obstructs the electrolyte access. This difficulty could be intercepted by reconciling the morphology of nanosheets. Different synthesis techniques were developed to produce nanosheets in leveled, twisty or wrinkled shapes, that when placed with current collectors ease the ion movements.

The 2D materials could be incorporated with other materials to form a basic building material for a series of hetro- or hybrid structures that could facilitate manufacturing of new materials with improved properties. This adaptable nature of 2D materials offers the probability to engineer new frameworks that could be used for an expansive range of applications. Hexagonal boron nitride and other transition metal oxide nanomaterials could be integrated with electrically conducting graphene's to form a heterostructured composite to enhance the energy storage properties [10, 11]. Establishing new nanomaterials and intercalation of large molecules could overcome the problem of restacking.

In this review, we provide a summary on functionalised hexagonal boron nitride and its applications in energy storage fields. The properties of this 2D material will be highlighted and an elaboration of their use in energy storage applications will be focussed. The structural, morphological, electrical, mechanical and thermal studies are analysed from different research papers and are explained. A general survey of energy storage and conversion applications is evaluated and proof-read with newly penned review articles. Designing and fabrication of electrode structures, device layout, and manufacturing methods are considered. The different possibility of utilising this material in energy storage applications, future perspectives and challenges faced are identified and elaborated based on the literatures studied and recommendations are made for future research.

HEXAGONAL BORON NITRIDE (H-BN)

Hexagonal boron nitride is a compound included under group III-V elements comprised of an equal number of nitrogen and boron atom layers in a honeycomb lattice arrangement under sp^2 hybridisation, where the layers are organised by van der Waals forces.

The arrangement of boron and nitrogen atoms in 2D h-BN is depicted below (Fig. 1). It is also identified as "white graphene" and is isotypic to graphene.

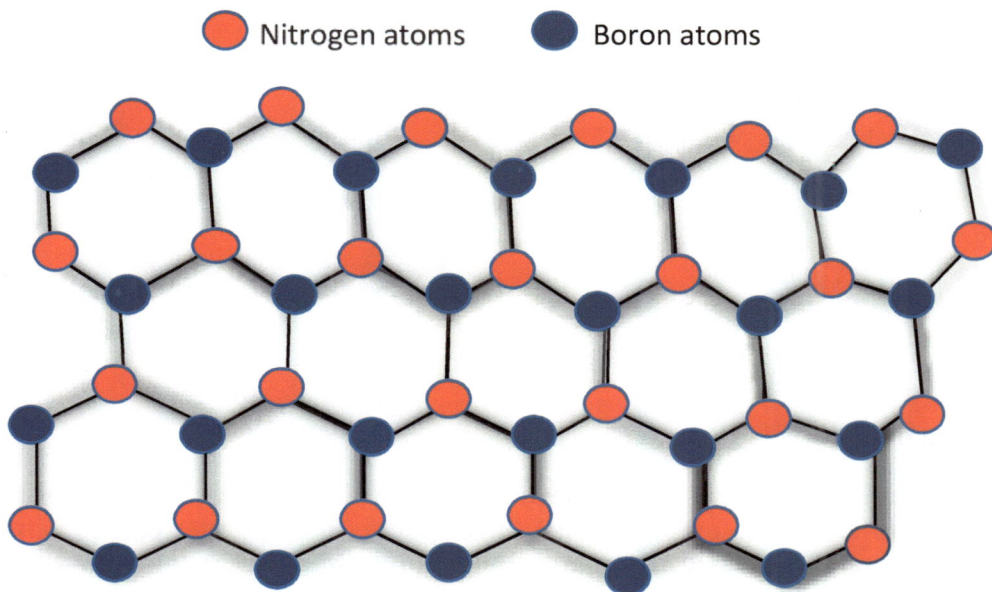

Fig. (1). The arrangement of boron and nitrogen atoms in 2D h-BN.

The nanosheets of h-BN (h-BNNS) depict a layered structure, exhibit good chemical and thermal stability and could be used in many applications. The synthesis approaches of bulk h-BN to h-BNNS is illustrated (Fig. 2). The energy band gap of pristine h-BN is equal to 5.9 eV, and thus, it could emit deep ultraviolet light (200 nm) [12]. The bandgap energy of h-BN is considered to be wide, and hence, h-BN acts as an insulator consequently retarding its energy storage and conversion applications. Many research activities have been carried out further on the modification of h-BN to enhance its properties. Thermal conductivity studies of epoxy nanosheets of h-BN were carried out to find out the dispersion rate of the particles as well as their conducting nature [13]. The two different methods that are found to be productive are doping of heteroatoms and grafting of functional groups. Many functional groups and dopants such as Hydrogen, Fluorine, -OH groups, $-CH_3$, -CHO, -CN and NH_2 are introduced to lower the energy gap of h-BN. The theoretical values indicate the lowering of bandgap from 0.3 eV to 3.1 eV [14] and experimental results of carbon doped h-BNNS reveal a decreased energy band gap width from 5.9 eV to 2.6 eV consequently indicating it as a favourable non-metallic photocatalyst [15]. h-BN modified by carbon and oxygen doping show good optical, electronic, thermal and surface properties. Thus, it could be used in many major applications, including electrochemical energy storage [16 - 18].

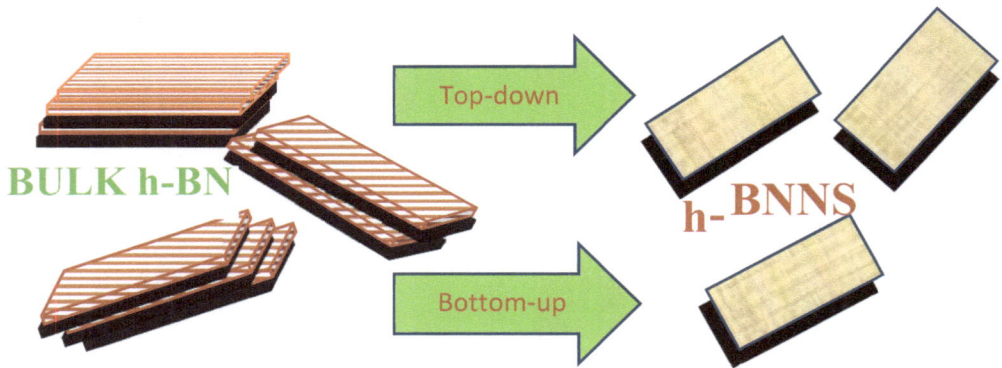

Fig. (2). Demonstration of synthesis approach of bulk h-BN to h-BNNS.

Doping of 2D h-BN Nanosheets

P type or N type materials are doped with 2D-hBN nanosheets to improve the optical and electronic properties and have specific control over charge carriers (electrons and holes) and determine the band gap movement. C atom with an appropriate structural arrangement and atomic size is the most promising material used in certain application fields of 2D h-BN nanosheets [19]. The research studies of C doped h-BN revealed that the theoretical values showed that C-C bonds are more specific and stable than other bonds like C-N and C-B bonds, thus forming isolated domains of C due to the segregation of C-C bonds. The work function and band gap of a semiconducting h-BN could be altered easily by creating a change in the composition mixture of C-B-N bonds, while doping even substitutional C atoms and doping of odd C atoms could result in the metallic properties [20]. The replacement of B and N atoms could be done successfully by inducing the acceptor and donor states by C doping. Tensile and compressive strains are applied to ensure control of the donor and acceptors' ionisation energies [21]. A new method of intercalation of lithium, sodium and fluorine atoms on h-BN was proposed by Zhang *et al.*. [22], known as bipolar doping, which showed that the donor and acceptor states are shallow and more effective with ionisation energies considered to be at low levels. Doping of many atoms considerably helped to transform the properties of h-BN. Xue *et al.*. [23] have investigated Fluorine doped h-BN nanosheets to examine the electrical properties, and help in changing the h-BN in insulator category to a material with high electrical conductivity. A superior electrical conductivity was obtained at 15.854 mA. Doping of h-BN with transition metal oxides and chalcogenides improved the storage capacity and doped h-BN could be readily used for energy storage applications [24].

Functionalised 2D h-BN in Energy Storage, Conversion and Utilisation

Energy storage is a method of storing the generated energy produced by different methods and using it later on, and devices used for this are called as energy storage devices. Renewable energy sources such as solar energy, wind energy *etc.*, should be stored carefully and used productively for the future. Good carrier mobility rates [25], wide and large specific surface area [26], high thermal and mechanical strength [27] and distinct chemical and structural characteristic features, have also made 2D materials a promising candidate in energy storage applications. Recent developments depicted that h-BNNS is a widely accepted 2D material with high chemical stability and excellent mechanical and thermal strength, and many other improved properties. Functionalised h-BNNS offers wide range of peculiar properties and could be effectively used in many energy storage applications such as in supercapacitors, solar energy conversion, hydrogen storage and rechargeable batteries. Various energy storage applications are made using functionalised h-BN incorporated with other 2D materials (Fig. 3). In this section, the theoretical insights and experiments of functionalised h-BN structures and their potential applications are discussed.

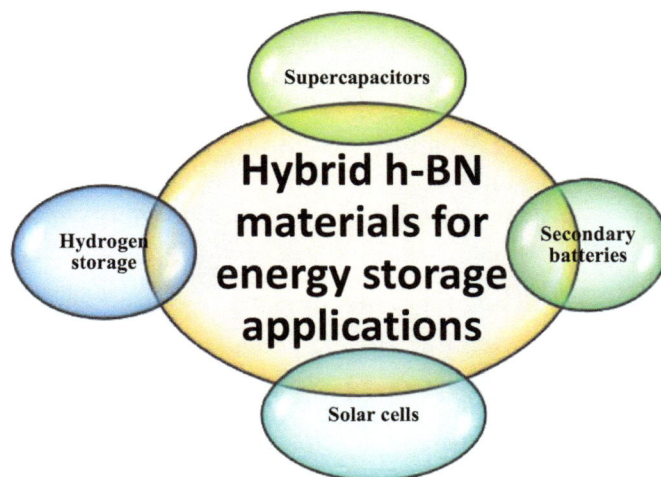

Fig. (3). Different energy storage applications using functionalised h-BN incorporated with other 2D materials.

Theoretical and Experimental Studies

2D h-BN is considered as an insulator with a wide band gap, and hence, dose not possess electrochemical catalytic activity by itself. By undergoing various treatments and chemical reactions they can be used for certain applications. The studies by Uosaki *et al..* [28], identified that gold covered by h-BNNS shows enhanced ORR (oxygen reduction rates). The electronic properties of h-BN could be improved by blending up of p and d metal orbitals. The resultant reaction between the gold surface and h-BN depicted that O_2 formed bond with two B atoms comparatively nearer to N atom above the gold atoms. The gold atoms incorporated h-BNNS (h-BNNS/Au) showed that the oxygen atoms are preferentially adsorbed on the edges of 2D h-BNNS at many highly activated states [29]. h-BN/Pt showed some peculiar properties that are theoretically and experimentally proved [30]. The comparison of h-BNNS/Cu and h-BNNS/Au revealed that the electrocatalytic activity of 2D h-BN completely depends on the metal supporting it. The materials thermal, electronic and catalytic activities could be maintained by doping a single metal atom or by defect engineering [31, 32].

A single Mo atom incorporated at the B vacancy areas in a defective h-BN and graphene-h-BN nanosheets displays good catalytic properties for many reactions [33, 34]. The enzymatic mechanism which deals with the activation of N_2 molecules could be done by implanting Mo atoms. In a nitrogen reduction reaction (NRR), the electrocatalyst used is exfoliated liquid h-BNNS [35]. The following NRR is done through an enzymatic route same as that by using a modified single Mo atom h-BN as catalyst [36]. The studies proved that instead of using Pt/C, hetero doping of many carbon based nanostructures free from metals could be used for electrocatalysis process [37, 38]. Especially, Han *et al.* [39], fabricated B and N co-doped graphene with excellent catalytic properties. C-doped h-BNNS could be synthesised from B, N co-doped graphene by the increase in concentrations of B and N co-doping. Chen *et al.* [40] carried out the DFT calculations on various sorting arrangements of C-doped h-BNNS and discovered -0.81 eV as the moderate adsorption energy and 0.61eV as the activation barrier.

Experimental studies revealed the usage of h-BN based electrodes in many industrial applications. Limani *et al.* [41] analysed many electrodes based on carbon, fitted with pristine h-BN to study the electrocatalyst nature. They discovered that the substrate on which h-BN underlies paves a route to find out the catalytic behaviour. Narayanan *et al.* [42] explored the process of electrocatalysis by h-BN with vertically stacked graphene hetero structures. An improved hydrogen evolution reaction (HER) activities are shown by h-BN/Graphene vertical van der Waals structures manifesting high catalytic behaviour. Many experimental studies have been carried out to intensify the properties of h-BN usually by carbon doping. Rafiq *et al.* [43] designed standard h-BN nanoreactors that shows high oxygen reduction reaction (ORR), hydrogen oxidation reaction (HOR) and oxygen evolution reactions (OER) rates with high stability and excellent catalytic reaction and are surrounded by catalyst nanoparticles. Therefore, future research activities should be highly concentrated on fabricating new advanced materials using modern technologies to attain experimental understanding of theoretical ideas. Many methods including doping of heteroatoms, defect engineering, decorating the material surface using functional groups, *etc.* should be implemented precisely to tune the different properties and to gain high performance results.

Supercapacitors

In an energy storage system, supercapacitors play a vital role analogous to rechargeable batteries. It is also a widely used technology for electrochemical storage. Certain properties exhibited by supercapacitors have made them more superior than rechargeable batteries. Excellent cyclic stability, enhanced catalytic reactions, improved power density and quick charge and discharge capabilities have made them most accustomed than other energy storage applications [44 - 46]. Hexagonal boron nitride nanosheets (h-BNNS) and its functionalisation have made them more accurate material for supercapacitor applications. Doping carbon atoms (heteroatom doping) is more efficient method to fabricate devices and applications with superior capacitive properties. By making certain adjustments in the precursor carrying carbon, it is possible to vary the carbon doping level in h-BNNS [47]. Fabrication of film-shaped and fibre-shaped supercapacitors from functionalised h-BN nanosheets are depicted in the recent studies (Fig. **4**) [48, 49].

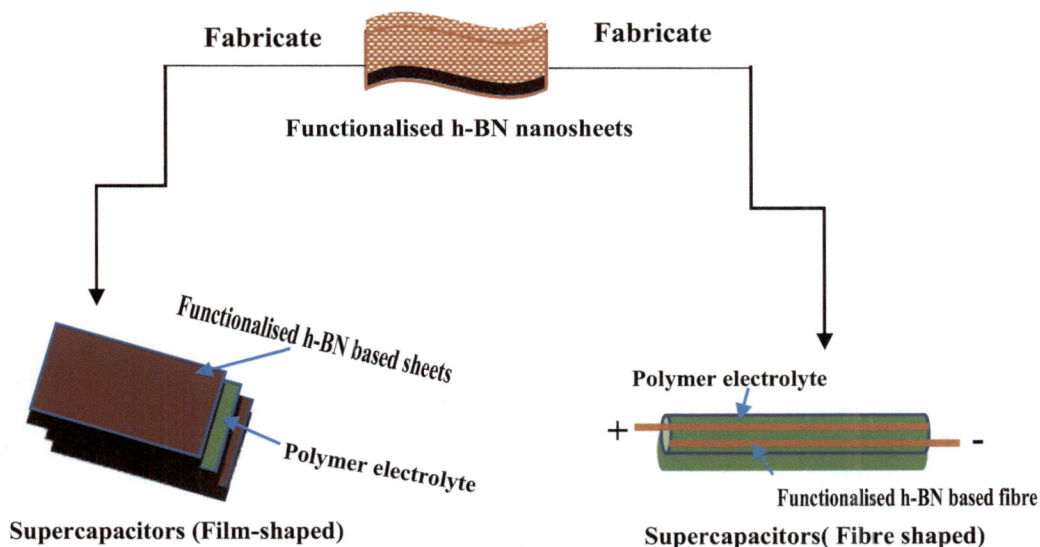

Fig. (4). Two type of supercapacitors designed from hybridised h-BN nanosheets and polymer electrolytes.

The outstanding chemical stability behaviour of h-BN revealed that the machine highlighted a capacitive decay of only 4% at elevated current densities even after 3000 cycles. Hydrothermal treatment of h-BN/rGO could increase the specific capacitance by surging of the electrical conductivity and charge transfer properties and can be readily used in asymmetric supercapacitor (ASC) devices [50]. The restacking process of graphene oxide sheets are repressed by h-BN that is harmful to the capacitance assured but ensures a good charge storage capacity. h-BN based materials and composites for supercapacitor applications are still an ongoing research area. The properties of h-BNNS including good surface area, superlative chemical stability and high thermal and mechanical properties makes them the most promising 2D material for future electrochemical applications. The vacant p orbital of B atoms and alteration of B and N atoms oxidation states results in better electronic properties bringing about a rise in the charge storage capacities. Even-though, h-BN indicates all these superior properties, the small electrical conductance of h-BN has remained as an obstacle for its usage as electrode materials in supercapacitors with superlative performances. Thus, functionalised h-BNNS and h-BNNS incorporated with other 2D materials will result in enhanced electrochemical properties and could be thus used to promote its usage in supercapacitor applications.

Hydrogen Storage

Fuel cell consumption produces H_2O that forms the basis of hydrogen storage. Hence, this could be otherwise called as a substitute to fossil fuels that generate harmful and toxic materials under consumption. The major challenge faced is how to store large amounts of hydrogen and produce it safely and systematically. It is known that hydrogen gas is comparatively lighter and is very difficult to compress and liquify. Many research methods are followed as hydrogen storage provides exceptional gravimetric as well as volumetric densities [51]. The peculiar structure of h-BN and its varying properties which includes B-N polar bonds, marvellous thermal and chemical stability have made them more appropriate material for applications under hydrogen storage. Many speculative researches are carried out based on this and one of the study suggested that the B-N bond in h-BN offers a good ionic character, and thus, increasing its dipole moment helps in the fastest adsorption of H_2 [51, 52]. The hydrogen storing characteristics credibly occurs by the functionalization of h-BN, and generating some changes in its porosity and electronic structure. Advanced investigations have been carried out to find out the application of functionalised h-BN in energy storage and reserving applications. Wang *et al.* [53 - 55], investigated on h-BN use as sorbent material for hydrogen and correlated it with graphite based materials (hydrogen storage). h-BN based materials for hydrogen storage are considered to be of low cost compared to that of borohydride and amine borane [56 - 60]. h-BN based nanomaterials present high reversible capacities [61]. Ball milling technique is used for the breaking down of large ordered materials and to reduce their crystal size. The so obtained nanostructured h-BN material expressed ~3 nm as its lateral size and also 2.6 wt.% of hydrogen was absorbed. The material could maintain its defective nanostructure structure even at high temperatures (1173K) and also at 950K nanostructured graphite underwent recrystallisation process [62]. The simple techniques involved during the preparation and excellent stability features possessed by these nanomaterials make them more suitable for hydrogen storage applications. The major factors involved in hydrogen storage capacity is porosity with regards to the size and volume. Many efforts are carried out to synthesize porous h-BN with a large surface area. Microbelts of h-BN accompanied by specific surface area values extensively ranging between 1161 to 1488 m^2/g were fabricated using a combination of melamine in ammonia and boric acid by the method of pyrolysis [63]. Preparation of porous h-BN under a temperature of 77K and pressure of 1MPa is obtained with an uptake of about 2.3 wt.% of hydrogen. In another work, holding the same conditions, specific surface area of 1900 m^2g^{-1} micro-sponges of h-BN are produced with a hydrogen adsorption (reversible) of only 2.57 wt% [64]. So it is clear from above that bulk form of h-BN illustrates infinitesimal adsorption rates.

h-BN based materials and their hydrogen storage capacities could exceed even at low-temperature conditions. Large surface area supports hydrogen storage. The electronic and chemical structural properties are also a vital factor in the determination of hydrogen storage capacity [65]. The improvement of their hydrogen storage properties is done by doping with hetero atoms and by functionalisation, thus altering the chemical and electronic properties. Lei *et al.* [66] initiated work on h-BNNS doped with oxygen resulting in a 5.7 wt% capacity at room temperature and under a pressure of 5 MPa. They also declared that that 2D structure not only enhanced the cycling stability but also eased the site doped b oxygen regulating the H_2 adsorption. Various other dopants like platinum, palladium, nickel and many more are used with h-BN to improve H_2 adsorption [67 - 69]. The low price and simple methods help to produce a porous structure and makes them appropriate for hydrogen storage applications. Functionalisation such as doping using hetero atoms, vacancy or defect engineering and grafting of functional groups can generate different adsorption sites with high binding energies, thus, providing diverse strategic methods to boost up their hydrogen storage capacity. The robust nature of interaction between the adsorbent materials and H_2 creates a deleterious effect triggering a difficulty in the release of hydrogen that needs to be also considered.

Secondary Batteries

h-BN blended with other conducting materials could be used as electrolytes, electrodes and many more in secondary batteries. Many investigations have been conducted to find out the applications of h-BN and its related materials. Li *et al.* [70], proposed a study on h-BN by actively depositing LiF on the vacant sites of h-BN by atomic layer deposition method. The so formed film showed high cycling stability rates and good battery performance. LIB electrodes free from binders are produced with h-BN and rGO composites [71]. A reversible capacity of 278 mAh/g at 100 mA/g is observed with only 2 wt.% of BN. The researchers even demonstrated that the excellent battery execution and performance of this synthesised film is due to the electrostatic and van der Waals interaction between the incorporated 2D materials. Monajjem *et al.* [72] have represented that mixing of h-BN with two doped graphene layers could improve the reversible cycling capacity of LIBs and enhance good cycling stability with improved performance rates. This also revealed that the electrochemical performance is very high and is stable compared to that of graphene-based materials. The method that could be used for augmenting the battery performance is by the addition of certain elements as dopants. High ionic conductivity of about 3 mS/cm is displayed by BN-based electrolytes and separators. Aydin *et al.* [73] introduced polyacrylonitrile (PAN) incorporated with hexagonal boron nitride to form a composite (nanofiber) with respect to an organic-based electrolyte (1 M LiPF6 in EC/DMC (1:1)) for lithium ion batteries. Electro-spinning method was used for the synthesis of this composite. The outcomes revealed Li-formulated cells with outstanding thermal, electrochemical and chemical stability, high ionic conductivity rate and excellent electrolyte uptakes. The experimental results also confirmed that with 10 wt.% composite loading, the Li-based cell showed good thermal stability, elevated ionic conductivity, outstanding electrochemical stability and high electrolyte uptake. In another noval study by Shim *et al.* [74], piperidinium-based liquid ionic electrolyte with h-BN was synthesised for LIBs. They also depicted large ionic and thermal conductivity rates, with an increase in their electrochemical stabilities.

Solar Energy Transformation and Utilisation

Solar energy is a renewable form of energy. Conversion of energy from sun to chemical energy by various methods may involve many chemical reactions which in turn produces electricity. Many research works are based on the conversion of solar energy and its usage in further applications. Band gap tuning and the excellent chemical stability of h-BN have made them more suitable for solar energy conversion applications. h-BN should be functionalised for this process. The new advanced technologies use functionalised h-BN as catalysts in photocatalytic reactions, reduction of CO_2 and in photovoltaic cells. Various materials are incorporated with h-BN (functionalised) and their heterostructures offer superior thermal and chemical stability. h-BN possesses the ability to accept or transport photo-generated charges ensuring the separation of charges with its recombination [75]. Wang *et al.* [76], investigated on the usage of h-BN materials for photocatalytic activities. In finding out deep space, pristine h-BNNS could be used as it possesses a broad band gap. The sandwich structure arrangement of graphene/h-BN/GaAs also exhibited an improved power transformation efficiency in solar cell due to the insertion of h-BN between graphene and gallium arsenide [77]. The band gap of h-BN could be tuned by enhancing its properties by the addition of functional groups. The small and narrowed band gap could be seen when doped by sulphur is due to the decrease in the potential of the conduction band as sulphur dopants are used and the position of the valence band remains unaltered. This in-turn provides more photogenerated charge separation manifested by a decrease in the photoluminescence intensity. The tuned-up properties of functionalised h-BN offers high optical absorption, excellent separation of charges and outstanding reacting surface properties. All these peculiar features bestow the good performance in photocatalytic reactions.

RECENT TECHNICAL ADVANCEMENTS OF FUNCTIONALISED 2D H-BN AND ITS VARIOUS APPLICATIONS

Many recent techniques and methods are implemented for the usage of 2D h-BN in various fields. Goel *et al.* [78], studied the usage of 2D h-BN, its synthesis processes, functionalisation methods and its wide utilisation in the field of gas sensing. Due to the decreased surface reactivity levels and low conductivity nature of 2D h-BN, they further undergo the process of functionalisation and doping, thus inheriting enhanced properties suitable for gas sensors. The new and interesting research area is on regenerative medicine, which helps in overcoming the defects evolved in tissues [79]. Graphene, functionalised 2D h-BN and many other nanomaterials are developed with improved properties and are applicable in tissue engineering fields. Yuanyuan Zheng *et al.*. [80] studied various aspects on how to use these nanomaterials in the field of tissue engineering to replicate tissues and to restore the defects in tissues, skin and cartilages. The new possibilities showed the designing and fabrication of 2D h-BN materials in nano- and micro-scale. The various interesting properties have made them widely accepted two dimensional nanomaterials. Functionalised and doping of 2D h-BN made them more suitable in photonic applications and in energy storage fields [81]. Different studies were performed before to understand the physical as well as chemical features of polymer electrolytes and their further implementations in energy storage applications. h-BN was incorporated with sulfonated polysulfone and an ionic liquid, blended in numerous fractions to synthesise a polymer electrolyte and characterised to figure out the proton conductivity values. The so prepared membranes depict high thermal and chemical stability with excellent conductivity, capacitance and energy density, thus making them more suitable for supercapacitor energy applications [82]. 2D h-BN nanomaterials (nanoflakes- h-BNNFs) are used as additives to fabricate gel polymer electrolytes, as shown in Fig. (**5**) with good cycling stability and high performance rates, thus, enabling their usage in lithium ion batteries [83].

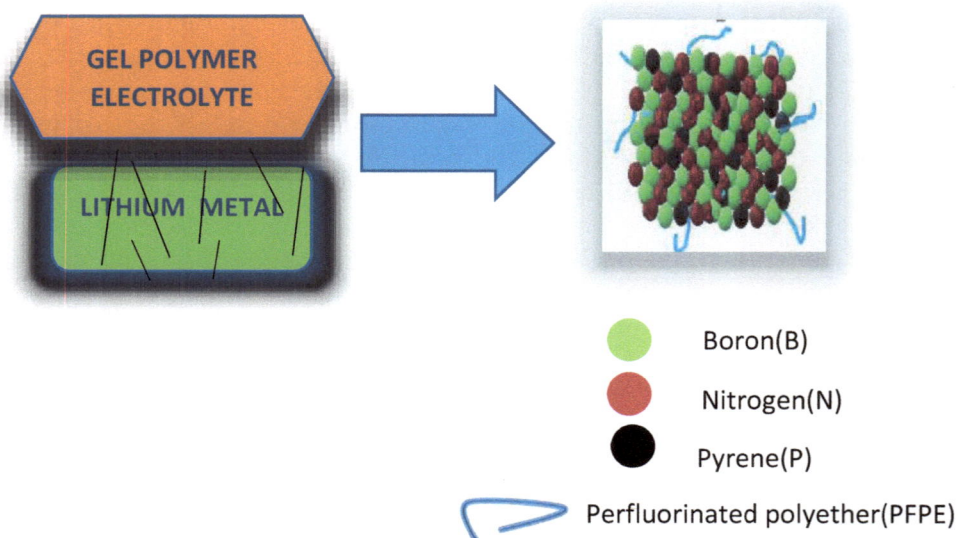

Fig. (5). Illustration of h-BN nanoflakes as additives in the synthesis of gel polymer electrolytes for Lithium-ion batteries.

The various research studies also provide an insight to the different methods executed for the preparation of polymer electrolytes using functionalised h-BN as fillers. Xin Liu *et al.* [84] explored the hexagonal boron nitride functionalised using amines to synthesise ion electrolytes for lithium ion battery applications. These electrolytes (ion-gel) depict various peculiar features. Often, they reveal less transference number values in lithium-ion batteries. But the latter studies also proved that, the addition of amine group functionalised h-BN could enhance the movement of lithium ions, exhibits extraordinary performance and capacity rates by decreasing the mobility rates of anions. Another work relates to the use of 2D h-BN with stearic acid to form composite materials for thermal energy storage applications [85]. The synthesis of SA/hBN-E composites is shown in (Fig. **6**). h-BN undergoes exfoliation process and when incorporated with stearic acid (SA), this composite material in-turn showed high thermal conductivity (more than 70%) than pure SA which enhanced its usage in many applications.

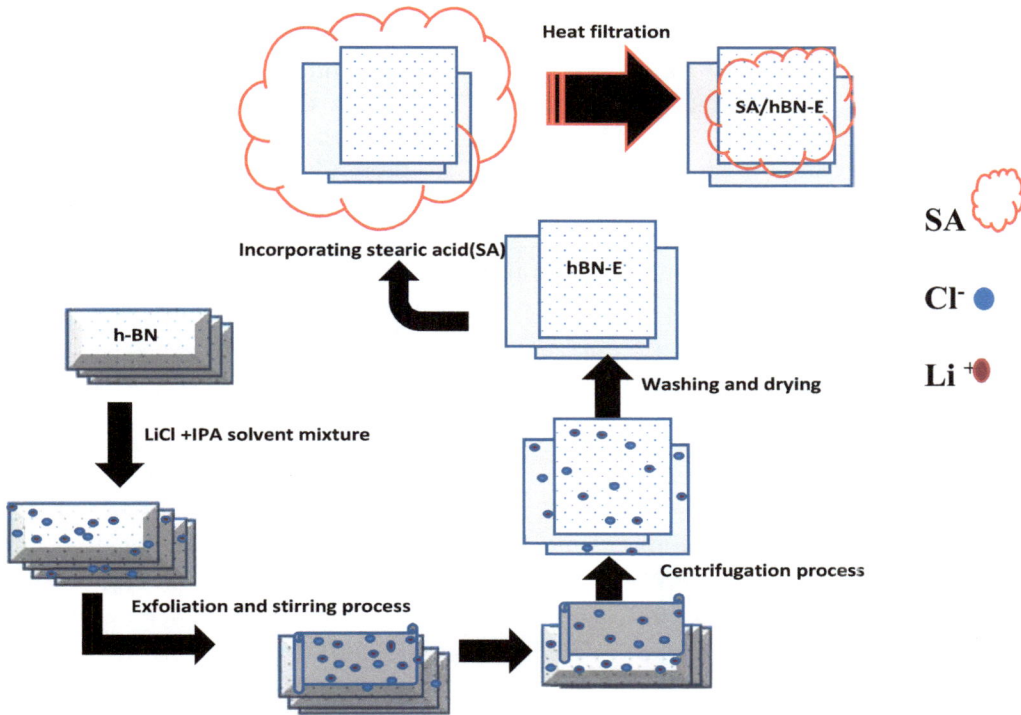

Fig. (6). Synthesis of SA/hBN-E composites.

CONCLUDING REMARKS

With a new and arising significance in the development of 2D materials, 2D-h-BN nanosheets have been diligently explored as a group of distinct and eccentric substances with interesting peculiar characteristic features different from that of bulk BN. There is an enormous outbreak in the development of new materials incorporated with 2D h-BN and different strategic methods involved in the synthesis of 2D h-BN are successfully depicted throughout this chapter. Functionalised h-BN with innumerable characteristic features are thus further made beneficial in electrical, mechanical, biomedical, and other application related areas. h-BN nanosheets show good mechanical properties and outstanding thermal conductivity. They could be incorporated with certain materials to enhance their properties. Considering the phonon Boltzmann transport equation and its numerical solution at room temperature, thermal conductivity k of about 600 W m^{-1} K^{-1} is estimated theoretically for single layered h-BN, whereas the reasoned value for bulk h-BN is 400 W m^{-1} K^{-1}. It is also revealed that the decrease in layer numbers is due to the reduced phonon–phonon scattering in 2D h-BN. The determination of the k values for multi-layered h-BN depends on the intrinsic phonon–phonon scattering that exhibits a foremost role in providing vast contributions. The studies further proved that the exfoliated h-BN contains impurities of polymer residues. This leads to the deterioration of k value with an increase in the scattering of phonons (low frequency) by these residues. The surface modulations altered the physical as well as chemical behaviour and produced h-BN with extraordinary properties, including bandgap tuning and high porosity nature, making them more appropriate for energy storage applications. The high thermal and chemical stability often made h-BN a promising material for catalytic reactions. 2D h-BN nanosheets when compared with 3D h-BN (centrosymmetric), show piezoelectric effect as they are ionic crystals. The sub-lattices are inhabited by discrete elements and the lattice points do not act as center of symmetry, and thereby, ascending to a splintered inversion symmetry depicting piezoelectric effect. The identification of piezoelectric effects in 2D h-BN authorise their usage in manufacturing many device applications. h-BN nanosheets possess wide bandgap exhibiting ultraviolet luminescence property, and thus, act as emerging widely accepted materials for many optical applications such as lasing using ultraviolet radiations, emission of phonons and in deep ultraviolet detectors. h-BN has shown outstanding energy reserving capacities, but despite of its low electrical conductivity, h-BN is being incorporated with other materials with superior conductive properties. Deep penetration properties with reduced photo damage effect, high 3D resolution and good hyper polarizability revealed by h-BN nanosheets make them more appropriate for applications in optical fields. Heterostructures of h-BN with other conductive materials exhibit improved electrochemical performance, and thus, could be used as an electrode, an electrolyte and separators in energy storage fields. Theoretical and experimental studies are reviewed that gave an insight onto the structure of 2D h-BN and its bonding nature. Functionalisation of h-BN helps in tuning the properties and characterisation techniques further help in the designing and preparation of 2D h-BN for energy storage and conversion applications.

CONSENT FOR PUBLICATION

Not applicable.

CONFLICT OF INTEREST

The authors declare no conflict of interest, financial or otherwise.

ACKNOWLEDGEMENTS

Declared none.

REFERENCES

[1]   Yu, Z.; Tetard, L.; Zhai, L.; Thomas, J. Supercapacitor electrode materials: nanostructures from 0 to 3 dimensions. *Energy Environ. Sci.,* **2015**, *8*(3), 702-730.
[http://dx.doi.org/10.1039/C4EE03229B]

[2]   Butler, S.Z.; Hollen, S.M.; Cao, L.; Cui, Y.; Gupta, J.A.; Gutiérrez, H.R.; Heinz, T.F.; Hong, S.S.; Huang, J.; Ismach, A.F.; Johnston-Halperin, E.; Kuno, M.; Plashnitsa, V.V.; Robinson, R.D.; Ruoff, R.S.; Salahuddin, S.; Shan, J.; Shi, L.; Spencer, M.G.; Terrones, M.; Windl, W.; Goldberger, J.E. Progress, challenges, and opportunities in two-dimensional materials beyond graphene. *ACS Nano,* **2013**, *7*(4), 2898-2926.
[http://dx.doi.org/10.1021/nn400280c] [PMID: 23464873]

[3]   Peng, X.; Peng, L.; Wu, C.; Xie, Y. Two dimensional nanomaterials for flexible supercapacitors. *Chem. Soc. Rev.,* **2014**, *43*(10), 3303-3323.
[http://dx.doi.org/10.1039/c3cs60407a] [PMID: 24614864]

[4]   Chhowalla, M.; Shin, H.S.; Eda, G.; Li, L.J.; Loh, K.P.; Zhang, H. The chemistry of two-dimensional layered transition metal dichalcogenide nanosheets. *Nat. Chem.,* **2013**, *5*(4), 263-275.
[http://dx.doi.org/10.1038/nchem.1589] [PMID: 23511414]

[5]   Naguib, M.; Mashtalir, O.; Carle, J.; Presser, V.; Lu, J.; Hultman, L.; Gogotsi, Y.; Barsoum, M.W. Two-dimensional transition metal carbides. *ACS Nano,* **2012**, *6*(2), 1322-1331.
[http://dx.doi.org/10.1021/nn204153h] [PMID: 22279971]

[6]   Naguib, M.; Bentzel, G. W.; Shah, J.; Halim, J.; Caspi, E. N.; Lu, J. & Barsoum, M. W.. New Solid Solution MAX Phases:(Ti0. 5, V0. 5) 3AlC2, (Nb0. 5, V0. 5) 2AlC,(Nb0. 5, V0. 5) 4AlC3 and (Nb0. 8, Zr0. 2) 2AlC. *Mater. Res. Lett.,* **2014**, *2*(4), 233-240.
[http://dx.doi.org/10.1080/21663831.2014.932858]

[7]   Naguib, M.; Kurtoglu, M.; Presser, V.; Lu, J.; Niu, J.; Heon, M.; Hultman, L.; Gogotsi, Y.; Barsoum, M.W. Two-dimensional nanocrystals produced by exfoliation of Ti3 AlC2. *Adv. Mater.,* **2011**, *23*(37), 4248-4253.
[http://dx.doi.org/10.1002/adma.201102306] [PMID: 21861270]

[8]   Naguib, M.; Mochalin, V.N.; Barsoum, M.W.; Gogotsi, Y. 25th anniversary article: MXenes: a new family of two-dimensional materials. *Adv. Mater.,* **2014**, *26*(7), 992-1005.
[http://dx.doi.org/10.1002/adma.201304138] [PMID: 24357390]

[9]   Lukatskaya, M.R.; Mashtalir, O.; Ren, C.E.; Dall'Agnese, Y.; Rozier, P.; Taberna, P.L.; Naguib, M.; Simon, P.; Barsoum, M.W.; Gogotsi, Y. Cation intercalation and high volumetric capacitance of two-dimensional titanium carbide. *Science,* **2013**, *341*(6153), 1502-1505.
[http://dx.doi.org/10.1126/science.1241488] [PMID: 24072919]

[10]   Mendoza-Sánchez, B.; Coelho, J.; Pokle, A.; Nicolosi, V. A 2D graphene-manganese oxide nanosheet hybrid synthesized by a single step liquid-phase co-exfoliation method for supercapacitor applications. *Electrochim. Acta,* **2015**, *174*, 696-705.
[http://dx.doi.org/10.1016/j.electacta.2015.06.030]

[11]   Peng, L.; Peng, X.; Liu, B.; Wu, C.; Xie, Y.; Yu, G. Ultrathin two-dimensional MnO_2/graphene hybrid nanostructures for high-performance, flexible planar supercapacitors. *Nano Lett.,* **2013**, *13*(5), 2151-2157.
[http://dx.doi.org/10.1021/nl400600x] [PMID: 23590256]

[12]   Watanabe, K.; Taniguchi, T.; Kanda, H. Direct-bandgap properties and evidence for ultraviolet lasing of hexagonal boron nitride single crystal. *Nat. Mater.,* **2004**, *3*(6), 404-409.
[http://dx.doi.org/10.1038/nmat1134] [PMID: 15156198]

[13]   Tsekmes, I.A.; Kochetov, R.; Morshuis, P.H.F.; Smit, J.J. Measuring and modeling the thermal conductivity of epoxy-Boron nitride nanocomposites. *Proceedings of 2014 International Symposium on Electrical Insulating Materials,* **2014**, pp. 26-29.
[http://dx.doi.org/10.1109/ISEIM.2014.6870711]

[14]   Bhattacharya, A.; Bhattacharya, S.; Das, G.P. Band gap engineering by functionalization of BN sheet. *Phys. Rev. B,* **2012**, *85*(3), 035415.
[http://dx.doi.org/10.1103/PhysRevB.85.035415]

[15]   Huang, C.; Chen, C.; Zhang, M.; Lin, L.; Ye, X.; Lin, S. Carbon-doped BN nanosheets for metal-free photoredox catalys. *Nat. Commun.,* **2015**, *6*(1), 1-7.

[16]   Liu, F.; Nattestad, A.; Naficy, S.; Han, R.; Casillas, G.; Angeloski, A. Fluorescent Carbon☐and Oxygen☐Doped Hexagonal Boron Nitride Powders as Printing Ink for Anticounterfeit Applications. *Adv. Opt. Mater.,* **2019**, *7*(24), 1901380.
[http://dx.doi.org/10.1002/adom.201901380]

[17]   Lei, W.; Qin, S.; Liu, D.; Portehault, D.; Liu, Z.; Chen, Y. Large scale boron carbon nitride nanosheets with enhanced lithium storage capabilities. *Chem. Commun. (Camb.),* **2013**, *49*(4), 352-354.
[http://dx.doi.org/10.1039/C2CC36998B] [PMID: 23108161]

[18]   Weng, Q.; Wang, X.; Wang, X.; Liu, D.; Jiang, X.; Zhi, C.; Bando, Y.; Golberg, D. Preparation and hydrogen sorption performances of BCNO porous microbelts with ultra-narrow and tunable pore widths. *Chem. Asian J.,* **2013**, *8*(12), 2936-2939.
[http://dx.doi.org/10.1002/asia.201300940] [PMID: 24115448]

[19]   Thomas, S.; Ajith, K.M.; Valsakumar, M.C. Directional anisotropy, finite size effect and elastic properties of hexagonal boron nitride. *J. Phys. Condens. Matter,* **2016**, *28*(29), 295302.
[http://dx.doi.org/10.1088/0953-8984/28/29/295302] [PMID: 27255345]

[20]   Novoselov, K. S.; Geim, A. K.; Morozov, S. V.; Jiang, D.; Zhang, Y.; Dubonos, S. V. Electric field effect in atomically thin carbon films. *Science,* **2004**, *306*(5696), 666-669.

[21]   Jayaramulu, K.; Dubal, D.P.; Nagar, B.; Ranc, V.; Tomanec, O.; Petr, M.; Datta, K.K.R.; Zboril, R.; Gómez-Romero, P.; Fischer, R.A. Ultrathin hierarchical porous carbon nanosheets for high-performance supercapacitors and redox electrolyte energy storage. *Adv. Mater.,* **2018**, *30*(15), e1705789.
[http://dx.doi.org/10.1002/adma.201705789] [PMID: 29516561]

[22]   Zhang, K.; Feng, Y.; Wang, F.; Yang, Z.; Wang, J. Two dimensional hexagonal boron nitride (2D-hBN): synthesis, properties and applications. *J. Mater. Chem. C Mater. Opt. Electron. Devices,* **2017**, *5*(46), 11992-12022.
[http://dx.doi.org/10.1039/C7TC04300G]

[23]   Xue, Y.; Liu, Q.; He, G.; Xu, K.; Jiang, L.; Hu, X.; Hu, J. Excellent electrical conductivity of the exfoliated and fluorinated hexagonal boron nitride nanosheets. *Nanoscale Res. Lett.,* **2013**, *8*(1), 49.
[http://dx.doi.org/10.1186/1556-276X-8-49] [PMID: 23347409]

[24]   Uosaki, K.; Elumalai, G.; Dinh, H.C.; Lyalin, A.; Taketsugu, T.; Noguchi, H. Highly efficient electrochemical hydrogen evolution reaction at insulating boron nitride nanosheet on inert gold substrate. *Sci. Rep.,* **2016**, *6*(1), 32217.
[http://dx.doi.org/10.1038/srep32217] [PMID: 27558958]

[25]   Guha, A.; Veettil Vineesh, T.; Sekar, A.; Narayanaru, S.; Sahoo, M.; Nayak, S. Mechanistic insight into enhanced hydrogen evolution reaction activity of ultrathin hexagonal boron nitride-modified Pt electrodes. *ACS Catal.,* **2018**, *8*(7), 6636-6644.
[http://dx.doi.org/10.1021/acscatal.8b00938]

[26]   Han, R.; Liu, F.; Wang, X.; Huang, M.; Li, W.; Yamauchi, Y. Functionalised hexagonal boron nitride for energy conversion and storage. *J. Mater. Chem. A Mater. Energy Sustain.,* **2020**, *8*(29), 14384-14399.
[http://dx.doi.org/10.1039/D0TA05008C]

[27]   Lyalin, A.; Nakayama, A.; Uosaki, K.; Taketsugu, T. Theoretical predictions for hexagonal BN based nanomaterials as electrocatalysts for the oxygen reduction reaction. *Phys. Chem. Chem. Phys.,* **2013**, *15*(8), 2809-2820.

[http://dx.doi.org/10.1039/c2cp42907a] [PMID: 23338859]

[28]   Uosaki, K.; Elumalai, G.; Noguchi, H.; Masuda, T.; Lyalin, A.; Nakayama, A.; Taketsugu, T. Boron nitride nanosheet on gold as an electrocatalyst for oxygen reduction reaction: theoretical suggestion and experimental proof. *J. Am. Chem. Soc.,* **2014**, *136*(18), 6542-6545.
[http://dx.doi.org/10.1021/ja500393g] [PMID: 24773085]

[29]   Han, R.; Liu, F.; Wang, X.; Huang, M.; Li, W.; Yamauchi, Y. Functionalised hexagonal boron nitride for energy conversion and storage. *J. Mater. Chem. A Mater. Energy Sustain.,* **2020**, *8*(29), 14384-14399.
[http://dx.doi.org/10.1039/D0TA05008C]

[30]   Deng, C.; He, R.; Shen, W.; Li, M.; Zhang, T. A single-atom catalyst of cobalt supported on a defective two-dimensional boron nitride material as a promising electrocatalyst for the oxygen reduction reaction: a DFT study. *Phys. Chem. Chem. Phys.,* **2019**, *21*(13), 6900-6907.
[http://dx.doi.org/10.1039/C9CP00452A] [PMID: 30863835]

[31]   Deng, C.; He, R.; Wen, D.; Shen, W.; Li, M. Theoretical study on the origin of activity for the oxygen reduction reaction of metal-doped two-dimensional boron nitride materials. *Phys. Chem. Chem. Phys.,* **2018**, *20*(15), 10240-10246.
[http://dx.doi.org/10.1039/C8CP00838H] [PMID: 29594280]

[32]   Zhang, C.; Zhang, X.; Zhao, W.; Zeng, C.; Li, W.; Li, B.; Luo, X.; Li, J.; Jiang, J.; Deng, B.; McComb, D.W.; Dong, Y. Chemotherapy drugs derived nanoparticles encapsulating mRNA encoding tumor suppressor proteins to treat triple-negative breast cancer. *Nano Res.,* **2019**, *12*(4), 855-861.
[http://dx.doi.org/10.1007/s12274-019-2308-9] [PMID: 31737223]

[33]   Zhao, J.; Chen, Z. Single Mo atom supported on defective boron nitride monolayer as an efficient electrocatalyst for nitrogen fixation: a computational study. *J. Am. Chem. Soc.,* **2017**, *139*(36), 12480-12487.
[http://dx.doi.org/10.1021/jacs.7b05213] [PMID: 28800702]

[34]   Huang, Y.; Yang, T.; Yang, L.; Liu, R.; Zhang, G.; Jiang, J. Graphene–boron nitride hybrid-supported single Mo atom electrocatalysts for efficient nitrogen reduction reaction. *J. Mater. Chem. A Mater. Energy Sustain.,* **2019**, *7*(25), 15173-15180.
[http://dx.doi.org/10.1039/C9TA02947H]

[35]   Zhu, Y.P.; Guo, C.; Zheng, Y.; Qiao, S.Z. Surface and interface engineering of noble-metal-free electrocatalysts for efficient energy conversion processes. *Acc. Chem. Res.,* **2017**, *50*(4), 915-923.
[http://dx.doi.org/10.1021/acs.accounts.6b00635] [PMID: 28205437]

[36]   Tang, J.; Liu, J.; Li, C.; Li, Y.; Tade, M.O.; Dai, S.; Yamauchi, Y. Synthesis of nitrogen-doped mesoporous carbon spheres with extra-large pores through assembly of diblock copolymer micelles. *Angew. Chem. Int. Ed. Engl.,* **2015**, *54*(2), 588-593.
[PMID: 25393650]

[37]   Zhao, J.; Chen, Z. Carbon-doped boron nitride nanosheet: an efficient metal-free electrocatalyst for the oxygen reduction reaction. *J. Phys. Chem. C,* **2015**, *119*(47), 26348-26354.
[http://dx.doi.org/10.1021/acs.jpcc.5b09037]

[38]   Khan, A.F.; Randviir, E.P.; Brownson, D.A.; Ji, X.; Smith, G.C.; Banks, C.E. 2D hexagonal boron nitride (2D-hBN) explored as a potential electrocatalyst for the oxygen reduction reaction. *Electroanalysis,* **2017**, *29*(2), 622-634.
[http://dx.doi.org/10.1002/elan.201600462]

[39]   Han, R.; Liu, F.; Wang, X.; Huang, M.; Li, W.; Yamauchi, Y.; Sun, X.; Huang, Z. Functionalised hexagonal boron nitride for energy conversion and storage. *J. Mater. Chem. A Mater. Energy Sustain.,* **2020**, *8*(29), 14384-14399.
[http://dx.doi.org/10.1039/D0TA05008C]

[40]   Chen, S.; Li, Y.; Zhang, Z.; Fu, Q.; Bao, X. The synergetic effect of h-BN shells and subsurface B in CoB x@ h-BN nanocatalysts for enhanced oxygen evolution reactions. *J. Mater. Chem. A Mater.*

Energy Sustain., **2018**, *6*(23), 10644-10648.
[http://dx.doi.org/10.1039/C8TA02312C]

[41]   Limani, N.; Boudet, A.; Blanchard, N.; Jousselme, B.; Cornut, R. Local probe investigation of electrocatalytic activity. *Chem. Sci. (Camb.)*, **2020**, *12*(1), 71-98.
[http://dx.doi.org/10.1039/D0SC04319B] [PMID: 34163583]

[42]   Rastogi, P.K.; Sahoo, K.R.; Thakur, P.; Sharma, R.; Bawari, S.; Podila, R.; Narayanan, T.N. Graphene-hBN non-van der Waals vertical heterostructures for four- electron oxygen reduction reaction. *Phys. Chem. Chem. Phys.*, **2019**, *21*(7), 3942-3953.
[http://dx.doi.org/10.1039/C8CP06155F] [PMID: 30706063]

[43]   Rafiq, M.; Hu, X.; Ye, Z.; Qayum, A.; Xia, H.; Hu, L.; Lu, F.; Chu, P.K. Recent Advances in Structural Engineering of 2D Hexagonal Boron Nitride Electrocatalysts. *Nano Energy*, **2021**, *91*, 106661.

[44]   Salunkhe, R.R.; Lee, Y.H.; Chang, K.H.; Li, J.M.; Simon, P.; Tang, J.; Torad, N.L.; Hu, C.C.; Yamauchi, Y. Nanoarchitectured graphene-based supercapacitors for next-generation energy-storage applications. *Chemistry*, **2014**, *20*(43), 13838-13852.
[http://dx.doi.org/10.1002/chem.201403649] [PMID: 25251360]

[45]   Karbhal, I.; Devarapalli, R.R.; Debgupta, J.; Pillai, V.K.; Ajayan, P.M.; Shelke, M.V. Facile green synthesis of bcn nanosheets as high-performance electrode material for electrochemical energy storage. *Chemistry*, **2016**, *22*(21), 7134-7140.
[http://dx.doi.org/10.1002/chem.201505225] [PMID: 27072914]

[46]   Ribeiro, H.; Trigueiro, J. P.; Woellner, C. F.; Pedrotti, J. J.; Miquita, D. R.; Silva, W. M. Higher thermal conductivity and mechanical enhancements in hybrid 2D polymer nanocomposites. *Polym. Test.*, **2020**, *87*, 106510.
[http://dx.doi.org/10.1016/j.polymertesting.2020.106510]

[47]   Yan, K.; Lee, H.W.; Gao, T.; Zheng, G.; Yao, H.; Wang, H.; Lu, Z.; Zhou, Y.; Liang, Z.; Liu, Z.; Chu, S.; Cui, Y. Ultrathin two-dimensional atomic crystals as stable interfacial layer for improvement of lithium metal anode. *Nano Lett.*, **2014**, *14*(10), 6016-6022.
[http://dx.doi.org/10.1021/nl503125u] [PMID: 25166749]

[48]   Khan, A.F.; Down, M.P.; Smith, G.C.; Foster, C.W.; Banks, C.E. Surfactant-exfoliated 2D hexagonal boron nitride (2D-hBN): role of surfactant upon the electrochemical reduction of oxygen and capacitance applications. *J. Mater. Chem. A Mater. Energy Sustain.*, **2017**, *5*(8), 4103-4113.
[http://dx.doi.org/10.1039/C6TA09999H]

[49]   Züttel, A.; Sudan, P.; Mauron, P.; Kiyobayashi, T.; Emmenegger, C.; Schlapbach, L. Hydrogen storage in carbon nanostructures. *Int. J. Hydrogen Energy*, **2002**, *27*(2), 203-212.
[http://dx.doi.org/10.1016/S0360-3199(01)00108-2]

[50]   Orimo, S.; Majer, G.; Fukunaga, T. Zü ttel A, Schlapbach L, Fujii H. Hydrogen in the mechanically prepared nanostructured graphite. *Appl. Phys. Lett.*, **1999**, *75*(20), 3093-3095.
[http://dx.doi.org/10.1063/1.125241]

[51]   Fukunaga, T.; Itoh, K.; Orimo, S.; Aoki, M.; Fujii, H. Location of deuterium atoms absorbed in nanocrystalline graphite prepared by mechanical alloying. *J. Alloys Compd.*, **2001**, *327*(1-2), 224-229.
[http://dx.doi.org/10.1016/S0925-8388(01)01420-7]

[52]   Orimo, S.; Matsushima, T.; Fujii, H.; Fukunaga, T.; Majer, G. Hydrogen desorption property of mechanically prepared nanostructured graphite. *J. Appl. Phys.*, **2001**, *90*(3), 1545-1549.
[http://dx.doi.org/10.1063/1.1385362]

[53]   Huang, W. Z., Wang, S., Dewhurst, R. D., Ignat'ev, N. V., Finze, M., & Braunschweig, H. Boron: Its Role in Energy-Related Processes and Applications. *Angew. Chem. Int. Ed.*, **2020**, *59*(23), 8800-8816.
[http://dx.doi.org/10.1002/anie.201911108]

[54]   Wang, P.; Orimo, S.; Matsushima, T.; Fujii, H.; Majer, G. Hydrogen in mechanically prepared

nanostructured h-BN: a critical comparison with that in nanostructured graphite. *Appl. Phys. Lett.,* **2002**, *80*(2), 318-320.
[http://dx.doi.org/10.1063/1.1432447]

[55]   Weng, Q.; Wang, X.; Zhi, C.; Bando, Y.; Golberg, D. Boron nitride porous microbelts for hydrogen storage. *ACS Nano,* **2013**, *7*(2), 1558-1565.
[http://dx.doi.org/10.1021/nn305320v] [PMID: 23301807]

[56]   Chen, W.; Yu, H.; Wu, G.; He, T.; Li, Z.; Guo, Z.; Liu, H.; Huang, Z.; Chen, P. Ammonium aminodiboranate: a long-sought isomer of diammoniate of diborane and ammonia borane dimer. *Chemistry,* **2016**, *22*(23), 7727-7729.
[http://dx.doi.org/10.1002/chem.201601375] [PMID: 27017580]

[57]   Staubitz, A.; Robertson, A.P.; Manners, I. Ammonia-borane and related compounds as dihydrogen sources. *Chem. Rev.,* **2010**, *110*(7), 4079-4124.
[http://dx.doi.org/10.1021/cr100088b] [PMID: 20672860]

[58]   Weng, Q.; Wang, X.; Wang, X.; Liu, D.; Jiang, X.; Zhi, C.; Bando, Y.; Golberg, D. Preparation and hydrogen sorption performances of BCNO porous microbelts with ultra-narrow and tunable pore widths. *Chem. Asian J.,* **2013**, *8*(12), 2936-2939.
[http://dx.doi.org/10.1002/asia.201300940] [PMID: 24115448]

[59]   Lei, W.; Zhang, H.; Wu, Y.; Zhang, B.; Liu, D.; Qin, S. Oxygen-doped boron nitride nanosheets with excellent performance in hydrogen storage. *Nano Energy,* **2014**, *6*, 219-224.
[http://dx.doi.org/10.1016/j.nanoen.2014.04.004]

[60]   Shevlin, S.A.; Guo, Z.X. Transition-metal-doping-enhanced hydrogen storage in boron nitride systems. *Appl. Phys. Lett.,* **2006**, *89*(15), 153104.
[http://dx.doi.org/10.1063/1.2360232]

[61]   Leybo, D.; Tagirov, M.; Permyakova, E.; Konopatsky, A.; Firestein, K.; Tuyakova, F. Ascorbic acid-assisted polyol synthesis of Iron and Fe/GO, Fe/h-BN composites for Pb2+ removal from wastewaters. *Nanomaterials (Basel),* **2020**, *10*(1), 37.
[http://dx.doi.org/10.3390/nano10010037]

[62]   Jhi, S.H.; Kwon, Y.K. Hydrogen adsorption on boron nitride nanotubes: a path to room-temperature hydrogen storage. *Phys. Rev. B,* **2004**, *69*(24), 245407.
[http://dx.doi.org/10.1103/PhysRevB.69.245407]

[63]   Zhang, L.P.; Wu, P.; Sullivan, M.B. Hydrogen adsorption on Rh, Ni, and Pd functionalized single-walled boron nitride nanotubes. *J. Phys. Chem. C,* **2011**, *115*(10), 4289-4296.
[http://dx.doi.org/10.1021/jp1078554]

[64]   Weng, Q.; Wang, X.; Bando, Y.; Golberg, D. One-step template-free synthesis of highly porous boron nitride microsponges for hydrogen storage. *Adv. Energy Mater.,* **2014**, *4*(7), 1301525.
[http://dx.doi.org/10.1002/aenm.201301525]

[65]   Portehault, D.; Giordano, C.; Gervais, C.; Senkovska, I.; Kaskel, S.; Sanchez, C.; Antonietti, M. High-surface-area nanoporous boron carbon nitrides for hydrogen storage. *Adv. Funct. Mater.,* **2010**, *20*(11), 1827-1833.
[http://dx.doi.org/10.1002/adfm.201000281]

[66]   Lei, W.; Zhang, H.; Wu, Y.; Zhang, B.; Liu, D.; Qin, S. Oxygen-doped boron nitride nanosheets with excellent performance in hydrogen storage. *Nano Energy,* **2014**, *6*, 219-224.
[http://dx.doi.org/10.1016/j.nanoen.2014.04.004]

[67]   Oshchepkov, A.G.; Braesch, G.; Bonnefont, A.; Savinova, E.R.; Chatenet, M. Recent advances in the understanding of nickel-based catalysts for the oxidation of hydrogen-containing fuels in alkaline media. *ACS Catal.,* **2020**, *10*(13), 7043-7068.
[http://dx.doi.org/10.1021/acscatal.0c00101]

[68]   Wu, X.; Yang, J.L.; Zeng, X.C. Adsorption of hydrogen molecules on the platinum-doped boron

nitride nanotubes. *J. Chem. Phys.,* **2006**, *125*(4), 44704.
[http://dx.doi.org/10.1063/1.2210933] [PMID: 16942171]

[69]   Venkataramanan, N.S.; Khazaei, M.; Sahara, R.; Mizuseki, H.; Kawazoe, Y. First-principles study of hydrogen storage over Ni and Rh doped BN sheets. *Chem. Phys.,* **2009**, *359*(1-3), 173-178.
[http://dx.doi.org/10.1016/j.chemphys.2009.04.001]

[70]   Li, Q.; Pan, H.; Li, W.; Wang, Y.; Wang, J.; Zheng, J. Homogeneous interface conductivity for lithium dendrite-free anode. *ACS Energy Lett.,* **2018**, *3*(9), 2259-2266.
[http://dx.doi.org/10.1021/acsenergylett.8b01244]

[71]   Li, H.; Tay, R.Y.; Tsang, S.H.; Liu, W.; Teo, E.H.T. Reduced graphene oxide/boron nitride composite film as a novel binder-free anode for lithium ion batteries with enhanced performances. *Electrochim. Acta,* **2015**, *166*, 197-205.
[http://dx.doi.org/10.1016/j.electacta.2015.03.109]

[72]   Monajjemi, M. Graphene/(h-BN) n/X-doped graphene as anode material in lithium ion batteries (X= Li, Be, B and N). *Maced. J. Chem. Chem. Eng.,* **2017**, *36*(1), 101-118.
[http://dx.doi.org/10.20450/mjcce.2017.1134]

[73]   Aydın, H.; Çelik, S.Ü.; Bozkurt, A. Electrolyte loaded hexagonal boron nitride/polyacrylonitrile nanofibers for lithium ion battery application. *Solid State Ion.,* **2017**, *309*, 71-76.
[http://dx.doi.org/10.1016/j.ssi.2017.07.004]

[74]   Shim, J.; Kim, H.J.; Kim, B.G.; Kim, Y.S.; Kim, D.G.; Lee, J.C. 2D boron nitride nanoflakes as a multifunctional additive in gel polymer electrolytes for safe, long cycle life and high rate lithium metal batteries. *Energy Environ. Sci.,* **2017**, *10*(9), 1911-1916.
[http://dx.doi.org/10.1039/C7EE01095H]

[75]   Lv, X.; Wang, J.; Yan, Z.; Jiang, D.; Liu, J. Design of 3D h-BN architecture as Ag₃VO₄ enhanced photocatalysis stabilizer and promoter. *J. Mol. Catal. Chem.,* **2016**, *418*, 146-153.
[http://dx.doi.org/10.1016/j.molcata.2016.03.036]

[76]   Wang, N.; Yang, G.; Wang, H.; Sun, R.; Wong, C.P. Visible light-responsive photocatalytic activity of boron nitride incorporated composites. *Front Chem.,* **2018**, *6*, 440.
[http://dx.doi.org/10.3389/fchem.2018.00440] [PMID: 30320071]

[77]   Li, X.; Lin, S.; Lin, X.; Xu, Z.; Wang, P.; Zhang, S.; Zhong, H.; Xu, W.; Wu, Z.; Fang, W. Graphene/h-BN/GaAs sandwich diode as solar cell and photodetector. *Opt. Express,* **2016**, *24*(1), 134-145.
[http://dx.doi.org/10.1364/OE.24.000134] [PMID: 26832245]

[78]   Goel, N.; Kumar, M. Recent advances in ultrathin 2D hexagonal boron nitride based gas sensors. *J. Mater. Chem. C Mater. Opt. Electron. Devices,* **2021**, *9*(5), 1537-1549.
[http://dx.doi.org/10.1039/D0TC05855F]

[79]   Zheng, Y.; Hong, X.; Wang, J.; Feng, L.; Fan, T.; Guo, R.; Zhang, H. 2D nanomaterials for tissue engineering and regenerative nanomedicines: recent advances and future challenges. *Adv. Healthc. Mater.,* **2021**, *10*(7), e2001743.
[http://dx.doi.org/10.1002/adhm.202001743] [PMID: 33511775]

[80]   Zheng, Y.; Hong, X.; Wang, J.; Feng, L.; Fan, T.; Guo, R.; Zhang, H. 2D nanomaterials for tissue engineering and regenerative nanomedicines: recent advances and future challenges. *Adv. Healthc. Mater.,* **2021**, *10*(7), e2001743.
[http://dx.doi.org/10.1002/adhm.202001743] [PMID: 33511775]

[81]   Gong, Y.; Xu, Z.Q.; Li, D.; Zhang, J.; Aharonovich, I.; Zhang, Y. Two-dimensional hexagonal boron nitride for building next-generation energy-efficient devices. *ACS Energy Lett.,* **2021**, *6*(3), 985-996.
[http://dx.doi.org/10.1021/acsenergylett.0c02427]

[82]   Fischer, M.; Caridad, J.M.; Sajid, A.; Ghaderzadeh, S.; Ghorbani-Asl, M.; Gammelgaard, L.; Bøggild, P.; Thygesen, K.S.; Krasheninnikov, A.V.; Xiao, S.; Wubs, M.; Stenger, N. Controlled generation of

luminescent centers in hexagonal boron nitride by irradiation engineering. *Sci. Adv.,* **2021,** *7*(8), eabe7138.
[http://dx.doi.org/10.1126/sciadv.abe7138] [PMID: 33597249]

[83]   Xu, L.; Zeng, J.; Li, Q.; Xia, L.; Luo, X.; Ma, Z. Defect-engineered 2D/2D hBN/g-C3N4 Z-scheme heterojunctions with full visible-light absorption: Efficient metal-free photocatalysts for hydrogen evolution. *Appl. Surf. Sci.,* **2021,** *547*, 149207.
[http://dx.doi.org/10.1016/j.apsusc.2021.149207]

[84]   Liu, X.; Zhang, R.; Li, M.; Li, L.; Wei, Z.; Jiao, F.; Geng, D.; Hu, W. The More, the Better–Recent Advances in Construction of 2D Multi-Heterostructures. *Adv. Funct. Mater.,* **2021,** 2102049.

[85]   Mirzaee, M.; Rashidi, A.; Zolriasatein, A.; Abadchi, M.R. Solid-state synthesis and characterization of two-dimensional hexagonal BCN nanosheet using a free template method. *Diamond Related Materials,* **2021,** *115*, 108350.
[http://dx.doi.org/10.1016/j.diamond.2021.108350]

CHAPTER 8

A Concise Summary of Recent Research on MOF-Based Flexible Supercapacitors

Ankita Mohanty[1,2] and **Ananthakumar Ramadoss**[1,*]

[1] *School for Advanced Research in Petrochemicals: Laboratory for Advanced Research in Polymeric Materials, Central Institute of Petrochemicals Engineering & Technology, Bhubaneswar-751024, India*

[2] *Department of Physics, Utkal University, Bhubaneswar, 751004, India*

Abstract: This book chapter elucidates the recent works accomplished in the platform of flexible/wearable supercapacitor devices based on metal-organic frameworks (MOFs) electrodes. Comprehensive insight into various types of supercapacitors, the advantage of MOF-based flexible supercapacitors among them, classifications of MOF-based flexible supercapacitors concerning their building blocks, and recent research accomplished with their pros and cons are illustrated. Finally, the performance assessment, strategies to improve efficiency, and future perspectives are briefed.

Keywords: Aspect ratio, Composites, Conducting polymers, Current collectors, Energy storage, Flexible, Gel-electrolyte, Metal-organic framework, Nano structure, Organic linker, Supercapacitor.

INTRODUCTION

The exponential growth of industrialization and the increasing population highly mandate the realization of sustainable energy storage devices. The intermittency of renewable energy sources and excessive utilization of fossil fuels align="center"has conceived a worldwide energy deficiency issue. Moreover, various electrochemical energy storage devices are being able to achieve this energy demand. Rechargeable batteries, fuel cells and capacitors are widely used as efficient energy storage devices; however, their poor power density, cycle life, and environmental hazard nature are the major bottlenecks in their application.

Supercapacitors are considered trending electrochemical energy storage systems owing to their special attributes like superior power density , excellent cycling

* **Corresponding author Ananthakumar Ramadoss:** School for Advanced Research in Petrochemicals: Laboratory for Advanced Research in Polymeric Materials, Central Institute of Petrochemicals Engineering & Technology, Bhubaneswar-751024, India; E-mail: ananth@larpm.in

Gaurav Manik and Sushanta Kumar Sahoo (Eds.)

stability, and nature safety [1]. Modern society is adopting wearable and flexible smart electronic devices as essential gadgets in day-to-day life.

Flexible supercapacitors are being proven to be quite reliable for the construction of efficient wearable or portable electronic devices [2, 3]. Light-weight, leakage-proof packaging, portable size, and high flexibility with good performance are the explicit features that make flexible supercapacitors stand out among other traditional electrochemical energy storage devices [4].

Predominantly, supercapacitors act upon two different types of charge storage behaviours, *i.e.*, pseudocapacitors and electric double-layer capacitors (EDLCs) [1]. The EDLC type supercapacitors undergo align="center"an arrangement of double layers of charges over the electrode surface. In contrast, pseudocapacitors include the faradaic oxidation-reduction reaction of the electrode and electrolyte ions [5]. Transition metal oxides (TMOs) [6, 7] and conducting polymers [8] obey pseudocapacitive behaviour while carbon-based materials show EDLC nature [9]. Novel materials like MXenes [10], layered double hydroxides [11], metal-organic frameworks (MOFs) [12], *etc.*., depict battery-like behaviour and are used in hybrid supercapacitor devices. A hybrid supercapacitor device reflects both sustainable energy density and optimized power density by exploiting the merits of both supercapacitor and battery like electrodes in the same system [1]. MOFs are treated as excellent hybrid supercapacitor electrode materials because of their structural tunability, huge surface area, high aspect ratio with excellent porosity [12]. They can be of different structures and dimensions (1D, 2D and 3D) depending on the selection of organic linkers and metal ions [13]. Usually, pristine MOFs reflect poor electrical conductivity due to the existence of insulating organic linker chains [12, 13]. Unlimited assignment of organic linkers and metal ions avails MOFs with various structures, varying porosity, and huge internal surface area [12, 14]. Therefore, mostly MOFs are used as templates to derive various metal oxides, metal sulfide, metal carbides, layered double hydroxides or carbon materials, *etc.* [15]. MOFs can also be converted into carbon materials when treated at very high temperatures under an inert atmosphere [16]. All these admirable properties of MOFs are utilized towards the application of design and assembly of wearable and smart electronic devices [12].

As yet, MOFs have not reached state of the art. However, much laboratory-scale research work is accomplished and proves the potentiality of MOFs in the assembly of flexible supercapacitor devices [17, 18]. The proper understanding of the requirements and mechanism of a flexible supercapacitor will pave the way towards real time application on an industrial scale. In this book chapter, recent research works based on MOFs for flexible supercapacitors are summarized systematically with their design, synthesis mechanism, performance evaluation, *etc..* A brief idea about flexible supercapacitor, its importance, and requirements are discussed in the below section.

FLEXIBLE SUPERCAPACITOR DEVICE

A flexible supercapacitor device consists of positive and negative electrodes, separated by a separator and sandwiched *via* an electrolyte between them. There is no significant difference between a flexible supercapacitor device and a conventional supercapacitor device in terms of charge storage mechanism. The only uniqueness of a flexible supercapacitor device is the total flexibility of the device. That means the device can work in all bending, twisting, rolling conditions retaining its original performance [4]. Another advantage of this device is very light-weight and align="center"portability. Therefore, these devices are highly preferred in fabricating portable and wearable smart electronic devices, including fitness bands, glucose sensors, smartwatches, wireless earphones *etc..*

A flexible supercapacitor device necessitates the flexibility of all of its components. Thus, a flexible supercapacitor device must have (i) a flexible current collector, (ii) free-standing active electrode and (iii) a sol-gel polymer electrolyte [4]. Various flexible substrates such as CC [19 - 22], nickel foil [23], copper foil [24], aluminium foil [25], stainless steel foil [26], titanium foil [27], metal tapes [28], graphite tapes [29] *etc..* are used as current collectors. Compositing all these components, a flexible supercapacitor device should retain confined volume, light-weight, appreciable stability, cost-effectiveness, swift charge-discharge ability, mechanical durability, resistance to expansion and compression and excellent efficiency [4].

Accordingly, flexible supercapacitor devices need to be designed perfectly so that they should be able to achieve good energy density with cycling stability and mechanical stability. Various kinds of active materials current collectors are being utilized for flexible supercapacitor device fabrication. In this book chapter, the prime focus is on MOF-based flexible supercapacitor devices. Hence, various flexible supercapacitors based on MOF and MOF-derived active electrodes are described hereafter.

MOF FOR FLEXIBLE SUPERCAPACITOR

Metal-organic frameworks (MOFs) constitute strong covalent bonds between various metal ions and organic linkers [12]. Hence, MOFs provide huge specific surface area, high porosity, good internal surface area, the excellent surface to volume ratio *etc.* [12]. Exploiting the special features of MOFs in designing flexible supercapacitors is quite welcoming and challenging. Most of the past research on MOF synthesis includes the binder method, *i.e.*, forming MOFs in powder form. However, a free-standing electrode film or a binder-free active electrode is highly desirable for the successful assembly of a flexible supercapacitor. The use of binder introduces the non-conducting binders in the electrode and thus increases the resistance. Recently, researchers have achieved MOF based supercapacitor electrodes and devices by using various methods such as hydrothermal synthesis [30], electrodeposition [31], chemical vapour deposition [32], plasma-enhanced chemical vapour deposition [33] *etc.*. These synthesis strategies can be adopted to design flexible supercapacitors also. In the following section, the MOF-based flexible supercapacitor device synthesis, design with performance is described with some particular classifications. They can be mentioned as (i) Based on the current collector, (ii) Based on active material, and (iii) Based on electrolytes employed in the full cell/three-electrode system.

MOF Based Flexible Supercapacitors: Various Current Collectors

Cloth or fibres are used as current collectors for the fabrication of flexible supercapacitor electrodes. A 2D cobalt MOF with nanosheet morphology was directly grown over carbon cloth (CC) using the solution method. Consequently, these MOFs were converted into nitrogen-doped cobalt sulfide, cobalt oxide and cobalt selenide. The N-doped cobalt selenide ($CoSe_2$) delivered highest gravimetric capacity of 106.7 mAh/g at 3 mA/cm^2 [22]. Similarly, Cao *et al.* synthesized a core-shell structured zinc oxide (ZnO) zeolitic imidazolate framework (ZIF-8) MOF hybrid *i.e.* ZnO@ZIF-8 over CC, and was composited with polyaniline (PANI) to achieve a flexible supercapacitor electrode [34]. With different bending situations, the fabricated symmetric supercapacitor device retained its electrochemical performance (Fig. **1**).

A hybrid array of nickel cobalt oxide ($NiCo_2O_4$) and nickel metal based MOF (Ni-MOF) was directly grown over CC *via* a two-step solution method [35]. Wang *et al.* electrochemically grown ZIF-67 (Ni based MOF based on zeolitic imidazolate framework) over CC and, consequently, coated PANI to design a flexible supercapacitor electrode [19]. Excellent conductivity, mechanical strength, flexibility, light-weight and pocket-friendliness of CC have made it quite famous in the field of flexible supercapacitor current collector [20]. Apart from all these special features, the tremendous stability of this CC make them highly resistant to any kind of wear and tear even after multiple chemical reactions or other treatments over that [21]. Exploiting this attribute, Yin *et al.* fabricated MOF-derived $NiCo_2O_4$ and silicon carbide nanowires (SiCNWs) over CC *via* a series of synthesis methods such as chemical vapour deposition (CVD), solution reaction followed by etc.hing and annealing [20]. Further, Zhang *et al.* implied a two-step hydrothermal treatment to fabricate cobalt oxide (Co_3O_4) and nickel MOF hybrid *i.e.* Co_3O_4/Ni-MOF over CC [21].

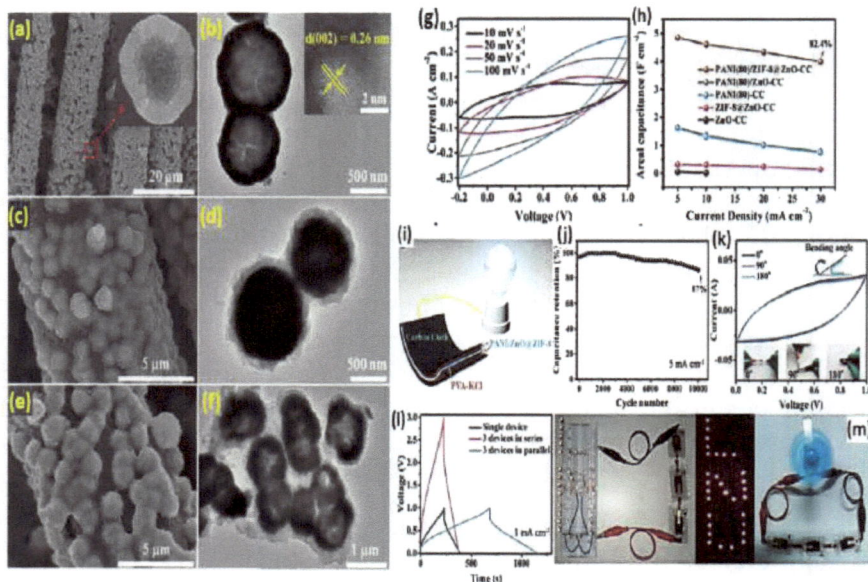

Fig. (1). Scanning electron microscopy (SEM) pictures of **(a)** ZnO-(CC), **(c)** ZnO@ZIF-8-CC, **(e)** PANI/ZnO@ZIF-8-CC; Transmission electron microscopy (TEM) image of **(b)** ZnO-CC with respective inset of High-resolution transmission electron microscopy (HRTEM) photo, **(d)** ZnO@ZIF-8-CC, **(f)** PANI/ZnO@ZIF-8-CC; **(g)** cyclic voltammetry (CV) at different scan rates and **(h)** specific capacitances for PANI(80)/ZnO@ZIF-8-CC; **(i)** pictorial depiction of the full cell; **(j)** cycling stability through 10,000 galvanostatic charge-discharge (GCD) cycles; **(k)** CV profiles of full cell at various bending states; **(l)** GCD profiles of the full cell in series and parallel connections; **(m)** digital photos of 3 full cells in series glowing a light emitting diode (LED) lighting matrix and a mini-fan. Reprinted with permission from ref [34]. Copyright ©The Royal Society of Chemistry 2019.

Likewise, CC [19], carbon paper, graphene fiber [36], graphite paper [37] *etc..*, are also being used and proved as reliable, current collectors for flexible supercapacitor electrode fabrication. Cheng *et al.* implied a polyester fabric as a substrate for coating rGO/Ni-MOF [38]. As the polyester fabric is non-conducting in nature, therefore nickel was coated over this polyester fabric. Meanwhile, carbon fiber (CF) was utilized to laminate MOF-derived Co_3O_4 composited with phosphorous and nitrogen co-doped carbon (PNC) by Liu *et al.* [39]. Compact volume, less weight, tremendous flexibility with mechanical strength proves graphene fiber as a potential conductive current collector for supercapacitor application [36]. Therefore, nanowire structured copper MOF was grown over reduced graphene oxide (rGO) fiber by Wang *et al.* [36]. Similarly, graphite paper of thickness 1 mm was adopted to grow MOF polypyrrole (PPy) composite as a flexible supercapacitor positive electrode [37].

Transparent, light-weight and highly flexible indium titanium oxide (ITO) and polyethylene terephthalate (PET) sheets are quite famous as a current collector for supercapacitor application [17, 18, 40]. To make these substrates conducting various electroactive materials are coated over them. Zhao *et al.* used ITO/PET (indium titanium oxide-polyethylene terephthalate) substrate to coat Ni-based MOF ($Ni_3(HITP)_2$) to fabricate a transparent and flexible supercapacitor electrode [40]. A PET sheet was utilized to assemble aluminium based MOF (MIL-53) using a layer-by-layer method [17]. However, PET is non-conducting in nature; less conductivity was surmounted by compositing MOF with rGO. In the sulfuric acid (H_2SO_4) electrolyte, the PET/MOF/rGO electrode showed a optimum specific capacitance value of 510 mF/cm² at 1 mV/s. Meanwhile, Dai *et al.* assigned PET sheets as the base to coat 3D bulk MOF and 2D MOF with nanosheet morphology [18]. These electrodes proved as very good flexible electrodes in an ionic electrolyte.

MOF Based Flexible Supercapacitors: Various Active Materials

Successful fabrication of high performance flexible supercapacitor device supremely needs a highly electrochemically active electrode material which is infirm attachment to the current collector and able to retain its performance even after continuous charge-discharge cycles and mechanical str*etc.*hing and bending. Various strategies are being adopted to design a high-performance active electrode material. MOFs are such novel materials which can actively compete with other active materials such as transition metal oxide, sulfide and phosphide *etc.*. However, pristine MOFs endow high specific surface area, structural tunability and good aspect ratio, but they have inferior conductivity due to the existence of insulating bridging linkers in them. Whence, generation of metal oxide, sulfide or phosphide from MOFs, hybridization of MOFs with metal oxide, conducting polymers and MOF composites can enhance the electrical conductivity, thus providing excellent electrochemical performance to the active electrode material. Apart from designing active electrode material, direct growth of active material over current collector or fabrication of free-standing electrode [41] reduces electrolyte ion path length and fast electrolyte ion diffusion. This section of the book chapter has analyzed recent research work based on MOFs as active electrode material for flexible supercapacitors.

A copper-based MOF with nanowire morphology was grown over graphene fibres. This highly flexible, light-weight and compact electrode was fabricated to use as a cloth. The three-electrode system furnished an optimal areal capacitance value of 44.6 mF/cm² at 5 mV/s [36]. A nickel MOF ($Ni_3(HITP)_2$) was used to establish a free-standing electrode film and was transferred to a ITO/PET transparent and flexible substrate [40]. The prepared electrode delivered highest specific capacitance value of 1.63 mF/cm² and 78.4% optical transmittance. Similarly, a 3D bulk MOF and 2D MOF nanosheet were grown over a PET sheet [18]. The ultrathin MOFs (less than 10 nm) with 2D nanosheet morphology reflected better electrochemical performance between the two electrodes.

MOFs hybridized with conducting polymers offer enhanced electrical conductivity with better ion intercalation/de-intercalation; thus, good pseudocapacitive property compared to pristine MOFs. Another conducting polymer *i.e.* PPy possesses a tubular structure with good electrical conductivity and porosity. Thus, Xu *et al.* assembled a 3D conductive network of ZIF-PPy, revealing the highest gravimetric capacitance value of 597.6 F/g in 1M Na_2SO_4 electrolyte [37]. Meanwhile, a cobalt based MOF (ZIF-67) was composited with PANI conducting polymer and coated over CC by Wang *et al.* [19] (Fig. **2**). This electrode exhibited superior specific capacitance 2146 mF/cm^2 at 10 mV/s. Similarly, MOF hybridized with rGO can multiply the electrical conductivity, structural stability and the specific surface area of the bare MOF. Accordingly, Cheng *et al.* used chemical plating synthesis over Ni-coated polyester substrate to coat rGO/Ni-MOF [38]. This flexible electrode provides the highest areal capacitance value 260 mF/cm^2 at 4 mA/cm^2. Meanwhile, for the sake of multiplying the ions and electrons' transportation nature, a 3D hierarchically porous (pore size 50-100 nm) MOF was composited with conducting polymer PANI [41]. Herein, PANI provided a reliable conductive base to the MOF. Additionally, to achieve flexibility and conducting network was availed by growing these MOF/PANI composite over carbon nanotube (CNT). This MOF/CNT/PANI electrode revealed a specific capacitance 342.5 F/g at 1 A/g.

MOF hybridized with metal oxide as a core-shell structure can supply highly electrochemically active metal oxide with excellent specific surface area, porosity and stability of MOF. Based on this concept, Cao *et al.* assembled zinc oxide and MOF hybrid *i.e.* ZnO@ZIF-8 core-shell hybrid over CC *via* a two-step hydrothermal method [34]. Likewise, Zhang *et al.* designed Co_3O_4/Ni-MOF core-shell electrodes over CC [21]. This flexible and binder-free electrode delivered a maximum specific capacity value 209 mAh/g at 1 A/g along with 90% capacitance retention after 3000 cycles. Further, as a conductivity improvement strategy, PANI was coated over the electrode. This flexible electrode furnished a superior specific capacitance value 4839 mF/cm^2 at 5 mA/cm^2.

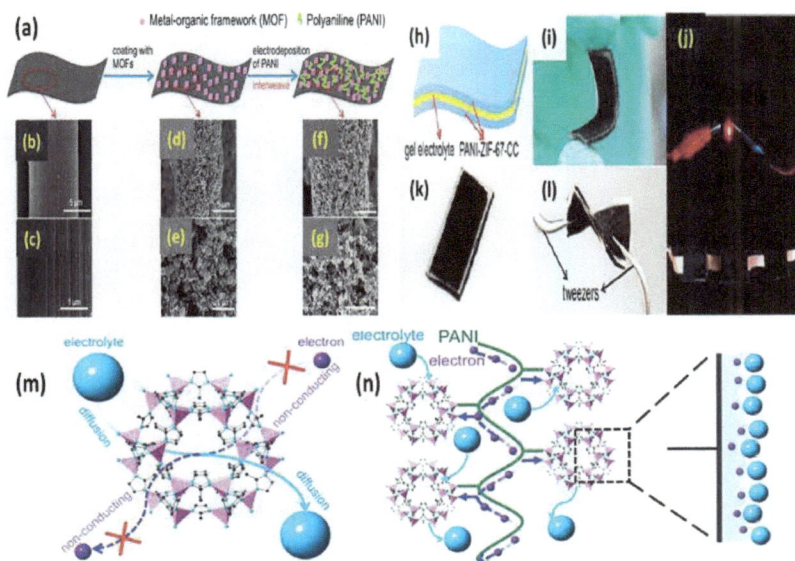

Fig. (2). (a) Pictorial depiction of the synthesis method of PANI-ZIF-67-CC electrode; SEM photographs of **(b, c)** CC fibers, **(d, e)** CC@ZIF-67, and **(f, g)** after electropolymerization of aniline respectively; **(h)** assembly design of PANI-ZIF-67-CC flexible SC full cell; Digital Photos of the as assembled device under (i) Flat, **(k)** bent, and **(l)** twisted state; **(j)** digital image three SCs connected in series glowing one red LED; **(m & n)** schematic illustration of the interaction of electrons through pristine MOF and after PANI composite formation respectively. Procreated with consent from ref [19] © 2015 American Chemical Society.

Further electrochemical performance can be enhanced *via* the fabrication of core-shell structured MOF hybridized with metal sulfide and carbon coating. Lee *et al.* designed a core-shell MOF@Co$_{1-x}$S@C electrode over Ni foam *via* a two-step hydrothermal, thus a binder-free method. This electrode revealed a maximum capacitance 13.1 F/cm^2 [42] (Fig. 3). Similarly, Li *et al.* fabricated bimetallic oxide (NiCo$_2$O$_4$) over CC and Ni-MOF over that to form a core-shell structure [35]. Herein, the MOF furnishes a huge surface area and high porosity to the core NiCo$_2$O$_4$ (Fig. 4). This bi-metallic oxide and MOF composite allow better electrochemical efficiency due to the added oxidation states and better ion intercalation and de-intercalation scope compared to single metal oxide. Often, metal sulfides tend to exhibit better electrochemical performance than their oxide counterparts due to their better electrical conductivity. On account of this, Liu *et al.* assembled a cobalt sulfide (Co$_9$S$_8$) nanoarray and Co-MOF core-shell 3D structure over nickel foam [43]. The as-fabricated flexible electrode offered a maximum specific capacitance value of 416 mF/cm^2 and was stable up to 500 GCD cycles.

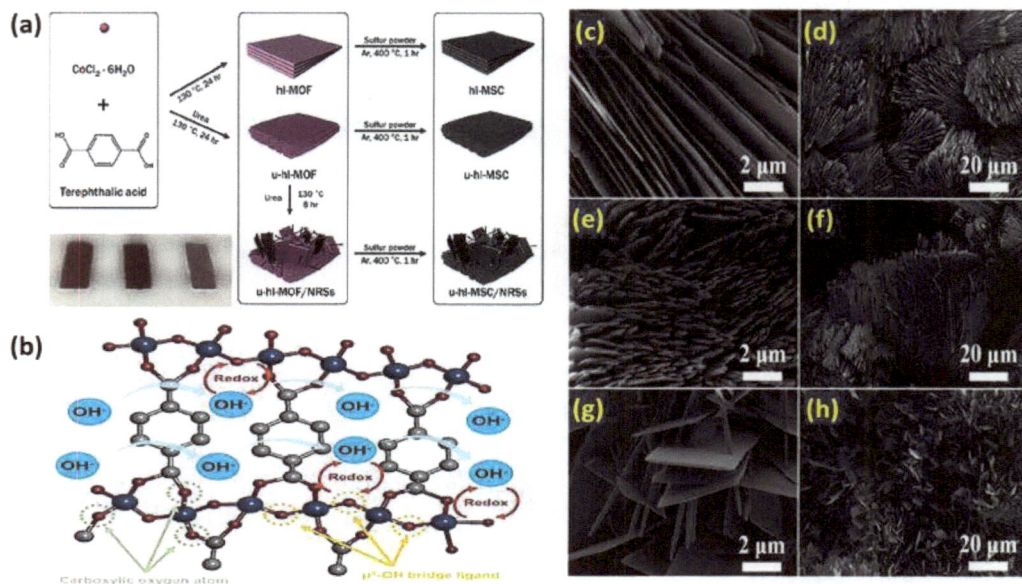

Fig. (3). (a) Pictorial depiction of the fabrication method of hl-MSC, u-hl-MSC, and u-hl-MSC/NRS; **(b)** Diagram displaying the intercalation and redox reaction between hydroxyl ions and hl-MOF or u-hl-MOF; FE-SEM images of the **(c, d)** hl-MOF, **(e, f)** u-hl-MOF, and **(g, h)** u-hl-MOF/NRSs on Ni foam. Procreated with consent from ref [42] © 2020 American Chemical Society.

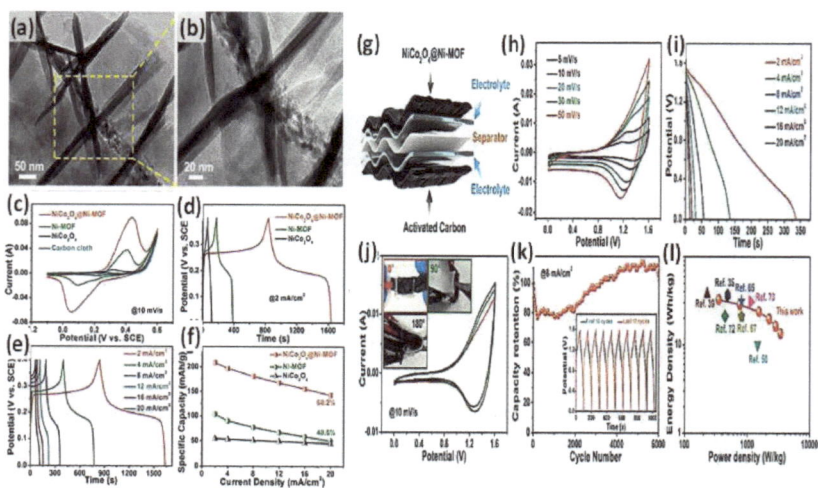

Fig. (4). (a & b) TEM pictures of $NiCo_2O_4$@Ni-MOF core/shell nanostructures; **(c)** CV and **(d)** GCD profiles of the CC, $NiCo_2O_4$, Ni-MOF (36 h) and $NiCo_2O_4$@Ni-MOF (12 h) electrodes; **(e)** GCD profiles of the $NiCo_2O_4$@Ni-MOF (12 h), **(f)** gravimetric capacity vs. current density of these electrodes; **(g)** graphical view of hybrid supercapacitor device; **(h)** CV and **(i)** GCD, **(j)** CV profiles at 10 mV/s under various bending conditions (inset displaying the respective photos of the device); **(k)** stability test at 8 mA/cm^2 (inset displaying the initial and final 10 cycles, (l) Ragone plot for full cell, compared various related researches. Procreated with permission from ref [35] Copyright © 2019 American Chemical Society.

Free-standing electrode films have the advantage of excellent ion diffusion capacity due to reduced electrolyte and ion diffusion path length, reduced mass, and cost due to the absence of any current collector [44]. Chen *et al.* fabricated a MOF/rGO free-standing paper electrode for flexible and editable microsupercapacitor application [45]. This rGO paper retains its electrochemical performance even after rigorous mechanical changes such as cutting, bending, and heavy loading (Fig. **5**).

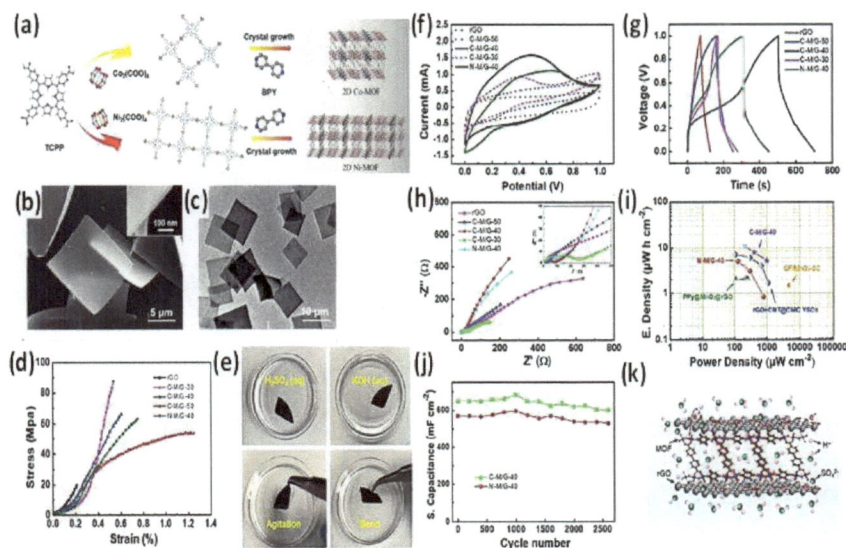

Fig. (5). (a) Pictorial overview of detailed synthesis of MOF nanosheets, **(b)** SEM picture and **(c)** TEM photo of 2D Co-MOF; **(d)** stress-strain profile for different rGO composites, **(e)** mechanical stability test of C-M/-
-40 paper under 6M H_2SO_4 and 6M KOH for 7 days, followed by water after agitation and 15 bending conditions, **(f)** CV at 10 mV s⁻¹, **(g)** GCD profiles at 1 mA cm⁻², **(h)** Nyquist plots, **(i)** Ragone plot, **(j)** cyclic test using 2500 GCD cycles for rGO, C-M/G-50, C-M/G-40, C-M/G-30, N-M/G-40 paper based flexible symmetric device, and **(k)** graphical illustration of reaction mechanism in electrode in 2D-MOF/rGO hybrid papers. Reprinted with consent from ref [45] Copyright © The Royal Society of Chemistry 2018.

Free-standing hybrid sponge and aerogels are also novel binder free electrodes with very good mechanical strength and flexibility [41]. MXenes are trending electroactive materials with excellent specific surface areas with 2D structures, which can be utilized to assemble free-standing films. MOF-derived cobalt ferrite ($CoFe_2O_4$) was coated over MXene films ($Ti_3C_2T_x$) to achieve a high performance flexible supercapacitor electrode by Xie *et al.* [44]. Herein, $CoFe_2O_4$ adds spacing between the MXene nanosheets, which usually tend to stack easily. This enhanced ion transportation behaviour. This $CoFe_2O_4/Ti_3C_2T_x$ paper electrode attached with PET substrate obtained a maximum volumetric capacitance value of 2467.3 F/cm³ in 1M LiCl electrolyte.

MOFs have structural adjustability due to the large possibilities of the selection of metal ions and linkers. Therefore, mostly MOFs are exploited as templates to grow metal oxides, sulfides and carbon material *etc.*. Liu *et al.* utilize this phenomenon for flexible supercapacitor application, as this group derived P-doped Co_3O_4 and PNC from Co-MOF [39]. This strategy successfully achieved both a positive (P-Co_3O_4@PNC) and a negative electrode (PNC). These self-supported active materials were grown over CF, thus availing excellent flexibility and unchanged supercapacitor performance under various bending conditions [39]. Another group, Yin *et al.*, derived hollow nanocage morphology $NiCo_2O_4$ from MOF and composited it with silicon carbide nanowire to achieve a 3D sugar gourd shape [20]. Chemical vapour deposition followed by the solution reaction method of synthesis was utilized to obtain this novel electrode over CC. This electrode CC/SiCNWS@$NiCo_2O_4$ exhibited the highest specific capacitance value of 1377.6 F/g at 1 A/g, with 68.8% rate capability at 20 A/g.

Certainly, selecting a reliable and high performance active electrode material and assembling it over a flexible substrate has become effortless for researchers. However, achieving sustainable electrochemical efficiencies like superior specific capacitance, reliable rate capability, and cyclic stability in a flexible supercapacitor electrode is still a great challenge. The successful fabrication of a competent flexible supercapacitor electrode demands good active material and qualitative electrolytes. The next section of the book chapter shades light upon this aspect briefly.

MOF Based Flexible Supercapacitors: Various Electrolytes

As the predominant necessity of a flexible supercapacitor device is flexibility with leakage-proof ability hence, the electrolyte must be in a sol-gel state. However, MOF-based flexible supercapacitors are majorly focused on various aqueous electrolytes such as potassium hydroxide (KOH), sodium sulphate (Na_2SO_4), lithium chloride (LiCl), potassium chloride (KCl) *etc.*. Thus, for device fabrication, these electrolytes are composited with a highly soluble, bio-degradable and hydrophilic polymer such as polyvinyl alcohol (PVA). The optimization of polymer and electrolyte concentration in the final polymer-gel electrolyte is of great importance. Otherwise, the dominance of polymer in the polymer-gel electrolyte may reduce the electrical conductivity of the overall electrolyte. Additionally, the improper mixing of the polymer network in the electrolyte may lead to obstruction in the electron and ion diffusion, thus providing poor ion intercalation/de-intercalation in the device. Meanwhile, the electrolyte transparency is an indicator of the successful fabrication of the polymer-gel electrolyte in a flexible supercapacitor device. Many research works have been accomplished based on polymer-gel electrolytes like PVA/KOH,

PVA/KCl, PVA/H$_2$SO$_4$, PVA/Na$_2$SO$_4$ for MOF-based flexible supercapacitors. The major drawback of these aqueous electrolytes is their compact potential window (~ 1V for three-electrode) due to the water decomposition point. A free-standing electrode-based supercapacitor electrode was fabricated by Dai *et al.* in an ionic electrolyte named EMIMBF$_4$ (1-ethyl-3-methylimidazolium tetrafluoroborate), which exhibited a potential window value 0 to 2V [18]. Thus, further research can be extended to design various MOF-based flexible supercapacitors out of organic and ionic electrolytes, retaining their gel consistency, transparency, and electrical conductivity.

Finally, the real-time electrochemical performance and mechanical stability can be analysed by assembling a flexible supercapacitor full cell. Thus, in the following segment of the book chapter, various MOF-based flexible supercapacitors are explained with their electrochemical performance, pros, cons and improvement strategies.

MOF BASED FLEXIBLE SUPERCAPACITORS DEVICES

Several flexible supercapacitor devices are successfully fabricated and used for various smart electronics applications such as, smartwatch, glucose sensors, fitness trackers, smart displays, smart cloth *etc.*. However, MOF-based flexible supercapacitors are yet to reach from laboratory scale to industry level. The high material cost, inadequacy, complex synthesis mechanism are the bottlenecks in the scale-up of MOF-based flexible supercapacitors. At the laboratory level, many research works related to this field have been executed successfully.

Some of the MOF-based flexible supercapacitors are symmetric. That means the same electrode acts as a negative and positive one. An all-solid-state flexible symmetric supercapacitor was designed using PET/MOF/rGO electrode and PVA-H$_2$SO$_4$ gel-electrolyte. The fabricated device revealed a high volumetric capacitance value of 3.5 F/cm^3, highest energy density and power density value of 64 µWh/cm^3 and 0.6 mW/cm^3, respectively. This device retained 85% of initial capacitance after 1000 GCD cycles [17]. Another flexible symmetrical supercapacitor device PANI/ZnO@ZIF-8-CC// PANI/ZnO@ZIF-8-CC was assembled by Cao *et al.* [34]. In PVA/KCl gel electrolyte, this device exhibited a maximum specific capacitance value of 226.9 mF/cm^2 at 0.5 mA/cm^2 with the highest energy density of 137 µWh/cm^3 and maximum power density value of 1.421 W/cm^3. Similarly, PANI-ZIF-67-CC//PANI-ZIF-67-CC flexible device 80% capacitance retention after 2000 cycles in PVA/H$_2$SO$_4$ electrolyte [19]. Excellent cyclic stability was achieved by the symmetrical flexible supercapacitor CoFe$_2$O$_4$/Ti$_3$C$_2$T$_X$//CoFe$_2$O$_4$/Ti$_3$C$_2$T$_X$ *i.e.* 88.2% even after 10,000 cycles [44].

Compared to symmetric flexible supercapacitors, asymmetric supercapacitors result in better electrochemical performance due to the exploitation of two different electrode materials. Asymmetric flexible supercapacitors provide wide operation potential windows, high energy and power density. Usually, carbon based materials like rGO, activated carbon *etc..*, are used as supercapacitor negative electrodes. However, MOFs have been utilized to derive novel high-performing negative electrodes. Co-MOF derived Co-NC@Fe_2O_3 proved as a potential flexible negative electrode grown over CC. The as-fabricated NC@$CoSe_2$//Co-NC@Fe_2O_3 delivered an energy density value of 15 Wh/kg at a power density of 23.5 kW/kg in PVA-KOH gel-electrolyte [22]. Another flexible MOF-based supercapacitor device was $Ni_3(HITP)_2$//PEDOT:PSS in PVA-KCl electrolyte was designed and fabricated by Zhao *et al.* [40]. This device furnished an optimum specific energy value of 240 μWh/cm^3 at a specific power value of 2.7 mW/cm^3. Under various bending angles ranging from 0°, 30°, 60° up to 150°, this device retained its electrochemical performance. Cheng *et al.* synthesized 2D Co-MOF/rGO//2D Ni-MOF/rGO flexible asymmetric supercapacitor device using rGO papers as substrate in PVA/H_2SO_4 electrolyte [45] (Fig. **5**). This microsupercapacitor delivered optimized volumetric energy value 1.87 mWh/cm^3 at power density value 270 mW/cm^3 and maximum power density value 250 mW/cm^3. Using polyester fabric coated PPy as negative electrode, an all-soli-state fibre supercapacitor device rGO/Ni-MOF//polyester fabric/PPy was fabricated by Cheng *et al.* [38]. This device offered the highest specific capacitance value, 95 mF/cm^2. PNC was directly grown over CF and was a negative electrode for a flexible supercapacitor device. The full cell P-Co_3O_4@PNC//PNC delivered excellent gravimetric energy of 69.6 Wh/kg at gravimetric power 750 W/kg and very good cycle life, *i.e.*, 96.8% even after 10,000 continuous GCD cycles [39].

In summary, binder-free approach to fabricate electrodes, choice of flexible substrates such as fibres, cloths, metal foils, sol-gel electrolyte with appropriate electrical conductivity, consistency, transparency and solubility and leakage proof packing are the primary criteria to attain the goal of successful fabrication of a flexible supercapacitor device. Many research works have already been accomplished using the MOF-based flexible supercapacitor device. However, various electrolytes can still be explored. The specific capacitance and cyclic stability need to be further optimized to increase the lifetime of these devices.

CONCLUDING REMARKS

This book chapter briefly summarizes the identity of supercapacitors and their specialty among other energy storage devices. Consequently, the advantages and use of flexible supercapacitors are explained. Meanwhile, MOFs are introduced as potential and novel electroactive materials for flexible supercapacitor application.

Fundamental requirements and elements of a flexible supercapacitor device are described, and various MOF-based flexible supercapacitors are narrated according to each component of a flexible supercapacitor. A detailed description is provided regarding the advantages and disadvantages of various current collectors, active materials, electrolytes and device designing with their improvement strategies. Indeed, it is believed that this book chapter will be quite informative to the community of researchers working in the platform of MOF-based energy storage devices.

- Indeed, a flexible supercapacitor necessitates all its constituents to be flexible in nature. The formation of flexible supercapacitors demands full flexibility, light-weight, confined volume, and goodresistance to various mechanical deformations.
- The performance of the active materials is usually poor in the case of flexible supercapacitors, compared to conventional supercapacitors, due to the low mass loading *i.e.* thin film. To overcome this difficulty MOF-derived 3D hierarchical porous nanostructures electrodes can be availed to achieve high electrochemical performance.
- MOF based self-standing electrodes must be designed in future research for flexible supercapacitors.
- The magnificent traits of MOFs like structural tailorability, high specific surface area, huge porosity should be exploited to explore various novel materials for *e.g.* MOF@MXene, MOF@COF (COF = covalent metal-organic framework), MOF@LDH (LDH = layered double hydroxide), Ternary metallic MOF derived ternary metal sulfides/phosphides, MOF derived hierarchical nanoporous carbons *etc.*.
- The search for negative electrodes for supercapacitors are limited. Thus, MOF can be utilized to derive various carbon materials with suitable morphology and higher performance compared to traditional EDLC type materials.
- The electrolytes used for flexible supercapacitors must be in semi-solid state with high ion conductivity, non-align="center"reactivity, stable under a wide range of temperature and harmless. Additionally, the electrolyte concentration can be regulated to attain optimal potential window for the flexible supercapacitor device.
- In conclusion, the electrochemical efficiency of the MOF based flexible supercapacitors can be even magnified in future research to make them competitive with conventional supercapacitors taking into account the above-mentioned points.

CONSENT FOR PUBLICATION

Not applicable.

CONFLICT OF INTERESTS

The authors declare no conflict of interest, financial or otherwise.

ACKNOWLEDGEMENTS

This research was funded by DST-INSPIRE (IF170869) and Indo-German Project (scheme no. INT/FRG/DAAD/P-09/2018), Department of Science and Technology (DST), New Delhi, India.

REFERENCES

[1]   Balasubramaniam, S.; Mohanty, A.; Balasingam, S.K.; Kim, S.J.; Ramadoss, A. Comprehensive Insight into the Mechanism, Material Selection and Performance Evaluation of Supercapatteries. *Nano-Micro Lett.,* **2020**, *12*(1), 85.
[http://dx.doi.org/10.1007/s40820-020-0413-7] [PMID: 34138304]

[2]   Ramadoss, A.; Yoon, K-Y.; Kwak, M-J.; Kim, S-I.; Ryu, S-T.; Jang, J-H. Fully flexible, lightweight, high performance all-solid-state supercapacitor based on 3-Dimensional-graphene/graphite-paper. *J. Power Sources,* **2017**, *337*, 159-165.
[http://dx.doi.org/10.1016/j.jpowsour.2016.10.091]

[3]   Ramadoss, A.; Saravanakumar, B.; Kim, S.J. Thermally reduced graphene oxide-coated fabrics for flexible supercapacitors and self-powered systems. *Nano Energy,* **2015**, *15*, 587-597.
[http://dx.doi.org/10.1016/j.nanoen.2015.05.009]

[4]   Xue, Q.; Sun, J.; Huang, Y.; Zhu, M.; Pei, Z.; Li, H.; Wang, Y.; Li, N.; Zhang, H.; Zhi, C. Recent Progress on Flexible and Wearable Supercapacitors. *Small,* **2017**, *13*(45), 1701827.
[http://dx.doi.org/10.1002/smll.201701827] [PMID: 28941073]

[5]   Chen, J.; Lee, P.S. Electrochemical supercapacitors: From mechanism understanding to multifunctional applications. *Adv. Energy Mater.,* **2021**, *11*(6), 2003311.
[http://dx.doi.org/10.1002/aenm.202003311]

[6]   Kumar, S.A.; Mohanty, A.; Saravanakumar, B.; Mohanty, S.; Nayak, S.K.; Ramadoss, A. Three-dimensional Bi_2O_3/Ti microspheres as an advanced negative electrode for hybrid supercapacitors. *Chem. Commun. (Camb.),* **2020**, *56*(85), 12973-12976.
[http://dx.doi.org/10.1039/D0CC04057F] [PMID: 32996474]

[7]   Arun, T.; Mohanty, A.; Rosenkranz, A.; Wang, B.; Yu, J.; Morel, M.J.; Udayabhaskar, R.; Hevia, S.A.; Akbari-Fakhrabadi, A.; Mangalaraja, R.V.; Ramadoss, A. Role of electrolytes on the electrochemical characteristics of Fe_3O_4/MXene/RGO composites for supercapacitor applications. *Electrochim. Acta,* **2021**, *367*, 137473.
[http://dx.doi.org/10.1016/j.electacta.2020.137473]

[8]   Meng, Q.; Cai, K.; Chen, Y.; Chen, L. Research progress on conducting polymer based supercapacitor electrode materials. *Nano Energy,* **2017**, *36*, 268-285.
[http://dx.doi.org/10.1016/j.nanoen.2017.04.040]

[9]   Chen, G.Z. Understanding supercapacitors based on nano-hybrid materials with interfacial conjugation. *Prog. Nat. Sci.,* **2013**, *23*(3), 245-255.
[http://dx.doi.org/10.1016/j.pnsc.2013.04.001]

[10]   Gogotsi, Y.; Anasori, B. The Rise of MXenes. *ACS Nano,* **2019**, *13*(8), 8491-8494.
[http://dx.doi.org/10.1021/acsnano.9b06394] [PMID: 31454866]

[11]   Li, X.; Du, D.; Zhang, Y.; Xing, W.; Xue, Q.; Yan, Z. Layered double hydroxides toward high-performance supercapacitors. *J. Mater. Chem. A Mater. Energy Sustain.,* **2017**, *5*(30), 15460-15485.
[http://dx.doi.org/10.1039/C7TA04001F]

[12]   Mohanty, A.; Jaihindh, D.; Fu, Y-P.; Senanayak, S.P.; Mende, L.S.; Ramadoss, A. An extensive review on three dimension architectural Metal-Organic Frameworks towards supercapacitor application. *J. Power Sources,* **2021**, *488*, 229444.
[http://dx.doi.org/10.1016/j.jpowsour.2020.229444]

[13]   Wang, R.; Jin, D.; Zhang, Y.; Wang, S.; Lang, J.; Yan, X.; Zhang, L. Engineering metal organic framework derived 3D nanostructures for high performance hybrid supercapacitors. *J. Mater. Chem. A Mater. Energy Sustain.,* **2017**, *5*(1), 292-302.
[http://dx.doi.org/10.1039/C6TA09143A]

[14]   Sundriyal, S.; Kaur, H.; Bhardwaj, S.K.; Mishra, S.; Kim, K-H.; Deep, A. Metal-organic frameworks and their composites as efficient electrodes for supercapacitor applications. *Coord. Chem. Rev.,* **2018**, *369*, 15-38.
[http://dx.doi.org/10.1016/j.ccr.2018.04.018]

[15]   Chen, D.; Wei, L.; Li, J.; Wu, Q. Nanoporous materials derived from metal-organic framework for supercapacitor application. *J. Energy Storage,* **2020**, *30*, 101525.
[http://dx.doi.org/10.1016/j.est.2020.101525]

[16]   Wang, C.; Kim, J.; Tang, J.; Kim, M.; Lim, H.; Malgras, V.; You, J.; Xu, Q.; Li, J.; Yamauchi, Y. New Strategies for Novel MOF-Derived Carbon Materials Based on Nanoarchitectures. *Chem,* **2020**, *6*(1), 19-40.
[http://dx.doi.org/10.1016/j.chempr.2019.09.005]

[17]   Barakzehi, M.; Montazer, M.; Sharif, F.; Norby, T.; Chatzitakis, A. MOF-modified polyester fabric coated with reduced graphene oxide/polypyrrole as electrode for flexible supercapacitors. *Electrochim. Acta,* **2020**, *336*, 135743.
[http://dx.doi.org/10.1016/j.electacta.2020.135743]

[18]   Dai, F.; Wang, X.; Zheng, S.; Sun, J.; Huang, Z.; Xu, B. Toward high-performance and flexible all-solid-state micro-supercapacitors: MOF bulk vs. MOF nanosheets. *Chem. Eng. J.,* **2020**, *413*, 127520.

[19]   Wang, L.; Feng, X.; Ren, L.; Piao, Q.; Zhong, J.; Wang, Y.; Li, H.; Chen, Y.; Wang, B. Flexible solid-state supercapacitor based on a metal-organic framework interwoven by electrochemically-deposited pani. *J. Am. Chem. Soc.,* **2015**, *137*(15), 4920-4923.
[http://dx.doi.org/10.1021/jacs.5b01613] [PMID: 25864960]

[20]   Yin, X.; Li, H.; Yuan, R.; Lu, J. Hierarchical self-supporting sugar gourd-shape MOF-derived $NiCo_2O_4$ hollow nanocages@SiC nanowires for high-performance flexible hybrid supercapacitors. *J. Colloid Interface Sci.,* **2021**, *586*, 219-232.
[http://dx.doi.org/10.1016/j.jcis.2020.10.086] [PMID: 33158557]

[21]   Zhang, L.; Zhang, Y.; Huang, S.; Yuan, Y.; Li, H.; Jin, Z.; Wu, J.; Liao, Q.; Hu, L.; Lu, J.; Ruan, S.; Zeng, Y-J. Co3O4/Ni-based MOFs on carbon cloth for flexible alkaline battery-supercapacitor hybrid devices and near-infrared photocatalytic hydrogen evolution. *Electrochim. Acta,* **2018**, *281*, 189-197.
[http://dx.doi.org/10.1016/j.electacta.2018.05.162]

[22]   Zhu, C.; Guan, C.; Cai, D.; Chen, T.; Wang, Y.; Du, J.; Wu, J.; Xu, X.; Yu, H.; Huang, W. Carbon nanoarrays embedded with metal compounds for high-performance flexible supercapacitors. *Batter. Supercaps,* **2020**, *3*(1), 93-100.
[http://dx.doi.org/10.1002/batt.201900174]

[23]   Wen, J.; Xu, B.; Zhou, J. Toward Flexible and Wearable Embroidered Supercapacitors from Cobalt Phosphides-Decorated Conductive Fibers. *Nano-Micro Lett.,* **2019**, *11*(1), 89.
[http://dx.doi.org/10.1007/s40820-019-0321-x] [PMID: 34138049]

[24]   Lin, J.; Zhong, J.; Bao, D.; Reiber-kyle, J.; Wang, W.; Vullev, V., Eds. Electrochemical supercapacitor based on flexible pillar graphene nanostructures. *69th Device Research Conference,* **2011**.
[http://dx.doi.org/10.1109/DRC.2011.5994429]

[25]   Ghai, V.; Chatterjee, K.; Agnihotri, P. *Vertically aligned carbon nanotubes-coated aluminium foil as*

flexible supercapacitor electrode for high power applications; Carbon Letters, **2020**, pp. 1-9.

[26]   Dong, L.; Xu, C.; Li, Y.; Huang, Z-H.; Kang, F.; Yang, Q-H.; Zhao, X. Flexible electrodes and supercapacitors for wearable energy storage: a review by category. *J. Mater. Chem. A Mater. Energy Sustain.,* **2016**, *4*(13), 4659-4685.
[http://dx.doi.org/10.1039/C5TA10582J]

[27]   Xie, Y.; Fang, X. Electrochemical flexible supercapacitor based on manganese dioxide-titanium nitride nanotube hybrid. *Electrochim. Acta,* **2014**, *120*, 273-283.
[http://dx.doi.org/10.1016/j.electacta.2013.12.103]

[28]   Kang, K-N.; Ramadoss, A.; Min, J-W.; Yoon, J-C.; Lee, D.; Kang, S.J.; Jang, J.H. Wire-Shaped 3D-Hybrid Supercapacitors as Substitutes for Batteries. *Nano-Micro Lett.,* **2020**, *12*(1), 28.
[http://dx.doi.org/10.1007/s40820-019-0356-z] [PMID: 34138068]

[29]   Yu, N.; Xiong, R.; Wang, Y.; Zhou, C.; Li, Y.; Pang, C.; Li, Z.; Zou, L.; Guo, K. Facile fabrication of low-cost and scalable graphite tape as novel current collectors for flexible supercapacitors. *J. Alloys Compd.,* **2021**, *861*, 158476.
[http://dx.doi.org/10.1016/j.jallcom.2020.158476]

[30]   Amo-Ochoa, P.; Givaja, G.; Miguel, P.J.S.; Castillo, O.; Zamora, F. Microwave assisted hydrothermal synthesis of a novel CuI-sulfate-pyrazine MOF. *Inorg. Chem. Commun.,* **2007**, *10*(8), 921-924.
[http://dx.doi.org/10.1016/j.inoche.2007.04.024]

[31]   Zhang, X.; Wan, K.; Subramanian, P.; Xu, M.; Luo, J.; Fransaer, J. Electrochemical deposition of metal–organic framework films and their applications. *J. Mater. Chem. A Mater. Energy Sustain.,* **2020**, *8*(16), 7569-7587.
[http://dx.doi.org/10.1039/D0TA00406E]

[32]   Stassin, T.; Rodríguez-Hermida, S.; Schrode, B.; Cruz, A.J.; Carraro, F.; Kravchenko, D.; Creemers, V.; Stassen, I.; Hauffman, T.; De Vos, D.; Falcaro, P.; Resel, R.; Ameloot, R. Vapour-phase deposition of oriented copper dicarboxylate metal-organic framework thin films. *Chem. Commun. (Camb.),* **2019**, *55*(68), 10056-10059.
[http://dx.doi.org/10.1039/C9CC05161A] [PMID: 31369024]

[33]   DeCoste, J.B.; Peterson, G.W. Preparation of hydrophobic metal-organic frameworks *via* plasma enhanced chemical vapor deposition of perfluoroalkanes for the removal of ammonia. *J. Vis. Exp.,* **2013**, (80), 51175.
[http://dx.doi.org/10.3791/51175] [PMID: 24145623]

[34]   Cao, X-M.; Han, Z-B. Hollow core-shell ZnO@ZIF-8 on carbon cloth for flexible supercapacitors with ultrahigh areal capacitance. *Chem. Commun. (Camb.),* **2019**, *55*(12), 1746-1749.
[http://dx.doi.org/10.1039/C8CC09847F] [PMID: 30663750]

[35]   Li, G.; Cai, H.; Li, X.; Zhang, J.; Zhang, D.; Yang, Y.; Xiong, J. Construction of Hierarchical NiCo$_2$O$_4$@Ni-MOF Hybrid Arrays on Carbon Cloth as Superior Battery-Type Electrodes for Flexible Solid-State Hybrid Supercapacitors. *ACS Appl. Mater. Interfaces,* **2019**, *11*(41), 37675-37684.
[http://dx.doi.org/10.1021/acsami.9b11994] [PMID: 31532185]

[36]   Wang, Y-F.; Yang, S-Y.; Yue, Y.; Bian, S-W. Conductive copper-based metal-organic framework nanowire arrays grown on graphene fibers for flexible all-solid-state supercapacitors. *J. Alloys Compd.,* **2020**, *835*, 155238.
[http://dx.doi.org/10.1016/j.jallcom.2020.155238]

[37]   Xu, X.; Tang, J.; Qian, H.; Hou, S.; Bando, Y.; Hossain, M.S.A.; Pan, L.; Yamauchi, Y. Three-dimensional networked metal-organic frameworks with conductive polypyrrole tubes for flexible supercapacitors. *ACS Appl. Mater. Interfaces,* **2017**, *9*(44), 38737-38744.
[http://dx.doi.org/10.1021/acsami.7b09944] [PMID: 29082737]

[38]   Cheng, C.; Xu, J.; Gao, W.; Jiang, S.; Guo, R. Preparation of flexible supercapacitor with RGO/Ni-MOF film on Ni-coated polyester fabric. *Electrochim. Acta,* **2019**, *318*, 23-31.
[http://dx.doi.org/10.1016/j.electacta.2019.06.055]

[39] Liu, S.; Kang, L.; Zhang, J.; Jung, E.; Lee, S.; Jun, S.C. Structural engineering and surface modification of MOF-derived cobalt-based hybrid nanosheets for flexible solid-state supercapacitors. *Energy Storage Mater.,* **2020**, *32*, 167-177.
[http://dx.doi.org/10.1016/j.ensm.2020.07.017]

[40] Zhao, W.; Chen, T.; Wang, W.; Jin, B.; Peng, J.; Bi, S.; Jiang, M.; Liu, S.; Zhao, Q.; Huang, W. Conductive Ni3(HITP)2 MOFs thin films for flexible transparent supercapacitors with high rate capability. *Sci. Bull. (Beijing),* **2020**, *65*(21), 1803-1811.
[http://dx.doi.org/10.1016/j.scib.2020.06.027]

[41] Gong, J.; Xu, Z.; Tang, Z.; Zhong, J.; Zhang, L. Highly compressible 3-D hierarchical porous carbon nanotube/metal organic framework/polyaniline hybrid sponges supercapacitors. *AIP Adv.,* **2019**, *9*(5), 055032.
[http://dx.doi.org/10.1063/1.5109042]

[42] Lee, C.S.; Moon, J.; Park, J.T.; Kim, J.H. Highly interconnected nanorods and nanosheets based on a hierarchically layered metal–organic framework for a flexible, high-performance energy storage device. *ACS Sustain. Chem.& Eng.,* **2020**, *8*(9), 3773-3785.
[http://dx.doi.org/10.1021/acssuschemeng.9b06999]

[43] Liu, X.F.; He, Q.; Wang, Y.; Wang, J.; Xiang, Y.; Blackwood, D.; Wu, R.; Chen, J S. MOF-reinforced Co9S8 self-supported nanowire arrays for highly durable and flexible supercapacitor. *Electrochim. Acta,* **2020**, *346*, 136201.
[http://dx.doi.org/10.1016/j.electacta.2020.136201]

[44] Xie, W.; Wang, Y.; Zhou, J.; Zhang, M.; Yu, J.; Zhu, C.; Xu, J. MOF-derived CoFe2O4 nanorods anchored in MXene nanosheets for all pseudocapacitive flexible supercapacitors with superior energy storage. *Appl. Surf. Sci.,* **2020**, *534*, 147584.
[http://dx.doi.org/10.1016/j.apsusc.2020.147584]

[45] Cheng, J.; Chen, S.; Chen, D.; Dong, L.; Wang, J.; Zhang, T.; Jiao, T.; Liu, B.; Wang, H.; Kai, J-J.; Zhang, D.; Zheng, G.; Zhi, L.; Kang, F.; Zhang, W. Editable asymmetric all-solid-state supercapacitors based on high-strength, flexible, and programmable 2D-metal–organic framework/reduced graphene oxide self-assembled papers. *J. Mater. Chem. A Mater. Energy Sustain.,* **2018**, *6*(41), 20254-20266.
[http://dx.doi.org/10.1039/C8TA06785F]

CHAPTER 9

Advanced Batteries and Charge Storage Devices based on Nanowires

Sunil Kumar[1,2], **Ravi Prakash**[1] and **Pralay Maiti**[1,*]

[1] *School of Materials Science and Technology, Indian Institute of Technology (Banaras Hindu University), Varanasi-221005, India*

[2] *Department of Chemistry, L.N.T.College (B.R.A. Bihar University), Muzaffarpur-842002, India*

Abstract: Compositional designed electrodes exhibiting high specific capacities are of great interest towards align="center"high performance charge storage devices. Electrode surface can store charge or guest ions due to structural confinement effect. Ion storage capacity depends on the structural integrity of electrode (anode) materials of batteries. Electrolyte selection also decides the storage capacity of batteries and other charge storage devices. Volume expansion or variation can be minimized through structural variation of the electrode. align="center"The charging phenomenon proceeds through the continuous ion destruction process of adsorbed ions into semipermeable align="center"pores. Dimension controlled electrode materials possess superior ion storage capacity. The contemporary design is an effective way to improve the charge storage capacity of electrodes. Low dimension materials exhibit better charge storage capacity due to high surface density (surface to volume ratio) and efficient charge confinement. The confined dimensions (quantum confinement) play important roles in orienting the desired kinetic properties of nanomaterials, such as charge transport and diffusion. This chapter emphasizes critical overviews of the state-of-the-art nanowires-based align="center"electrodes for energy storage devices, such as lithium-ion batteries, lithium-ion capacitors, sodium-ion batteries, and supercapacitors. Ions or charges can be percolated easily through nanowire networks due to fast adsorption and diffusion. High-rate capability is intensified align="center"over large electroactive surface in align="center"an ordered nanowire electrode.

Keywords: Active electrodes, Batteries, Capacitors, Charge storage capacity, Capacity retention, Composite storage surface, Energy density, Electrolyte wettability, Electrode strain accommodation, Functionalized phase, Ion adsorption, Interfacial ion exchange, align="center"Metal-organic framework, Nanowire, Nanowire array, Nanowire nucleation sites, Redox active sites, Redox reaction, Surface penetration.

---

* **Corresponding author Pralay Maiti:** School of Materials Science and Technology, Indian Institute of Technology (BHU), Varanasi 221005, India; E-mail: pmaiti.mst@itbhu.ac.in

Gaurav Manik and Sushanta Kumar Sahoo (Eds.)

INTRODUCTION

Today's fast-growing generations require renewable and sustainable energy to drive the life cycle. The unique shaped electrodes realize intense research to develop energy storage devices with high energy density (~100 – 200 Whkg⁻¹), power density (≥5 kWkg⁻¹) and long cycle life (≥105 cycles) [1, 2]. Batteries and electrochemical capacitors (ECs) align="center"explore intense interest in the configurational design of align="center"electrodes to assemble highly functional and smart energy-storage devices with their high energy densities and long sustainability [3]. The devices can store energy *via* electrostatic accumulation (non-Faradic) or electrochemical reaction (Faradic reaction) [4]. The electrochemical energy storage behavior strongly depends on the structural and compositional design of the electrode materials [5]. Traditional electrode fabrication scheme requires a large quantity of binding and conducting chemicals for sustainable development of energy and device [6]. Materials with one-dimensional nanostructures have kindled great ialign="center"nterest in the development of energy storage electrodes because of their unique properties, such as shorter diffusion distance of ions along the diameter, lesser charge/discharge time and greater electrode-electrolyte interfacial contact [6, 7]. Battery and capacitor salign="center"ystems are composed of anode, cathode, and electrolyte [8]. Anode and cathode are bridged by align="center"electrolytes for efficient and reversible charge transport [9]. The critical theme to improve energy storage capacity is to optimize electrical and ionic conductivities and consequently maximize the active material utilization with minimum strain in electrode [10]. align="center"The electrolyte is a source of ions align="center"that can be intercalated at the surface of the electrode. Superior electrode-electrolyte wettability offers enhanced kinetics of ion transport. The degree of reaction reversibility determines the energy efficiency and columbic efficiency of battery materials. Capacitive materials with composite structure (d-block metal oxides/sulfides and conjugated functional polymers) display higher capacitance or energy density, primarily due to continuous transmission of surface redox reactions through multioxidation state metal center. Redox-active chromophore improves storage capacity due to faster reaction reversibility. Thus, functionalities can improve redox activity on the surface of align="center"the electrode. Anodes are susceptible to storing ions or electronic align="center"charges during the charging process. The sufficient spaces between the nanowires arrays can ease electrolyte penetrating the integrated anode to increase the interfacial contact areas between electrolyte and electrode [11]. The nanoscale dimension of building blocks promotes the kinetics of ion storage due to the shorter diffusion paths [9]. In general, nanomaterials bring important improvements in energy storage devices because of their reduced size and shape, which widens the surface and interfacial contact to minimize transport and transmission path [12]. Pseudo-capacitors store charge on the surface layer where electrolyte ions can penetrate the matrix of active materials. Size reduction can improve the charge storage capability. The oriented nanowire align="center"has vacant space to retain the large capacity variations without destroying the configuration, shape, and size of active content [13]. Nanowire arrays align="center"exhibit faster ion diffusion than the nanowire network. This is why arrays structure leads to higher capacitance with more structural balance in the continuous phase of nanowire electrode. Li-ion capacitors, typically align="center"combining the align="center"benefits of high-capacity battery type anodes and high-power capacitor type cathodes, have been realized as the most functioning devices. Nanowire serves as a structural resistance to accommodate the volume change during charging and discharging. The nanowire array structure has a tendency to segregate interfacial vacancy within neighboring 1D nanostructure. The array spaces confine electrolyte absorption causing enhancement of internal conductance and high-power performance characteristics. Moreover, the porous structure strengthens the absorption of electrolyte to facilitate reversibility of the redox reaction. Nanowires improve rate capability due to fast electron transport and insertion of ions along long dimensions [14, 15]. Nanowire arrays intensify the electrical and ionic conductivity of the electrode [16, 17]. Since, single nanowire electrode decreases its conductivity gradually, therefore, volume accommodation capacities are reduced during insertion and extraction of charges. But it resumes higher charging stability than that of other nanodimensions [3]. Metal organic frameworks (MOFs) such as zeolitic imidazolate framework-8 and zeolitic imidazolate framework-67 could easily be grown on different interfaces as a functional layer to enhance the charge storage properties [18, 19]. Surface reactions produce supercapacitance. The capacitive behavior is produced through ion exchange phenomenon at the electrode-electrolyte interface [20]. Metal oxides and hydroxides accumulate specific energy by polar sites and redox configuration. Complex metal oxides with variable valence states have a tendency to be reduced and oxidized potentially and thus better suits as pseudocapacitive electrode materials [21]. Thus, the chapter is focused on review of the structural stability of nanowire based electrodes for efficient designing of energy storage devices.

CHARGE STORAGE DEVICES

Advanced charge storage devices are Li-ion, Na-ion and Li-S batteries, which differ with each other due to differing storage ions (charge) [5]. The capacitor and supercapacitor are efficient charge storage devices. Capacitors are charge storage devices that have traditional design, and consist of two or pairs of conductors (electrodes) separated by an insulating surface. The conductors have similar and opposite charges extent on their surface with respect to the dielectric. The interfacial charge accumulation involves no change in chemical composition, which confirms that the charge storage process is bidirectional (reversible), and the charge-discharge process exhibits a longer life time and sustainable cycle life. Interfacial exposure implicates fast reversible redox reaction, which connects the faradic charge storage process in pseudocapacitors. Asymmetric capacitors are characterized through integration of different electrochemical activities on different electrodic align="center"surfaces having dissimilar configuration and chemical environment. The asymmetric nature of electrodes generates capacitance due to non-identical potential drop. Charging rate depends on electrode designing *i.e.*, size, shape, dimensions and intercalating ions like Li^+ or Na^+, which can maintain structural stability in acidic or basic electrolytes during charge-storage process.

Nanowire Electrodes as Charge Storage Medium

Nanowire could drive electron transport along the length with a high confinement effect across the diameter. The one-dimensional (1D) nanostructure provides a shorter ion diffusion path due to the narrow diameter combined with long continuous paths. The porous architecture accommodates volume shrinkage [5]. *Zhan et al.* [17] developed an electrode with grass-like Co_3O_4 nanowire arrays for Li-ion battery which exhibits a specific capacity of 1031 $mAhg^{-1}$ with a current density of 500 mAg^{-1} and retains stable capacity after 500 cycles. Co_3O_4 nanowires crosslink each other because of the volume variation. *Liu et al.* reported an anode composed of carbon encapsulated Ge/C nanowires having large exposed surface (223 m^2g^{-1}) and high specific capacity 1200 $mAhg^{-1}$ at 0.2 C. The electrode delivered a stable discharge capacity of 770 $mAhg^{-1}$ after 500 cycles at 10C [22]. *Li et al.* [23] developed an anode with lithium-reactive zinc phosphide (Zn_3P_2) nanowires arrays on carbon fabrics as a grafted and binder free structure for advanced lithium ion batteries (LIBs). The integrated anode exhibits an initial and reversible Li^+ storage capacity of 1200 $mAhg^{-1}$. Thus, the interconnection and continuous surface morphology of nanowire increases the structural stability for better electronic conduction and further lowers the impedance of the battery (Fig. **1 a,b**). *Ko et al.* [24] developed a high power Li-ion battery with self-supported SnO_2 nanowire anode having excellent rate capability as a renewable discharge capacity of 600 $mAhg^{-1}$ at 3 C. *Gao et al.* [25] reported highly aligned WO_3 nanowire arrays on conductive carbon cloth (WNCC) for supercapacitor and LIBs. The supercapacitors demonstrate a high specific capacitance of 521 Fg^{-1} at a current density of 1 $Ag^{-1,}$ while LIBs exhibit a high capacity of 662 $mAhg^{-1}$ after 140 cycles at a 0.28 C rate.

(a)   (b)

Fig. (1). (a) SEM images CoO – MnO_2– $MnCo_2O_4$ hybrid nanowires (HNWs) electrode for the application of asymmetric supercapacitor **(b)** TEM image of $CuGeO_3$ nanowire anode for Li-ion battery. (Reproduced with copyright permission from ref [26, 27].

Function and Mechanism of Charge Storage Devices

Electrochemical capacitors (*i.e.* supercapacitors) contain electrochemical double-layer capacitance, first due to charge storage of ion adsorption and second involving fast surface redox reactions to account for pseudo-capacitive nature [27]. Ion intercalation is promoted on the surface of the electrode *via* the oxidation and reduction phenomenon of electrode active species. The charge storage mechanism based on pseudocapacitance arises due to reversible reactions in redox active centers (having multioxidation states) available on the surface of the electrode materials. Asymmetric supercapacitors (ASCs) consisted of the cathode (a battery type Faradic electrode) and anode (a capacitor-type electrode) to the interlink energy source and power source separately [28]. Rechargeable batteries and pseudocapacitors store charges through the redox phenomenon that occurs due to redox active metal ions present in electrode materials. Crystalline structure can control the cation diffusion and charge storage rate in the battery. The composition and nature of electrolyte determine the functions of charge storage and ion adsorption on the surface of electrode architecture [29].

Role of Nanowire based Separator in Energy Storage Devices

Nanowire based separators protect internal short circuit. Separators play a critical role in determining cell performances. A highly porous structure offers suitability towards electrolyte adsorption and retention improvement, which in turn helps to improve cyclability and rate capability of batteries. The separator not only prevents the physical contact between the anode and cathode but also works as an electrolyte reservoir to facilitate ion transportation across electrodes [30]. Proper electrolyte wettability and permeability can drive fast ion transport. Separators having high thermal stability can enhance the battery lifetime and the operating temperature of the cells. Polyolefin based separators offer low porosity which in turn hinders electrolyte adsorption and retention, thereby, degrading the battery performance [30].

Functionalized Nanowires and its Integrated Storage Capacity

Functionalization influences the electronic properties of nanowires. It can reduce the surface charge trap states, thereby enhancing ion insertion/extraction capacity. It lowers the activation energy of electrode-electrolyte interface [31]. Surface functionalization introduces a unique host for the intercalation of ions. Negatively charged functional group can absorb or bind the electrolyte cation rapidly during charging of the battery. The oxygenated functional groups grafted over carbonaceous nanowire favour high energy storage towards battery-type negative electrodes (anode) and capacitor-type positive electrodes respectably. *Wang et al.* [32] developed LiCl/PVA gel electrolyte stabilized V_2O_3 nanowire electrode for pseudocapacitor. Gel electrolyte prevents degradation of electrode active sites by minimizing hydrophilic (aqueous) content. Thus, the gel electrolyte works as the matrix to stabilize the nanowire structure for sustainable charge transport. The pseudocapacitor exhibited 85% capacity retention after 5000 cycles of the charging/discharging process. Thus, functionalities improve the capacity loss.

Surface Chromophore (Active Sites) and its Energy Retention Capacity

Charge storage capacity can be intensified through grafting various chromophoric active sites on the surface dimensions of nanowires. Electron withdrawing chromophore enhances the kinetics of outer circuit electron extraction which facilitates ion storage during charging of electrochemical device. Redox active sites adsorb on the electrolyte surface, which leads to improvement of specific capacitance. Supercapacitor stores charges on the surface of redox active functional groups materials through adsorption phenomenon. On the other hand, LIBs store charges by the intercalation of Li^+ ions on the surface and in the bulk of the active materials. Therefore, LIBs deliver a much higher energy density than that of other charge storage devices. Storage capacity retention can be intensified through functionalization of nanowire. The storage energy (E_g) of electrodes can be enhanced by increasing the specific capacitance (C_G) and the potential window (V), which further invites structural variation towards electrode designing. Redox active grafted functional groups can store electric charge through highly redox reaction, and thus, maximize the capacitance of electrode materials [33]. The potential window in the supercapacitor can be extended by fabricating an asymmetric device configuration having two electrode materials with complementary potential (Eq. **1**). Capacitance enhancement depends on the quantity of electrochemically active sites in the electrodes accessed by the solvated ions (*n*) in the electrolyte [26]. Organic electrolytes can enhance the voltage window [34].

$$E_g = 0.5 \times C_G \times V^2 \tag{1}$$

$$n = \frac{C_S \times m \times \Delta V}{F \times Z} \tag{2}$$

Where, physical parameters *m*, ΔV, *F* and *Z* demonstrate molecular weight, redox potential, the Faraday's constant and oxidation state of electrode active substance, respectively. Surface reversible redox reaction depends on the nature of functionalities that aid the capacitance in the electrode. The surface functionalities exhibit different behaviour in different electrolytes. Redox active chromophores are susceptible towards additional storage due to fast reversible redox reaction during charging/discharging process of devices. The electrode potential is expressed as an electrochemical conversion of the electroactive substance. In other words, the electrode potential is a measure of the utilization index of unoccupied sites on the active surface (Eq. **3**) [27].

$$E = E^\circ + \frac{RT}{nF} \ln\frac{X}{1-X} \tag{3}$$

Where, parameter *E* denotes potential produced on the electrode. Other entities (*R*, *T*, and *F*) address universal gas constant, temperature (*K*) and Faraday's constant, respectively. The numeric *n* stands for the number of electrons transferred electrochemically while *X* corresponds to characteristics of materials such as occupant density of the surface or structural dimensions. The occupancy factor depends on the electrode dimensions.

STORAGE PERFORMANCE DEPENDENT ON HYBRID AND COMPOSITE PHASE OF NANOWIRE ELECTRODE

Hybrid or composite phase creates a nanochannel for ion transportation. Nanowire dimensions have a high porosity and favorable orientations which intensifies electrode/electrolyte interface and transport efficiency for ion as well as electrons within electrodes [26, 35]. Hybrid configuration stores charges simultaneously by fast surface ion adsorption/desorption and slow faradaic intercalation. Graphene oxide has a large number of oxygen functionalities confined as a –O-, -OH located on the surface while edges are occupied with carbonyl and carboxyl groups. The negative functional groups adsorb electroactive ions and thus, intensify the additional storage capacity of composite nanowire [35]. *Nair et al.* [36] reported superior electrochemical performances of the nitrogen doped graphene -Ag nanowire, which creates a cooperative effect because the doped phase notably ameliorates the space or volume for Li ion since it creates more surface imperfections which serve as a green atmosphere for lithium ions. *Shen et al.* [37] showed a lithium ion capacitor device that consists of Li_3VO_4 and N-doped carbon integrated as nanowires and that conveys a high energy density of 136.4 Whkg^{-1} at a power density of 532 Wkg^{-1}. *Chen et al.* [38] developed $Ni(OH)_2$ nanowire /graphite foam type complex (mixed) electrode for super capacitor. The capacitive behaviour was characterized by measuring specific capacity in highly concentrated strong electrolytes as a basic medium. The specific capacity value 2144 Fg^{-1} supports unique nanostructure and easy diffusion (penetration) of electrolyte into the internal core of the electrode. *Fu et al.* [39] developed a $CuGeO_3$ nanowire as anode for Na ion battery. The sodium storage process of $CuGeO_3$ is featured with the merging of chemical constituents and amalgamation reaction. Electrode delivered an initial charge capacity of 306.7 mAhg^{-1} along with an amicable cycling function.

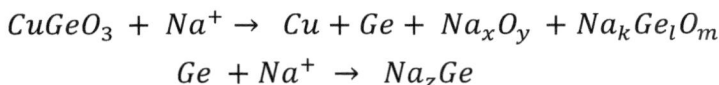

$$CuGeO_3 + Na^+ \rightarrow Cu + Ge + Na_xO_y + Na_kGe_lO_m$$
$$Ge + Na^+ \rightarrow Na_zGe$$

Yao et al. [40] reported that the unification of soft graphene nanosheets and spinel type Co_3O_4 nanowires causes operational structure spanning, which essentially intensifies the electrical and redox properties of the mixed character. Lithium ion battery delivers high capacity performance even at high current density with mixed configurational design. *O'Neill et al.* [41] reported enhanced charge storage performance of Fe_3O_4/FeOOH nanowire electrode in the presence of aqueous electrolyte having Na_2SO_3 species. The pseudocapacitor exhibits additional storage behaviour because of adsorbed SO_3^{2-} which shows redox activity on the surface of nanowires. *Xie et al.* [35] developed an asymmetric solid state supercapacitor with N-doped graphene-coated complex structure of cobalt carbonate hydroxide nanowire and absorptive and electronegative nitrogen-doped carbonaceous graphene electrodes. The structure offers an electrochemically denser active region with a capacitance value of 1358 Fg^{-1} at a current density of 10 Ag^{-1}, which further depicts enhancement of redox reaction.

$$\text{Specific volumetric capacitance } (Cv) = \frac{I\Delta t}{V_{device}\,\Delta V} \tag{4}$$

Where, I and Δt denote applied current and discharge time, respectively. Other parameters ΔV and V_{device} dictate voltage drop and volume within the device. 1D electrode nanomaterial exhibits higher volumetric energy than 2D or 3D nanomaterials [35]. Hence, the capacitance C_V is reliant on the nanowire configuration and active electrode materials (Eq. **4**). The specific capacitance increases in a composite phase of the nanowire. The surface defects and dimensions intensify the storage behaviour.

Storage Behaviour of Core Shell and Porous Structures

Capacity retention depends on the integrity of electrode composition [42]. Directional charge transport properties arise from the quantum confinement effect in 1D nanowire. Nanowire electrodes having core-shell structures are susceptible to protecting active sites from environmental changes, which leads to intensifying new chemical or physical capability, *i.e.*, percolation of ions or molecules onto the core selectively by maintaining the structural integrity [43]. The core-sheath structured nanowire increases the contact area with electrolytes and thus, greatly improves the redox activity and capacitance of electrode materials [3]. Porous electrodes can accommodate volume change during charging, and hence, improve capacitive retention. Redox active group grafted nanowire capture charge due to fast reversible electrochemical reaction. The SEM and TEM images display the successful formation of the nanowire (Figs. **1a** and **1b**). *Han et al.* [44] reported Co_3O_4@PPy@MnO_2 core shell nanowire with a high energy density of 34.3 Whkg^{-1} at a power density of 80.0 Wkg^{-1} and remarkable long-term cycling stability. Porous and nano-sized electrode materials exhibit improved capacitance. *Wu et al.* [45] developed a bendable asymmetric supercapacitor with ($NiCo_2S_4$@PPy NWs//AC-Ni foam) electrodes which produced maximum energy of 86.6 Whkg^{-1} at a power density of 1046 Wkg^{-1} with a high capacitance 1925 Fg^{-1}. Thus, asymmetric device exhibited high energy density than the symmetric device and presumably due to higher specific capacitance [35].

Synergistic Interaction of Nanowire with Conjugated Polymer and Metal Oxide/Sulfide

The charge storage originates from the redox or interconversion reaction. Transition metal oxide exhibits high capacity due to the conversion into multi-oxidation state during ion storage. $MO_x + 2xLi + 2xe^- \rightarrow M + Li_2O$ [2, 46]. In another context, extra capacity can be achieved by maximizing the oxygen content on the surface of electrode materials. Conducting polymer improves the redox activity of electrode during charging of the device. The encapsulation of a flexible and conducting polymer can enhance the structural and functional properties of the nanowire electrode [34, 47]. Composite metal and binary metal oxide nanowires have high Li storage capacity. *Huo et al.* [48] developed the 3D Co_3S_4 nanophase coated with polypyrrole and structurally supported on nickel foam as a hybrid electrochemical composite electrode with bi-polar conductive frameworks. The conductive layer can facilitate smooth electron transmission by minimizing the diffusion pathway of electrolytic ions. These hybrid electrodes exhibit a battery-type storage behavior with ultrahigh specific capacity of 723 Cg^{-1} and 98.6% retention capacity after 1000 cycles. Materials interaction increases the capacity retention of charge storage electrodes.

Influence of Doping, Coating and Redox Active Group on Nanowire Electrode

Energy barrier of ion migration can be minimized by doping of electron rich species [46]. Lower charge transfer resistance improves specific capacity *via* fast storage kinetics. Coating effectively accommodates the huge volume change of composite nanowire during cycling and maintains perfect electrical conductivity throughout the electrode [49]. The carbon coating offered additional and efficient nanochannels on the surface of the electrode for electron transport. The mechanism of enhancement in the charging rate is shown in Fig. (2a). Conductive materials generate heterogeneous structures on the nanowire surface that enhance additional conductivity, which in turn helps to store high charge density [28]. Composite surface has a synergistic interface that effectively boosts the energy storage from both electrode and electrolyte. *Sun et al.* [33] reported that using $[Fe(CN)_6]^{4-}$ as a functional layer grafted on the surface of a Co_3O_4 nanowire electrode leads to the creation of a synergistic interface. The energy density of aqueous batteries or pseudo capacitors is increased due to functionalized interface, which is suitable to achieve a double gain in charge storage from the redox reactions occurring at solid as well as liquid sides [50]. *Xu et al.* [28] developed asymmetric supercapacitor composed of sulfur-doped Co_3O_4 (S-Co_3O_4) as a negative electrode and MnO_2 as a positive electrode to achieve an energy density of 0.86 mWh /cm^3 and power density of 0.79 W/cm^3. *Yuan et al.* [51] reported dodecanethiol-passivated Ge nanowires as an efficient anode for LIB which exhibits an excellent electrochemical performance with a reversible specific capacity of 1130 $mAhg^{-1}$ at 0.1 C rates after 100 cycles. Charge storage in a redox-active nanowire is driven by the concentration gradient of the redox active species. *Wang et al.* [46] synthesized surface modified electrode with N-doped carbon layer over CoO nanowire arrays on Ni foam followed by encapsulation with metal organic frameworks such as zeolitic imidazolate frameworks-67 (ZIF-67), which provides better cyclability of 1884.1 $mAhg^{-1}$ at current density 1 Ag^{-1} after 100 cycles.

(a)   (b)

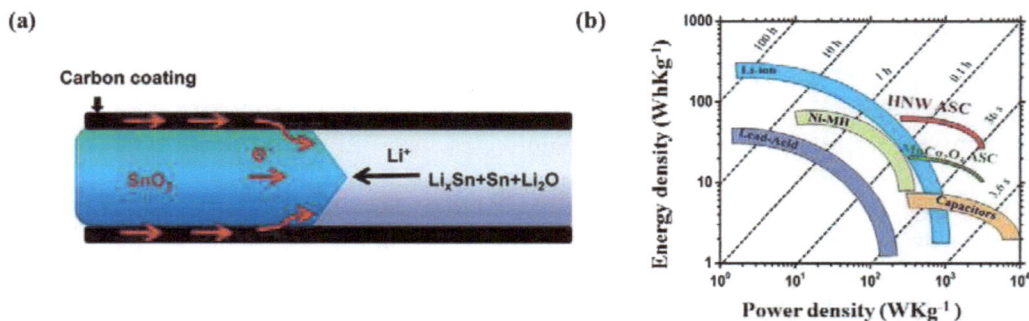

Fig. (2). (a) Lithium ion battery and its lithiation (storage) mechanism. The lithiation rate was 6.6 on carbon coated SnO_2 nanowires comparatively higher than 1.2 on non coated SnO_2 nanowires. (b) Ragone plot of materials in different charge storage devices integrated with commercial activated carbon (AC). (Reproduced with copyright permission from ref [26, 49].

Phase Dependent Functional Properties and Electrochemical Storage Correlation

The function and specific capacities strongly rely on sizes, morphologies, phases and chemical structures. Electrochemical properties and specific capacity vary with the number of operating cycles during charging/discharging. One-dimensional morphologies are highly interesting because of directional charge transport behaviour. Electrochemical properties and specific capacity vary with the number of cycles operated during charging/discharging. Phase variation shifts the specific capacitance of nanomaterials. *Ko et al.* [24] reported phase dependent specific capacitance for Li-ion battery. The SnO_2 nanowire electrode exhibited specific discharge capacity of 510 mAhg^{-1} with better cyclic stability at 50th cycle. The discharge capacity was comparatively higher than that of electrodes composed of SnO_2 nanopowder and Sn nanopowder. Structural stability helped maintain the capacity degradation. This demonstrates that 1D nanowire electrodes with a high aspect ratio of length to diameter efficiently enhance the charge-transfer properties compared with the powder form and other phases. Nano dimension (nanowire) maintains almost the constancy of capacity loss. Hybrid phase exhibits high-capacity retention than single or any other shape nanowire. Doped nanowire electrode exhibits high capacitance than the normal nanowire. *Nair et al.* reported Nitrogen doped graphene embedded Ag nanowire (NG-Ag NW) offers 724 mAhg^{-1} capacitance value, which is higher than the capacitance value (53 mAhg^{-1}) of normal Ag nanowire for the 50 cycle [36]. Rangone plot depicts increasing order of energy density for different devices composed of a different phase of nanowire electrode (Fig. **2b**). This clearly indicates that the materials integrity and surface nanostructures play a key role in extending the energy density and charge storage capacity. The nano dimension (nanowire) maintains almost constancy in capacity loss.

The hybrid phase exhibits high-capacity retention than single or any other shape nanowires.

Integrated Storage based on 1D Metal - Organic Frameworks (MOFs) Anchored Nanowire

Metal organic frameworks (MOFs) enhance the storage properties of nanowire due to the synergistic effect [19]. Conductive MOFs have a tendency to deliver outperforming redox activity because of smooth conduction towards ion and electron transport. Metal–organic frameworks (MOFs) are composed of metal ions and organic ligands with suitable functionalities. They have tunable properties, nano-porous network structure, high surface area, and a large number of active donor and acceptor sites that benefit electrolyte/electrode interfacial contact. Their superior performance is rationalized by the reversible structural deformation and the conversion reaction between two different metal oxidation states inside the organic framework. *Kim et al.* [42] developed a composite anode for LIB with Co_3O_4 nanowire and reduced graphene oxide followed by wrapping N-doped carbon layer. The Co_3O_4 NWs/rGO@N-doped carbon (NC) electrode delivered a high discharge capacity of 995 mAhg^{-1} at 0.1 C at 65 cycles and showed superiority to that of Co_3O_4NWs/rGO electrode, having a discharge capacity of 203 mAhg^{-1}. *Sun et al.* [52] developed lithium ion capacitor assembled by coupling $Co_3(HHTP)_2$ nanowire anode with a pore size of 13.1 nm and porous activated carbon spheres (ACS) as a cathode. The capacitor provided a large capacity value of 67 Fg^{-1} with energy density of 64 Whkg^{-1} under stable working voltage of 4.0 V.

CONCLUDING REMARKS

In summary, charge storage materials displayed capacitor or battery-like performance depending on the configurations and chemical composition of the electrode, host, and charge storage insertion ions. Charge storage capacity depends on the structure, geometry and shape of the electrode (anode). Charge or ion adsorption maintains the uniqueness of the surface of nanowire electrode without structural degradation. align="center"Charge storage capacity depends on the type of redox active group and surface defects available on the surface of nanowire designed architecture. Doped surfaces intensify electrical conductivity for fast transport and storage of ion. In this way, LIBs, lithium-ion capacitors and supercapacitors could be assembled through nanowire decorated electrodes for efficient energy storage. The density of the redox-active groups controls the storage capacity of nanowire. Reversible surface redox reaction drives ion storage. Multi-redox center integrated nanowires have a high charge retention capacity. Thus, surface confinement ignites the development of charge storage materials.

CONSENT FOR PUBLICATION

Not applicable.

CONFLICT OF INTERESTS

The authors declare no conflict of interest, financial or otherwise.

ACKNOWLEDGEMENTS

The author (Sunil Kumar) gratefully recognizes the cooperation of SMST, IIT (BHU) for availing any sort of support of research tools, management, and techniques during chapter design. The author also would like to acknowledge the authorities of L.N.T.College and R.B.B.M. College Muzaffarpur for promoting us academically.

REFERENCES

[1]   Vidhyadharan, B.; Misnon, I.I.; Aziz, R.A.; Padmasree, K.P.; Yusoff, M.M.; Jose, R. Superior Supercapacitive Performance in Electrospun Copper Oxide Nanowire Electrodes. *J. Mater. Chem. A Mater. Energy Sustain.,* **2014**, *2*(18), 6578-6588.
[http://dx.doi.org/10.1039/C3TA15304E]

[2]   Cao, K.; Jiao, L.; Liu, Y.; Liu, H.; Wang, Y.; Yuan, H. Ultra-High Capacity Lithium-Ion Batteries with Hierarchical CoO Nanowire Clusters as Binder Free Electrodes. *Adv. Funct. Mater.,* **2015**, *25*(7), 1082-1089.
[http://dx.doi.org/10.1002/adfm.201403111]

[3]   Zhang, C.; Wang, J.G.; Jin, D.; Xie, K.; Wei, B. Facile Fabrication of MnO/C Core-Shell Nanowires as an Advanced Anode Material for Lithium-Ion Batteries. *Electrochim. Acta,* **2015**, *180*, 990-997.
[http://dx.doi.org/10.1016/j.electacta.2015.09.050]

[4]   He, P.; Zhang, G.; Liao, X.; Yan, M.; Xu, X.; An, Q.; Liu, J.; Mai, L. Sodium ion stabilized vanadium oxide nanowire cathode for high-performance zinc-ion batteries. *Adv. Energy Mater.,* **2018**, *8*(10), 1-6.
[http://dx.doi.org/10.1002/aenm.201702463]

[5]   Li, F.; Ma, J.; Ren, H.; Wang, H.; Wang, G. Fabrication of MnO Nanowires Implanted in Graphene as an Advanced Anode Material for Sodium-Ion Batteries. *Mater. Lett.,* **2017**, *206*, 132-135.
[http://dx.doi.org/10.1016/j.matlet.2017.07.006]

[6]   Wang, X.; Li, G.; Seo, M.H.; Lui, G.; Hassan, F.M.; Feng, K.; Xiao, X.; Chen, Z. Carbon-coated silicon nanowires on carbon fabric as self-supported electrodes for flexible lithium-ion batteries. *ACS Appl. Mater. Interfaces,* **2017**, *9*(11), 9551-9558.
[http://dx.doi.org/10.1021/acsami.6b12080] [PMID: 27808493]

[7]   Yan, C.; Wang, X.; Cui, M.; Wang, J.; Kang, W.; Foo, C.Y.; Lee, P.S. Stretchable Silver-Zinc Batteries Based on Embedded Nanowire Elastic Conductors. *Adv. Energy Mater.,* **2014**, *4*(5), 1-6.
[http://dx.doi.org/10.1002/aenm.201301396]

[8]   Shen, L.; Che, Q.; Li, H.; Zhang, X. Mesoporous $NiCo_2O_4$ nanowire arrays grown on carbon textiles as binder-free flexible electrodes for energy storage. *Adv. Funct. Mater.,* **2014**, *24*(18), 2630-2637.
[http://dx.doi.org/10.1002/adfm.201303138]

[9]   Huang, L.; Wei, Q.; Sun, R.; Mai, L. Nanowire electrodes for advanced lithium batteries. *Front. Energy Res.,* **2014**, *2*(OCT), 1-13.

[10] Ferrara, G.; Damen, L.; Arbizzani, C.; Inguanta, R.; Piazza, S.; Sunseri, C.; Mastragostino, M. SnCo nanowire array as negative electrode for lithium-ion batteries. *J. Power Sources,* **2011**, *196*(3), 1469-1473.
[http://dx.doi.org/10.1016/j.jpowsour.2010.09.039]

[11] Yan, L.; Shu, J.; Li, C.; Cheng, X.; Zhu, H.; Yu, H.; Zhang, C.; Zheng, Y.; Xie, Y.; Guo, Z. W3Nb14O44 nanowires: ultrastable lithium storage anode materials for advanced rechargeable batteries. *Energy Storage Mater.,* **2019**, *16*, 535-544.
[http://dx.doi.org/10.1016/j.ensm.2018.09.008]

[12] Zhou, G.; Xu, L.; Hu, G.; Mai, L.; Cui, Y. Nanowires for electrochemical energy storage. *Chem. Rev.,* **2019**, *119*(20), 11042-11109.
[http://dx.doi.org/10.1021/acs.chemrev.9b00326] [PMID: 31566351]

[13] Chen, L.; Guo, X.; Lu, W.; Chen, M.; Li, Q.; Xue, H.; Pang, H. Manganese Monoxide-Based Materials for Advanced Batteries. *Coord. Chem. Rev.,* **2018**, *368*, 13-34.
[http://dx.doi.org/10.1016/j.ccr.2018.04.015]

[14] Li, L.; Zhang, Y.Q.; Liu, X.Y.; Shi, S.J.; Zhao, X.Y.; Zhang, H.; Ge, X.; Cai, G.F.; Gu, C.D.; Wang, X.L.; Tu, J.P. One-Dimension $MnCo_2O_4$ Nanowire Arrays for Electrochemical Energy Storage. *Electrochim. Acta,* **2014**, *116*, 467-474.
[http://dx.doi.org/10.1016/j.electacta.2013.11.081]

[15] Gowda, S.R.; Leela Mohana Reddy, A.; Zhan, X.; Ajayan, P.M. Building energy storage device on a single nanowire. *Nano Lett.,* **2011**, *11*(8), 3329-3333.
[http://dx.doi.org/10.1021/nl2017042] [PMID: 21755944]

[16] Park, M.S.; Wang, G.X.; Kang, Y.M.; Wexler, D.; Dou, S.X.; Liu, H.K. Preparation and electrochemical properties of SnO_2 nanowires for application in lithium-ion batteries. *Angew. Chem. Int. Ed.,* **2007**, *46*(5), 750-753.
[http://dx.doi.org/10.1002/anie.200603309] [PMID: 17163569]

[17] Zhan, L.; Wang, S.; Ding, L.X.; Li, Z.; Wang, H. Grass-like Co_3O_4 Nanowire Arrays Anode with High Rate Capability and Excellent Cycling Stability for Lithium-Ion Batteries. *Electrochim. Acta,* **2014**, *135*, 35-41.
[http://dx.doi.org/10.1016/j.electacta.2014.04.139]

[18] Yu, D.; Wu, B.; Ge, L.; Wu, L.; Wang, H.; Xu, T. Decorating Nanoporous ZIF-67-Derived $NiCo_2O_4$ Shells on a Co_3O_4 Nanowire Array Core for Battery-Type Electrodes with Enhanced Energy Storage Performance. *J. Mater. Chem. A Mater. Energy Sustain.,* **2016**, *4*(28), 10878-10884
[http://dx.doi.org/10.1039/C6TA04286D]

[19] Song, H.; Shen, L.; Wang, J.; Wang, C. Reversible Lithiation-Delithiation Chemistry in Cobalt Based Metal Organic Framework Nanowire Electrode Engineering for Advanced Lithium-Ion Batteries. *J. Mater. Chem. A Mater. Energy Sustain.,* **2016**, *4*(40), 15411-15419.
[http://dx.doi.org/10.1039/C6TA05925B]

[20] Krittayavathananon, A.; Pettong, T.; Kidkhunthod, P.; Sawangphruk, M. Insight into the Charge Storage Mechanism and Capacity Retention Fading of $MnCo_2O_4$ Used as Supercapacitor Electrodes. *Electrochim. Acta,* **2017**, *258*, 1008-1015.
[http://dx.doi.org/10.1016/j.electacta.2017.11.152]

[21] Pang, Q.; Sun, C.; Yu, Y.; Zhao, K.; Zhang, Z.; Voyles, P.M.; Chen, G.; Wei, Y.; Wang, X. $H_2V_3O_8$ nanowire/graphene electrodes for aqueous rechargeable zinc ion batteries with high rate capability and large capacity. *Adv. Energy Mater.,* **2018**, *8*(19), 1-9.
[http://dx.doi.org/10.1002/aenm.201800144]

[22] Liu, J.; Song, K.; Zhu, C.; Chen, C.C.; van Aken, P.A.; Maier, J.; Yu, Y. Ge/C nanowires as high-capacity and long-life anode materials for Li-ion batteries. *ACS Nano,* **2014**, *8*(7), 7051-7059.
[http://dx.doi.org/10.1021/nn501945f] [PMID: 24940842]

[23]   Li, W.; Gan, L.; Guo, K.; Ke, L.; Wei, Y.; Li, H.; Shen, G.; Zhai, T. Self-supported Zn_3P_2 nanowire arrays grafted on carbon fabrics as an advanced integrated anode for flexible lithium ion batteries. *Nanoscale,* **2016**, *8*(16), 8666-8672.
[http://dx.doi.org/10.1039/C5NR08467A] [PMID: 27049639]

[24]   Ko, Y.D.; Kang, J.G.; Park, J.G.; Lee, S.; Kim, D.W. Self-supported SnO_2 nanowire electrodes for high-power lithium-ion batteries. *Nanotechnology,* **2009**, *20*(45), 455701.
[http://dx.doi.org/10.1088/0957-4484/20/45/455701] [PMID: 19822930]

[25]   Gao, L.; Wang, X.; Xie, Z.; Song, W.; Wang, L.; Wu, X.; Qu, F.; Chen, D.; Shen, G. High-Performance Energy-Storage Devices Based on WO_3 Nanowire Arrays/Carbon Cloth Integrated Electrodes. *J. Mater. Chem. A Mater. Energy Sustain.,* **2013**, *1*(24), 7167-7173.
[http://dx.doi.org/10.1039/c3ta10831g]

[26]   Harilal, M.; Krishnan, S.G.; Yar, A.; Misnon, I.I.; Reddy, M.V.; Yusoff, M.M.; Dennis, J.O.; Jose, R. Pseudocapacitive charge storage in single-step-synthesized $CoO-MnO_2-MnCo_2O_4$ hybrid nanowires in aqueous alkaline electrolytes. *J. Phys. Chem. C,* **2017**, *121*(39), 21171-21183.
[http://dx.doi.org/10.1021/acs.jpcc.7b06630]

[27]   Liu, J.; Wang, J.; Xu, C.; Jiang, H.; Li, C.; Zhang, L.; Lin, J.; Shen, Z.X. Advanced energy storage devices: basic principles, analytical methods, and rational materials design. *Adv. Sci. (Weinh.),* **2017**, *5*(1), 1700322.
[http://dx.doi.org/10.1002/advs.201700322] [PMID: 29375964]

[28]   Xu, W.; Chen, J.; Yu, M.; Zeng, Y.; Long, Y.; Lu, X.; Tong, Y. Sulphur-Doped Co_3O_4 nanowires as an advanced negative electrode for high-energy asymmetric supercapacitors. *J. Mater. Chem. A Mater. Energy Sustain.,* **2016**, *4*(28), 10779-10785.
[http://dx.doi.org/10.1039/C6TA03153F]

[29]   Wang, Y.; Song, Y.; Xia, Y. Electrochemical capacitors: mechanism, materials, systems, characterization and applications. *Chem. Soc. Rev.,* **2016**, *45*(21), 5925-5950.
[http://dx.doi.org/10.1039/C5CS00580A] [PMID: 27545205]

[30]   Li, H.; Wu, D.; Wu, J.; Dong, L.Y.; Zhu, Y.J.; Hu, X. Flexible, high-wettability and fire-resistant separators based on hydroxyapatite nanowires for advanced lithium-ion batteries. *Adv. Mater.,* **2017**, *29*(44), 1-11.
[http://dx.doi.org/10.1002/adma.201703548] [PMID: 29044775]

[31]   Khandare, L.; Terdale, S. Gold nanoparticles decorated mno_2 nanowires for high performance supercapacitor. *Appl. Surf. Sci.,* **2017**, *418*, 22-29.
[http://dx.doi.org/10.1016/j.apsusc.2016.12.036]

[32]   Wang, G.; Lu, X.; Ling, Y.; Zhai, T.; Wang, H.; Tong, Y.; Li, Y. LiCl/PVA gel electrolyte stabilizes vanadium oxide nanowire electrodes for pseudocapacitors. *ACS Nano,* **2012**, *6*(11), 10296-10302.
[http://dx.doi.org/10.1021/nn304178b] [PMID: 23050855]

[33]   Sun, S.; Rao, D.; Zhai, T.; Liu, Q.; Huang, H.; Liu, B.; Zhang, H.; Xue, L.; Xia, H. Synergistic interface-assisted electrode-electrolyte coupling toward advanced charge storage. *Adv. Mater.,* **2020**, *32*(43), e2005344.
[http://dx.doi.org/10.1002/adma.202005344] [PMID: 32954557]

[34]   Duay, J.; Gillette, E.; Liu, R.; Lee, S.B. Highly flexible pseudocapacitor based on freestanding heterogeneous MnO2/conductive polymer nanowire arrays. *Phys. Chem. Chem. Phys.,* **2012**, *14*(10), 3329-3337.
[http://dx.doi.org/10.1039/c2cp00019a] [PMID: 22298230]

[35]   Xie, H.; Tang, S.; Zhu, J.; Vongehr, S.; Meng, X. A high energy density asymmetric all-solid-state supercapacitor based on cobalt carbonate hydroxide nanowire covered n-doped graphene and porous graphene electrodes. *J. Mater. Chem. A Mater. Energy Sustain.,* **2015**, *3*(36), 18505-18513.
[http://dx.doi.org/10.1039/C5TA05129K]

[36] Nair, A.K.; Elizabeth, I. S, G.; Thomas, S.; M.S, K.; Kalarikkal, N. nitrogen doped graphene – silver nanowire hybrids: an excellent anode material for lithium ion batteries. *Appl. Surf. Sci.,* **2018**, *428*, 1119-1129.
[http://dx.doi.org/10.1016/j.apsusc.2017.09.214]

[37] Shen, L.; Lv, H.; Chen, S.; Kopold, P.; van Aken, P.A.; Wu, X.; Maier, J.; Yu, Y. Peapod-like $Li_3 VO_4$ /n-doped carbon nanowires with pseudocapacitive properties as advanced materials for high-energy lithium-ion capacitors. *Adv. Mater.,* **2017**, *29*(27), 1-8.
[http://dx.doi.org/10.1002/adma.201700142] [PMID: 28466539]

[38] Chen, Y.; Zhang, Z.; Sui, Z.; Liu, Z.; Zhou, J.; Zhou, X. $Ni(OH)_2$ nanowires/graphite foam[composite as an advanced supercapacitor electrode with improved cycle performance. *Int. J. Hydrogen Energy,* **2016**, *41*(28), 12136-12145.
[http://dx.doi.org/10.1016/j.ijhydene.2016.05.104]

[39] Fu, L.; Zheng, X.; Huang, L.; Shang, C.; Lu, K.; Zhang, X.; Wei, B.; Wang, X. Synthesis and Investigation of $CuGeO_3$ Nanowires as Anode Materials for Advanced Sodium-Ion Batteries. *Nanoscale Res. Lett.,* **2018**, *13*(1), 193.
[http://dx.doi.org/10.1186/s11671-018-2609-z] [PMID: 29974272]

[40] Yao, X.; Guo, G.; Zhao, Y.; Zhang, Y.; Tan, S.Y.; Zeng, Y.; Zou, R.; Yan, Q.; Zhao, Y. Synergistic Effect of Mesoporous Co_3O_4 Nanowires Confined by N-Doped Graphene Aerogel for Enhanced Lithium Storage. *Small,* **2016**, *12*(28), 3849-3860.
[http://dx.doi.org/10.1002/smll.201600632] [PMID: 27283881]

[41] O'Neill, L.; Johnston, C.; Grant, P.S. Enhancing the Supercapacitor Behaviour of Novel Fe_3O_4/FeOOH Nanowire Hybrid Electrodes in Aqueous Electrolytes. *J. Power Sources,* **2015**, *274*, 907-915.
[http://dx.doi.org/10.1016/j.jpowsour.2014.09.151]

[42] Kim, Y.; Noh, Y.; Han, H.; Bae, J.; Park, S.; Lee, S.; Yoon, W.; Kim, Y.K.; Ahn, H.; Ham, M.H.; Kim, W.B. Effect of n-doped carbon layer on Co_3O_4 nanowire-graphene composites as anode materials for lithium ion batteries. *J. Phys. Chem. Solids,* **2018**, *2019*(124), 266-273.

[43] Lu, W.; Guo, X.; Luo, Y.; Li, Q.; Zhu, R.; Pang, H. Core-shell materials for advanced batteries. *Chem. Eng. J.,* **2018**, *2019*(355), 208-237.

[44] Han, L.; Tang, P.; Zhang, L. Hierarchical Co_3O_4 at PPy at MnO_2 core-shell-shell nanowire arrays for enhanced electrochemical energy storage. *Nano Energy,* **2014**, *7*, 42-51.
[http://dx.doi.org/10.1016/j.nanoen.2014.04.014]

[45] Wu, X.; Meng, L.; Wang, Q.; Zhang, W.; Wang, Y. A novel inorganic-conductive polymer core-sheath nanowire arrays as bendable electrode for advancedelectrochemical energy storage. *Chem. Eng. J.,* **2018**, *2019*(358), 1464-1470.

[46] Wang, D.; Yan, B.; Guo, Y.; Chen, L.; Yu, F.; Wang, G. N-doped carbon coated CoO nanowire arrays derived from zeolitic imidazolate framework-67 as binder-free anodes for high-performance lithium storage. *Sci. Rep.,* **2019**, *9*(1), 5934.
[http://dx.doi.org/10.1038/s41598-019-42371-y] [PMID: 30976045]

[47] Wang, K.; Wu, H.; Meng, Y.; Wei, Z. Conducting polymer nanowire arrays for high performance supercapacitors. *Small,* **2014**, *10*(1), 14-31.
[http://dx.doi.org/10.1002/smll.201301991] [PMID: 23959804]

[48] Huo, W.; Zhang, X.; Liu, X.; Liu, H.; Zhu, Y.; Zhang, Y.; Ji, J.; Dong, F.; Zhang, Y. Construction of advanced 3D Co_3S_4@PPy nanowire anchored on nickel foam for high-performance electrochemical energy storage. *Electrochim. Acta,* **2020**, *334*, 135635.
[http://dx.doi.org/10.1016/j.electacta.2020.135635]

[49] Zhang, L.Q.; Liu, X.H.; Liu, Y.; Huang, S.; Zhu, T.; Gui, L.; Mao, S.X.; Ye, Z.Z.; Wang, C.M.; Sullivan, J.P.; Huang, J.Y. Controlling the lithiation-induced strain and charging rate in nanowire electrodes by coating. *ACS Nano,* **2011**, *5*(6), 4800-4809.

[http://dx.doi.org/10.1021/nn200770p] [PMID: 21542642]

[50]   Yu, K.; Pan, X.; Zhang, G.; Liao, X.; Zhou, X.; Yan, M.; Xu, L.; Mai, L. Nanowires in energy storage devices: structures, synthesis, and applications. *Adv. Energy Mater.,* **2018**, *8*(32), 1802369.
[http://dx.doi.org/10.1002/aenm.201802369]

[51]   Yuan, F.W.; Yang, H.J.; Tuan, H.Y. Alkanethiol-passivated ge nanowires as high-performance anode materials for lithium-ion batteries: the role of chemical surface functionalization. *ACS Nano,* **2012**, *6*(11), 9932-9942.
[http://dx.doi.org/10.1021/nn303519g] [PMID: 23043347]

[52]   Sun, J.; Guo, L.; Sun, X.; Zhang, J.; Liu, Y.; Hou, L.; Yuan, C. Conductive co-based metal-organic framework nanowires: A competitive high-rate anode towards advanced li-ion capacitors. *J. Mater. Chem. A Mater. Energy Sustain.,* **2019**, *7*(43), 24788-24791.
[http://dx.doi.org/10.1039/C9TA08788E]

CHAPTER 10

Polymer Nanocomposite Membrane for Fuel cell Applications

Ratikanta Nayak[1,2,*], **Harilal** [3] and **Prakash Chandra Ghosh**[4]

[1] *Orson Resins & Coatings Pvt Ltd. Mumbai,400101, India*

[2] *NIST (AUTONOMOUS), Berhampur, Odisha, 761008, India*

[3] *School of Chemistry, University of Hyderabad, Hyderabad 500046, India*

[4] *Indian Institute Technology Bombay, Mumbai, 400076, India*

Abstract: Polymer nanocomposite is a new kind of material that offers to substitute traditionally filled polymers. The nanomaterial polymer matrix inter-phase area increases drastically due to the inherent high surface-to-volume ratio resulting in remarkably enhanced properties compared to the pristine polymers or their conventional counterpart filled nanocomposites. Nanocomposites have several novel properties such as nonlinear optical properties, electronic conductivity and luminescence. Therefore, their use has been projected in many areas like chemical sensors, polymer electrolyte membrane fuel cell (PEMFCs), electroluminescent devices, batteries, electrocatalysis, smart windows and memory devices. PEMFCs embody a potential candidate for electrochemical energy generation in the twenty-first century due to their better efficiency and environmentally friendly nature. Proton exchange/Polymer electrolyte membrane (PEM) plays a vital role in the PEMFCs. Currently, PEM like Nafion and Flemions are widely used in PEMFC, which have certain drawbacks such as fuel cross-over through the membrane, low operating temperature, and high cost. The researchers from several laboratories across the globe have put their extreme effort into preparing a novel polymer electrolyte membrane with high proton conductivity, better long-term stability, improved thermal stability, high peak power density (PPD), and less fuel crossover with minimum cost. The advent of nanotechnology has brought a new scope to this research area. The hybrid (organic polymer with inorganic nanoparticle) nanocomposite membrane has developed into an exciting alternative to the conventional polymer membrane applications. It provides an exclusive blend of inorganic and organic properties and helps to overcome the drawbacks of align="center"pristine polymer membranes. In this book chapter, we have focused on different nanomaterials and their effect is analyzed in polymer electrolyte nanocomposite membranes for PEMFC applications.

* **Corresponding author Ratikanta Nayak:** Orson Resins & Coatings Pvt Ltd. Mumbai,400101, India; E-mail: ratikantapolymer@gmail.com

Keywords: Nanocomposite, Proton conductivity, Acid functional group, PEMFC, Water absorption, Thermal Stability, Fuel cell Performance, Ionic medium, Inorganic particle, Oxidative stability, Functionalization.

INTRODUCTION

The demand for electric power is rapidly increasing throughout the world. The traditional way of power production generally depends upon centralized fossil fuel plants, which are now less preferred for a clean and green mode of power generation technologies. Emissions from coal and other fossil fuel-powered plants release huge amounts of carbon dioxide (CO_2), greenhouse gas (GHG), and pollutants that include nitrogen and sulfur oxides. Furthermore, all forms of unified power need a grid of high-voltage transmission lines to transfer energy to the consumer's home. These transmission lines create challenges to the infrastructure for the services, known for their perceived health threats and the enormous loss of about 20% of energy depending on the distance from the source to the consumer. Sunny days are essential for getting better solar energy, but the situation becomes pathetic during long days of cloudy skies. Another option of wind energy exists which however also depends on the factors that cannot be controlled. Similarly, a suitable speed of water flow is needed for producing hydroelectric power. Thus, it is difficult to think about the renewable energy sources without consistent sun, wind, and water flow. These sources are not align="center"under the control of human beings and may not be counted as suitable align="center"or reliable sources. On the contrary, fuel cell technology has advanced to a stage where it can be considered as the probable challenger to the traditional combustion-based engine and can produce a feasible means of generating clean and green power according to demand.

A fuel cell is like a factory, which produces electrical energy as output as long as input fuel is there. There is a different classification of fuel cells based on the electrolyte used in the cells. The five major types are:

a. Phosphoric acid (H_3PO_4) fuel cell (PAFC)
b. Polymer electrolyte (or Proton-exchange) membrane fuel cells (PEMFCs).
c. Alkaline fuel cell (AFC)
d. Molten carbonate fuel cell (MCFC)
e. Solid oxide fuel cell (SOFC)

PEMFC uses polymer electrolyte membrane as an electrolyte for proton-exchange purposes and is also called a proton exchange membrane fuel cell. However, other fuel cells use different types of electrolytes for proton transfer from anode to the cathode side of the PEM fuel cell.

POLYMER ELECTROLYTE MEMBRANE FUEL CELL

PEMFC is considered the field of interest in this chapter. PEMFC is nothing but an electrochemical device that produces electricity and water from hydrogen (H_2) and oxygen (O_2) fuels. The H_2 gas is oxidized to electron and proton (H^+) in the presence of the catalyst, whereas the electron passes through the external load and proton passes through the polymer electrolyte.

The electrochemical half reactions are shown as follows:

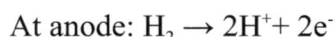

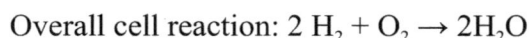

At anode: $H_2 \rightarrow 2H^+ + 2e^-$

At cathode: $1/2\ O_2 + 2H^+ + 2e^- \rightarrow H_2O$

Overall cell reaction: $2\ H_2 + O_2 \rightarrow 2H_2O$

Advantages of Fuel Cell

  i. It is more efficient than a combustion engine, as it produces direct electrical energy from chemical energy.
 ii. All parts of the cell are in static mode.
iii. Lack of moving align="center"parts implies that Fuel Cells are noiseless, highly reliable and long-lasting systems.
 iv. No emission of an undesirable product such as NO_2, SO_2 and other greenhouse gases.
  v. Battery scales poorly at large size, but fuel cells range from 1watt range (mobile phone) to power plant (1 Mw).
 vi. It offers higher power efficiencies compared to traditional batteries, rapidly recharges by refueling, whereas batteries need to be plugged in for recharge or need to be thrown away or appropriately disposed align="center' of.

Disadvantages of Fuel Cell

1. Cost is the major barrier.

2. Fuel availability and storage is a bigger problem.

3. Alternative fuels such as methanol (MeOH), formic acid (HCOOH) and gasoline are difficult to use directly.

4. Operational temperature compatibility.

5. The durability hampers during start-stop condition.

PEMFC is a promising technology for the 21st century, but there are some drawbacks that limit the application of fuel cells [1, 2], and that limitation can be solved by raising the operating temperature above 100°C. The PEMFC, which operates at 100-120 °C is called a high-temperature fuel Cell (HTPEMFC). In addition to high-temperature fuel cells, there is another type of fuel cell called Direct methanol fuel cell (DMFC). There are several components like Anode, Cathode, Gas diffusion electrode, Catalyst layer, Flow field, Proton exchange membrane (PEM) present in PEMFC, which are shown in Fig. (1). Though every component has its own importance, PEM is still considered the main part of the PEMF⌐

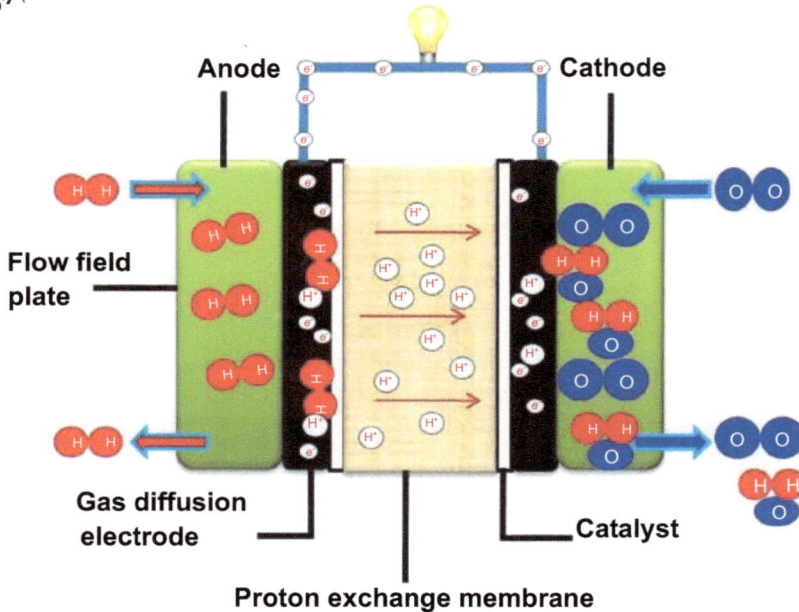

Fig. (1). Illustration of a PEM fuel cell.

POLYMER ELECTROLYTE (OR PROTON-EXCHANGE) MEMBRANE FUEL CELL

The essential properties of the PEM for fuel cells are high proton-conductivity, zero electronic conductivity, robust mechanical behavior, good thermal stability, excellent hydrolytic and chemical stability and good dimensional properties [3]. Nafion is used as a most successful commercial membrane in various applications, including fuel align="center"cells. Despite their many advantages, these membranes have some shortcomings like high cost, and less performance. Due to dehydration above 80°C, fuel permeability across the membrane decreases. Several studies have been conducted for alternative PEMs, which can work at high temperatures, and have overcome the fuel cell performance issues.

There are two ways to approach the solution, the first one is preparing thermally stable polymer, and the second one is to make an inorganic-organic nanocomposite membrane.

Thermally Stable Polymer

align="center"Developing membrane from a highly thermally stable polymer like PBI, ABPBI, SPEEK, SPSU, *etc.*, may overcome a few limitations [4 - 11]. The polymer electrolyte membrane mostly contains two parts. The first part includes the backbone and the side chain. The second part is the proton carrier; it may be water (H_2O) or an ionic medium such as an inorganic mineral aqueous H_3PO_4 or any ionic liquid. Membrane material needs to absorb the optimum amount of this medium; the excess absorbance of the medium may weaken the membrane mechanical strength; too small also results in less conductivity. The optimized proton conducting ionic medium absorbance should be defined for each medium and material operating from 100-120°C. For enhancing proton carrier uptake, the number of polar groups (acidic or base) on the polymer main chain must be maximized. The chemical structure of some polymers used for polymer electrolyte membrane materials is depicted in Table **1**.

Table 1. Structure of most common PEM polymers.

Name	Structure	References
Nafion		[12]
PBI		[13]
SPEEK		[14]
SPSU		[15]

The commonly used membranes are classified into the following categories:

Aromatic hydrocarbons (Ar-H) represent a huge number of polymer materials that are cheaper and abundantly available in the market. C-H bonds of the arene ring present in Ar-H possess higher characteristic bond energy (434 kJ.mol⁻1) than that of aliphatic hydrocarbons (351 kJ.mol⁻¹). Sulfonated Poly Ether Ether Ketone (SPEEK), Sulfonated Polysulfone (SPS), Sulfonated polyamide (SPA) sulfonated poly sulfide ketone (SPSK), *etc.* give good conductivity at 60-70°C but it sharply decreases with increasing temperature [16]. Interestingly, the nanocomposite shows a remarkable result with different inorganic fillers, which is discussed below.

NANOCOMPOSITE MEMBRANE

It is observed that the nano composite membrane exhibits good proton conduction, is highly thermally stable and offers better oxidative stabilities. The shape, size, dispersion and functionalization of nanoparticles also affect the final properties of the composite membrane [17]. The objective of this chapter is to highlight different types of nanofillers and their effects on the cell performance briefly illustrated in Table **2**.

Table 2. Different types of Nanomaterials and their effects on Fuel Cell application.

Type	PEM Example	References
Hygroscopic Oxides	TiO_2 filled Nafion upsurges water uptake, H^+ conductivity, and dimensional stability	[23, 24]
Clay	Sulfonated polysulfone /PTFE membrane loaded with montmorillonite clay showed better H_2O retention ability and mechanical durability	[25]
Zeolites	PVDF loaded with ETS-10 (SiO_4 tetrahedral and TiO_6 Octahedron) has increased proton-conductivity up to 150°C	[26]
Inorganic mineral Acid	Both PFSA and polybenzimidazole are successfully loaded with mineral H_3PO_4 increased proton conductivity under anhydrous condition.	[27]
Heteropoly acids (HPAs)	SPEEK with ZrO_2 and phosphomolybdic acid ($H_3Mo_{12}PO_{40} \cdot 12H_2O$) exhibit enhanced conductivity and improved thermal properties	[28]
Zirconium Phosphates	PTFE, sulphonated polyether ketone and Nafion loaded with ZrP showed higher proton-conductivity and oxidative stability	[29]
Graphene Oxide (GO)	Nafion loaded with Sulfonic acid functionalized GO showed increased proton-conductivity at 110°C and 30% relative humidity	[30]
Polymeric Capsule	Nafion loaded with sulfonated polystyrene nano-particles display protonic conductivity > 100°C	[31]

Along with perfluorosulfonic acid (PFSA) polymer, sulfonated hydrocarbon polymers-based nanocomposites are extensively used as a host matrix for PEMFC applications. The presence of inorganic nanofillers enhances the dimensional-mechanical properties and controls water management, further suppressing the fuel crossover by increasing the ion-transport highway.

The fabrication method of nanocomposite films containing inorganic proton conducting particles is significant as it will impact the membrane microstructure. The intrinsic characteristics of the particles such as shape, surface functionality, surface area, acidity, and their inter-electrostatic interactions with the base polymer microstructure are predominantly important and can bring huge variations in polymer membrane performance [18]. However, nano filler brought drastic changes in several properties [19]. The fillers can be classified into two main types (1) Solid non-porous filler such as aluminium oxide (Al_2O_3), titanium dioxide (TiO_2) and silicon dioxide (SiO_2) nanoparticles, and (2) Solid porous filler, such as Clay, porous metal oxides, zeolites, metal-organic frameworks (MOFs), carbon nanotubes (CNT) and graphene with organic polymer (Nafion, SPEEK, PVDF, SPSF *etc.*) shows better conductivity at high temperature. SPEEK with HPA gives proton conductivity of 5×10^{-1} S cm^{-1} > 100°C [20], PVDF with CsHSO4 gives conductivity of more than 0.01 Scm^{-1} >150 °C [21] and Nafion with ZrP shows 0.1 S cm^{-1} at 100 °C [22].

The nanocomposite membranes based on different fillers align="center"are explained in the following sections.

Graphene Oxide Nanocomposite

The graphene oxide is a two-dimensional layered structure, and the inter-planar distance can be increased to 1.2 nm in a saturated state. The hydrophilic oxygen-containing functional groups such as ether ($-O-$), hydroxy ($-OH$), and carboxylic($-COOH$) attract protons and pass-through hydrogen-bonded network from one end to other. The proton conductivity of Graphene oxide is high, and nearto 10^{-2} S cm^{-1}. Therefore, this two-dimensional proton conducting material has huge potential for proton conducting application in fuel cell. In addition to higher proton conductivity, it has large surface area, high mechanical properties, better chemical stability, and a great barrier for fuel cross-over, making graphene oxide as the best suitable filler for PEMFC [32 - 35]. Synthesis of sulfonated-graphene oxide (SGO) /SPEEK composite film and the effects of SGO on the properties of SPEEK such as conductivity and PEMFC performance were observed earlier by Kumar *et al.* (2014) [35]. The composite membranes exhibited a good proton conductivity (0.055 S cm^{-1} at 80 °C and 30% relative humidity (RH) and reasonable PEMFC performance (378 mW cm^{-2} at 80 °C and 30% RH), which were higher than that of re-formed SPEEK.

CNT Nanocomposite

Among inorganic fillers, cylindrically rolled graphene layered carbon nanotube (CNT) with its diameter in nano dimensions has caught the imagination of polymer composite community due to its high aspect ratio, low densities, extraordinary mechanical properties, hence CNT has been extensively explored as nanofiller for PEM application. Moreover, the performance and properties can be drastically increased by surface modification. Different functionalization like sulphonation [36 - 39], phosphonation [40], benzimidazole functionalized CNT, silica coated CNT, carboxylic functionalized CNT [41] have been investigated in the past. The surface of CNT was also modified with benzimidazole group and further inserted with Nafion by Asgari *et al.* (2013) [42]. They observed substantial improvement in the proton conductivity, power density, and reduced methanol permeability.

SiO$_2$ Silicate-based Nanocomposite

Several fillers have been extensively studied and nano silica found to be more suitable in terms of cost, electrical insulation properties, and improved water retention capacity than others. The surface modification of the silica nanoparticle by organic modifier improves the compatibility of modified filler with organic polymer matrix. The proton-conductivity of the silica-based composite is increased drastically owing to the hydrophilic nature of the nano silica. Further, the sulfonated and phosphonated modified silica composite enhanced the proton conductivity drastically along the well dispersed nanoparticle across the material. The fuel cell performance of Sulphonated polyamide comprising triazole group with silica nanoparticle-based membrane at 80°C is unusually enhanced due to the presence of silica nanoparticle (Sakamoto *et al.*, 2014) [43]. Sulphonated-graphene oxide-silica combination was used as a filler in Nafion matrix by Feng *et al.* (2014) for making membrane through solution casting method. This polymer electrolyte membrane-based fuel cell performed impressive proton transport properties, and the composite membrane also inhibited the methanol permeability compared with the recast Nafion membrane.

Titanium Dioxide (TiO$_2$) based Nanocomposite

The titanium dioxide is a hygroscopic material and even retains water at high temperature to provide unique thermal stability and water management system [44]. The functionalization of Titanium dioxide with phosphonic acid enhanced the compatibility of the inorganic filler with the polymer matrix. SPEEK with phosphonic acid modified with Titania nanocomposite membrane is synthesized through an in-situ method [45]. The proton conductivity of nanohybrid membrane-based fuel cell promoted 25% higher proton-conductivity and a 23%

decrease in MeOH permeability with better thermal and mechanical stabilities. Titanium nanotube surface was modified by amine-based compound for enhancing its compatibility with Nafion by Wu *et al.* (2014) [46]. The modified Titanium dioxide-based composite offered three times more conductivity than Nafion-unmodified titane nanotube composite and 4-5 times higher conductivity than Nafion.

Nanocomposite PEMs with Perovskite-type Oxides Protonic Conductors

Perovskite structured nano filler is considered as a suitable nanofiller for polymer electrolyte membrane due to its excellent chemical stability, better mechanical stability, better mechanical and thermal stability and cost effectiveness. Shabanikia *et al.* (2015) [47] and Hooshyari *et al.* (2015) [48] conducted the experiment with the Polybenzimidazole with perovskite nano composite membrane for high temperature fuel cell application. They prepared PBI nano composite membrane with $BaZrO_3$ & $SrCeO_3$ nano filler and observed greater water uptake, and higher conductivity as compared with pristine PBI membrane.

Nanocomposite PEMs with Zeolite

The presence of charged anionic framework and high pore volume produce extra surface area, contributing to its hydrophilic nature and highwater absorption capacity. To maintain higher proton conductivity, acid functionalized Zeolite has been made (Jones *et al.*, 1998) [49]. This acid functionalized zeolite is used due to its high proton-conductivity (0.02 S cm^{-1}), good hydrophilic, better acid stability and it was successfully used by (Holmberg *et al.*, 2005) [50]. Nafion was modified with Zeolite and the membrane was casted through solvent casting method; the resulting composite membrane demonstrated high proton conductivity with better water retention capacity [51].

Nanocomposite PEMs with Cellulose Whiskers (CWs)

Cellulose whisker (CW) is a biodegradable material that can be used as a filler in polymer electrolyte membrane due to its unique properties like high aspect ratio, high dispersibility in the polymer matrix and excellent capacity of absorption and retention of water. Also, it can reduce methanol crossover and increase proton conductivity. It was observed that 5 wt.% one dimensional cellulose whiskers (CWs) improve the proton conductivity at a temperature of more than 100°C. The reinforcement of CW highly suppressed the methanol permeability [52]. The formation of long-range order proton-conduction networks in the neighborhood of 1D cellulose nano-framework helps for better conductivity [53].

CONCLUDING REMARKS

There are lots of drawbacks observed in a conventional membrane-like Nafion, including the fact that this membrane cannot sustain at high temperatures. The inorganic filler has a high thermal stability and if the surface is organically modified, it could be compatible with the organic polymer. The inorganic fillers like oxides such as GO, MOFs, clays, CNTs, zeolites, ZrP, silane-based fillers and PTA are hygroscopic in nature. The inorganic filler component gives high mechanical stability conductivity; they inhibit fuel cross over and the organic backbone gives flexibility align="center"to organic-inorganic hybrid nanocomposite membrane. The final properties of the nanocomposite membrane can be manipulated in accordance with the concentration, their interaction, sizes and shape of the nanoparticle used. There are still lots of challenges during processing, like poor dispersion of nanoparticles in its respective solvent and polymer matrix, high chance of agglomeration of nanoparticles, weak interface or interaction between polymer and nanomaterials, difficulties in controlling the alignment of the one-dimensional nanotubes and less awareness about the nano waste management, *etc*. It may be concluded that nanocomposite membrane has huge potential to achieve desired properties for fuel cell application if several innovative approaches are additionally considered to avoid the challenges during processing.

CONSENT FOR PUBLICATION

Not applicable.

CONFLICT OF INTERESTS

The authors declare no conflict of interest, financial or otherwise.

ACKNOWLEDGEMENTS

All the supports provided by Mr. Narayan Goenka, Managing Director, Orson Resins & Coatings Pvt. Ltd, Mumbai for the completion of this project are highly acknowledged.

REFERENCES

[1]   Chandan, A.; Hattenberger, M.; El-Kharouf, A.; Du, S.; Dhir, A.; Self, V.; Pollet, B.G.; Ingram, A.; Bujalski, W. High temperature (HT) polymer electrolyte membrane fuel cells (PEMFC)- a review. *J. Power Sources,* **2013**, *231*, 264-278.
        [http://dx.doi.org/10.1016/j.jpowsour.2012.11.126]

[2]   Authayanun, S.; Im-orb, K.; Arpornwichanop, A. A review of the development of high temperature proton exchange membrane fuel cells. *Chin. J. Catal.,* **2015**, *36*(4), 473-483.
        [http://dx.doi.org/10.1016/S1872-2067(14)60272-2]

[3]   Kraytsberg, A.; Ein-Eli, Y. Review of advanced materials for proton exchange membrane fuel cells. *Energy Fuels*, **2014**, *28*(12), 7303-7330.
        [http://dx.doi.org/10.1021/ef501977k]

[4]   Zhengping, Z.; Oksana, Z.; Xiang-Fa, Wu.; Ted, A. Jivan, T.; John, H. *Energies*, **2021**, *14*(135), 1-27.

[5]   Gil, M.; Ji, X.; Li, X.; Na, H.; Hampsey, J.E.; Lu, Y. Direct synthesis of sulfonated aromatic poly (ether ether ketone) proton exchange membranes for fuel cell applications. *J. Membr. Sci.*, **2008**, *234*(1-2), 75-81.
        [http://dx.doi.org/10.1016/j.memsci.2003.12.021]

[6]   Ahmad, H.; Kamarudin, S.K.; Hasran, U.A.; Daud, W.R.W. Overview of hybrid membranes for direct-methanol fuel-cell applications. *Int. J. Hydrogen Energy*, **2010**, *35*(5), 2160-2175.
        [http://dx.doi.org/10.1016/j.ijhydene.2009.12.054]

[7]   Nayak, R.; Sundarraman, M.; Ghosh, P.C.; Bhattacharyya, A.R. Doped Poly (2,5-benzimidazole) membranes for high temperature polymer electrolyte fuelcell: Influence of various solvents during membrane casting on the fuel cell performance. *Eur. Polym. J.*, **2018**, *100*, 111-120.
        [http://dx.doi.org/10.1016/j.eurpolymj.2017.08.026]

[8]   Nayak, R.; Dey, T.; Ghosh, P.C.; Bhattacharyya, A.R. Phosphoric acid doped Poly (2,5-benzimidazole) based proton exchange membrane for high temp fuelcell application. *Polym. Eng. Sci.*, **2016**, *56*(12), 1366-1374.
        [http://dx.doi.org/10.1002/pen.24370]

[9]   Harilal, N.; Nayak, R.; Ghosh, P.C.; Jana, T. R.; Ghosh, P. C.; Jana, T. Crosslinked polybenzimidazole membrane for PEM Fuel Cells. *ACS Appl. Polym. Mater.*, **2020**, *2*(8), 3161-3170.
        [http://dx.doi.org/10.1021/acsapm.0c00350]

[10]   Nayak, R.; Ghosh, P. C. Mechanical and Impedance analysis of Poly (2, 5) benzimidazole proton exchange membrane for high temperature fuel cell Application; Materials Today Proceedings, 2018, 5, Issue-5, Part-5,13767-13775

[11]   Bhattacharyya, R.; Nayak, R.; Ghosh, P. C. HYPERLINK "javascript:void(0)" Study of ABPBI membrane as an alternative separator for vanadium redox flow batteries. *Energy Storage*, **2019**, *2-108*, 1-4.

[12]   Banerjee, S.; Curtin, D.E. Nafion® perfluorinated membranes in fuel cells. *J. Fluor. Chem.*, **2004**, *125*(8), 1211-1216.
        [http://dx.doi.org/10.1016/j.jfluchem.2004.05.018]

[13]   Asensio, J.A.; Sánchez, E.M.; Gómez-Romero, P. Proton-conducting membranes based on benzimidazole polymers for high-temperature PEM fuel cells. A chemical quest. *Chem. Soc. Rev.*, **2010**, *39*(8), 3210-3239.
        [http://dx.doi.org/10.1039/b922650h] [PMID: 20577662]

[14]   Iojoiu, C.; Chabert, F.; Maréchal, M.; Kissi, N.E.; Guindet, J.; Sanchez, J.Y. From polymer chemistry to membrane elaboration: A global approach of fuel cell polymeric electrolytes. *J. Power Sources*, **2006**, *153*(2), 198-209.
        [http://dx.doi.org/10.1016/j.jpowsour.2005.05.039]

[15]   Parvole, J.; Jannasch, P. Polysulfones Grafted with Poly(vinylphosphonic acid) for Highly Proton Conducting Fuel Cell Membranes in the Hydrated and Nominally Dry State. Macromolecules, 2008, 41-11, 3893-3903.

[16]   Rikukawa, M.; Sanui, K. Proton-conducting polymer electrolyte membranes based on hydrocarbon polymers. *Prog. Polym. Sci.*, **2000**, *25*(10), 1463-1502.
        [http://dx.doi.org/10.1016/S0079-6700(00)00032-0]

[17]   Di, Z.; Xie, Q.; Li, H.; Mao, D.; Li, M.; Zhou, D.; Li, L. Novel composite proton-exchange membrane based on proton conductive glass powders and sulfonated poly (ether ether ketone). *J. Power Sources*, **2015**, *273*, 688-696.

[http://dx.doi.org/10.1016/j.jpowsour.2014.09.122]

[18]   Laberty-Robert, C.; Vallé, K.; Pereira, F.; Sanchez, C. Design and properties of functional hybrid organic-inorganic membranes for fuel cells. *Chem. Soc. Rev.,* **2011**, *40*(2), 961-1005.
[http://dx.doi.org/10.1039/c0cs00144a] [PMID: 21218233]

[19]   Li, Y.; He, G.; Wang, S.; Yu, S.; Pan, F.; Wu, H.; Jiang, Z. Recent advances in the fabrication of advanced composite membranes. *J. Mater. Chem. A Mater. Energy Sustain.,* **2013**, *1*(35), 10058-10077.
[http://dx.doi.org/10.1039/c3ta01652h]

[20]   Li, Q.; He, R.; Gao, J.; Jensen, J.O.; Bjerrum, N. The CO Poisoning Effect in PEMFCs Operational at Temperatures up to 200°C, J. *J. Electrochem. Soc.,* **2003**, *150*(12), A1599-A1605.
[http://dx.doi.org/10.1149/1.1619984]

[21]   Olivia, B.; Huaneng, S.; Vladimir, L.; Bruno, G.; Sivakumar, P. CsHSO$_4$ as proton conductor for high-temperature polymer electrolyte membrane fuel cells. *J. Appl. Electrochem.,* **2014**, *44*(9), 1037-1045.
[http://dx.doi.org/10.1007/s10800-014-0715-x]

[22]   Costamagna, P.; Yang, C.; Bocarsly, A.B.; Srinivasan, S. Nafion® 115/zirconium phosphate composite membranes for operation of PEMFCs above 100 °C. *Electrochim. Acta,* **2002**, *47*(7), 1023-1033.
[http://dx.doi.org/10.1016/S0013-4686(01)00829-5]

[23]   Ye, G.; Li, K.; Xiao, C.; Chen, W.; Zhang, H.; Pan, M. Nafion®—titania nanocomposite proton exchange membranes. *J. Appl. Polym. Sci.,* **2011**, *120*(2), 1186-1192.
[http://dx.doi.org/10.1002/app.33031]

[24]   Zhengbang, W.; Tang, H.; Mu, P. Self-assembly of durable Nafion/TiO$_2$ nanowire electrolyte membranes for elevated-temperature PEM fuel cells. *J. Membr. Sci.,* **2011**, *369*(1-2), 250-257.
[http://dx.doi.org/10.1016/j.memsci.2010.11.070]

[25]   Xing, D.; He, G.; Hou, Z.; Ming, P.; Song, S. Preparation and characterization of a modified montmorillonite/sulfonated polyphenylether sulfone/PTFE composite membrane. *Int. J. Hydrogen Energy,* **2011**, *36*(3), 2177-2183.
[http://dx.doi.org/10.1016/j.ijhydene.2010.11.022]

[26]   Sancho, T.; Soler, J.; Pina, M.P. Conductivity in zeolite–polymer composite membranes for PEMFCs. *J. Power Sources,* **2007**, *169*(1), 92-97.
[http://dx.doi.org/10.1016/j.jpowsour.2007.01.079]

[27]   Dupuis, A.C. Proton exchange membranes for fuel cells operated at medium temperatures: Materials and experimental techniques. *Prog. Mater. Sci.,* **2011**, *56*(3), 289-327.
[http://dx.doi.org/10.1016/j.pmatsci.2010.11.001]

[28]   Luu, D.X.; Kim, D. sPEEK/ZPMA composite proton exchange membrane for fuel cell application. *J. Membr. Sci.,* **2011**, *371*(1-2), 248-253.
[http://dx.doi.org/10.1016/j.memsci.2011.01.046]

[29]   Kozawa, Y.; Suzuki, S.; Miyayama, M.; Okumiya, T.; Traversa, E. Proton conducting membranes composed of sulfonated poly(etheretherketone) and zirconium phosphate nanosheets for fuel cell applications. *Solid State Ion.,* **2010**, *181*(5-7), 348-353.
[http://dx.doi.org/10.1016/j.ssi.2009.12.017]

[30]   Zarrin, H.; Higgins, D.; Jun, Y.; Chen, Z.; Fowler, M. Functionalized graphene oxide nanocomposite membrane for low humidity and high temperature proton exchange membrane fuel cells. *J. Phys. Chem. C,* **2011**, *115*(42), 20774-20781.
[http://dx.doi.org/10.1021/jp204610j]

[31]   Pu, H.; Wang, D.; Yang, Z. Towards high water retention of proton exchange membranes at elevated temperature via hollow nanospheres. *J. Membr. Sci.,* **2010**, *360*(1-2), 123-129.
[http://dx.doi.org/10.1016/j.memsci.2010.05.012]

[32] Chien, H.C.; Tsai, L.D.; Huang, C.P.; Kang, C.Y.; Lin, J.N.; Chang, F.C. Sulfonated graphene oxide/Nafion composite membranes for high-performance direct methanol fuel cells. *Int. J. Hydrogen Energy,* **2013**, *38*(31), 13792-13801.
[http://dx.doi.org/10.1016/j.ijhydene.2013.08.036]

[33] Bayer, T.; Bishop, S.R.; Nishihara, M.; Sasaki, K.; Lyth, S.M. Characterization of a graphene oxide membrane fuel cell. *J. Power Sources,* **2014**, *272*, 239-247.
[http://dx.doi.org/10.1016/j.jpowsour.2014.08.071]

[34] Liu, K.L.; Lee, H.C.; Wang, B.Y.; Lue, S.J.; Lu, C.Y.; Tsai, L.D.; Fang, J.; Chao, C.Y. Sulfonated poly (styrene-block-(ethylene-ranbutylene)-block-styrene (SSEBS)-zirconium phosphate (ZrP) composite membranes for direct methanol fuel cells. *J. Membr. Sci.,* **2015**, *495*, 110-120.
[http://dx.doi.org/10.1016/j.memsci.2015.08.017]

[35] Kumar, R.; Mamlouk, M.; Scott, K. Sulfonated polyether ether ketone-sulfonated graphene oxide composite membranes for polymer electrolyte fuel cells. *RSC Advances,* **2014**, *4*(2), 617-623.
[http://dx.doi.org/10.1039/C3RA42390E]

[36] Yun, S.; Im, H.; Heo, Y.; Kim, J. Im, H.; Heo, Y.; Kim, J. Crosslinked sulfonated poly (vinyl alcohol)/sulfonated multi-walled carbon nanotubes nanocomposite membranes for direct methanol fuel cells. *J. Membr. Sci.,* **2011**, *380*(1-2), 208-215.
[http://dx.doi.org/10.1016/j.memsci.2011.07.010]

[37] Zhou, W.; Xiao, J.; Chen, Y.; Zeng, R.; Xiao, S.; Nie, H.; Li, F.; Song, C. Sulfonated carbon nanotubes/sulfonated poly (ether sulfone ether ketone ketone) composites for polymer electrolyte membranes. *Polym. Adv. Technol.,* **2011**, *22*(12), 1747-1752.
[http://dx.doi.org/10.1002/pat.1666]

[38] Yu, D.M.; Sung, I.H.; Yoon, Y.J.; Kim, T.H.; Lee, J.Y.; Hong, Y.T. Properties of sulfonated poly (arylene ether sulfone)/functionalized carbon nanotube composite membrane for high temperature PEMFCs. *Fuel Cells (Weinh.),* **2013**, *13*(5), 843-850.
[http://dx.doi.org/10.1002/fuce.201200105]

[39] Kannan, R.; Aher, P.P.; Palaniselvam, T.; Kurungot, S.; Kharul, U.K.; Pillai, V.K. Artificially designed membranes using phosphonated multiwall carbon nanotube-polybenzimidazole composites for polymer electrolyte fuel cells. *J. Phys. Chem. Lett.,* **2010**, *1*(14), 2109-2113.
[http://dx.doi.org/10.1021/jz1007005]

[40] Kannan, R.; Kagalwala, H.N.; Chaudhari, H.D.; Kharul, U.K.; Kurungot, S.; Pillai, V.K. Improved performance of phosphonated carbon nanotube-polybenzimidazole composite membranes in proton exchange membrane fuel cells. *J. Mater. Chem.,* **2011**, *21*(20), 7223-7231.
[http://dx.doi.org/10.1039/c0jm04265j]

[41] Thomassin, J.M.; Kollar, J.; Caldarella, G.; Germain, A.; Jérôme, R.; Detrembleur, C. Beneficial effect of carbon nanotubes on the performances of Nafion membranes in fuel cell applications. *J. Membr. Sci.,* **2007**, *303*(1-2), 252-257.
[http://dx.doi.org/10.1016/j.memsci.2007.07.019]

[42] Asgari, M.S.; Nikazar, M.; Molla-Abbasi, P.; Hasani-Sadrabadi, M.M. Nafion® /histidine functionalized carbon nanotube: High-performance fuel cell membranes. *Int. J. Hydrogen Energy,* **2013**, *38*(14), 5894-5902.
[http://dx.doi.org/10.1016/j.ijhydene.2013.03.010]

[43] Sakamoto, M.; Nohara, S.; Miyatake, K.; Uchida, M.; Watanabe, M.; Uchida, H. Effects of incorporation of SiO_2 nanoparticles into sulfonated polyimide electrolyte membranes on fuel cell performance under low humidity conditions. *Electrochim. Acta,* **2014**, *137*, 213-218.
[http://dx.doi.org/10.1016/j.electacta.2014.05.159]

[44] Bose, S.; Kuila, T.; Nguyen, T.X.H.; Kim, N.H.; Lau, K.T.; Lee, J.H. Polymer membranes for high temperature proton exchange membrane fuel cell: recent advances and challenges. *Prog. Polym. Sci.,* **2011**, *36*(6), 813-843.

[http://dx.doi.org/10.1016/j.progpolymsci.2011.01.003]

[45]   Wu, H.; Cao, Y.; Li, Z.; He, G.; Jiang, Z. Novel sulfonated poly (ether ether ketone)/phosphonic acid-functionalized titania nanohybrid membrane by an in situ method for direct methanol fuel cells. *J. Power Sources,* **2015**, *273*, 544-553.
[http://dx.doi.org/10.1016/j.jpowsour.2014.09.134]

[46]   Li, Z.; He, G.; Zhao, Y.; Cao, Y.; Wu, H.; Li, Y.; Jiang, Z. Enhanced proton conductivity of proton exchange membranes by incorporating sulfonated metal-organic frameworks. *J. Power Sources,* **2014**, *262*, 372-379.
[http://dx.doi.org/10.1016/j.jpowsour.2014.03.123]

[47]   Shabanikia, A.; Javanbakht, M.; Amoli, H.S.; Hooshyari, K.; Enhessari, M. Polybenzimidazole/strontium cerate nanocomposites with enhanced proton conductivity for proton exchange membrane fuel cells operating at high temperature. *Electrochim. Acta,* **2015**, *154*, 370-378.
[http://dx.doi.org/10.1016/j.electacta.2014.12.025]

[48]   Hooshyari, K.; Javanbakht, M.; Shabanikia, A.; Enhessari, M. Fabrication $BaZrO_3$/PBI-based nanocomposite as a new proton conducting membrane for high temperature proton exchange membrane fuel cells. *J. Power Sources,* **2015**, *276*, 62-72.
[http://dx.doi.org/10.1016/j.jpowsour.2014.11.083]

[49]   Jones, C.W.; Tsuji, K.; Davis, M.E. Organic-functionalized molecular sieves as shape-selective catalysts. *Nature,* **1998**, *393*(6680), 52-54.
[http://dx.doi.org/10.1038/29959]

[50]   Holmberg, B.A.; Hwang, S.J.; Davis, M.E.; Yan, Y. Synthesis and proton conductivity of sulfonic acid functionalized zeolite BEA nanocrystals. *Microporous Mesoporous Mater.,* **2005**, *80*(1-3), 347-356.
[http://dx.doi.org/10.1016/j.micromeso.2005.01.010]

[51]   Devrim, Y.; Albostan, A. Enhancement of PEM fuel cell performance at higher temperatures and lower humidities by high performance membrane electrode assembly based on Nafion/zeolite membrane. *Int. J. Hydrogen Energy,* **2015**, *40*(44), 15328-15335.
[http://dx.doi.org/10.1016/j.ijhydene.2015.02.078]

[52]   Mohammad, M.; Hasani, S.; Erfan, D.; Rasool, N.; Akbar, K.; Fatemeh, S.; Nassir, M.; Philippe, R.; Karl, J. Cellulose nanowhiskers to regulate the microstructure of perfluorosulfonate ionomers for high-performance fuel cells. *J. Mater. Chem.,,* **2014**, *2*, 11334-11340.

[53]   Hasani-Sadrabadi, M.M.; Dashtimoghadam, E.; Nasseri, R.; Karkhaneh, A.; Majedi, F.S.; Mokarram, N.; Renaud, P.; Jacob, K.I. Cellulose nanowhiskers to regulate the microstructure of perfluorosulfonate ionomers for high-performance fuel cells. *J. Mater. Chem. A Mater. Energy Sustain.,* **2014**, *2*(29), 11334-11340.
[http://dx.doi.org/10.1039/c4ta00635f]

CHAPTER 11

Graphene-based Nanocomposites for Electro-optic Devices

Monojit Bag[1,*], Jitendra Kumar[1] and Ramesh Kumar[1]

[1] *Advanced Research in Electrochemical Impedance Spectroscopy, Indian Institute of Technology Roorkee, Roorkee 247667, India*

Abstract: Graphene, the most exciting carbon allotrope, and its derivatives such as graphene oxide and graphene quantum dots have sparked a flurry of research and innovation owing to their unprecedented optoelectronic properties. Graphene and its nanocomposites have been widely used in a variety of opto-electronic devices such as photodetectors, transistors, actuators, biomedical aids, and membranes. Their sp^2 hybridization state provides some extraordinary opto-electronic and mechanical properties. Chemical exfoliation of graphite into graphene and graphene oxide allows us to mix graphene nanocomposites into various layers of organic solar cells and other organic semiconductor-based optoelectronic devices, especially for roll-to-roll fabrication of large-area devices at a lower cost. Recently, these nanocomposites have also been utilized as charge transport layers and surface modifiers in perovskite solar cells and perovskite light-emitting diodes. Researchers have found that the presence of graphene, even at very low loading, can significantly improve the device's performance. In this chapter, we have discussed the application of graphene oxide, reduced graphene oxide, and doped graphene oxide in various combinations in perovskite solar cells and perovskite light-emitting diodes; these nanomaterials can be utilized either in transport layers of a multilayered device or directly incorporated in the active layers of these optoelectronic devices. These nanocomposites generally improve the device efficiencies by improving the band alignment at heterojunctions in a multilayered device by substantially reducing the trap states and the charge transfer resistance. These nanocomposites are found to achieve significantly improved device power conversion efficiency and stability of perovskite-based optoelectronic devices.

Keywords: Band alignment, Graphene-oxide, Hybridization, Low cost, Nanocomposites, Opto-electronic, Perovskite optoelectronic devices, Power-conversion efficiency, Stability, Trap-states.

---

* **Corresponding author Monojit Bag:** Advanced Research in Electrochemical Impedance Spectroscopy, Indian Institute of Technology Roorkee, Roorkee 247667, India; E-mail: monojit.bag@ph.iitr.ac.in

Gaurav Manik and Sushanta Kumar Sahoo (Eds.)

INTRODUCTION

Graphene has emerged as one of the revolutionary materials and attracted tremendous attention from all the material scientists and device engineers around the world. The world's first 2D material, graphene, was first isolated and identified by Andre Geim and Kostya Novoselov in 2004. Graphene in its purest form has some exceptional opto-electronic and mechanical properties, which makes graphene a shining star among the list of other functional materials. Due to its extremely high mechanical strength, graphene has been utilized to functionalize a variety of materials to improve their mechanical strength. The sp^2 hybridization of the carbon atoms in graphene is arranged in a honeycomb structure attached by σ and π bonds with an interatomic distance of around 0.142 nm. Graphene has extremely high charge carrier mobility in the range of 15000 – 20000 cm^2/V.s at room temperature (RT) due to the delocalization of the p-electrons. In defect-free graphene sheets, electrons can move at an extraordinary speed as if there is no effective mass. In the pristine form of graphene, there is a zero-band gap in the material. However, introducing some defects or any functional group can open up a very small energy bandgap and therefore, in most applications, graphene nanocomposites such as graphene oxide (GO), reduced graphene oxide (rGO) or doped graphene is utilized in semiconducting industries. Apart from this, graphene is not used in its pure form because of its inadequate yield compared to graphene derivatives. It's easier to achieve a high production yield for the GO as compared to the purest form of graphene, subsequently, reduction of GO can lead us to the efficient production of graphene. Pristine GO is a wide bandgap, low conducting material; therefore, it cannot be useful in many applications. Several approaches have been utilized to reduce the graphene oxide and improve the electrical properties of GO; these methods include chemical, electrochemical reduction, and heat treatments. Hydrazine or dimethylhydrazine are usually used for the chemical reduction of GO. Hui-Lin *et al.* utilized an electrochemical reduction of the exfoliated GO at a graphite electrode to reduce GO in order to avoid the use of excessive reducing agents [1]. Apart from the reduction of GO, functionalization of graphene or GO could provide us with the desired optoelectronic properties of doped graphene. In this chapter, we will discuss the applications of graphene nanocomposites, namely GO, rGO and doped graphene quantum dots in perovskite solar cells (PeSCs) and perovskite light-emitting diodes (PeLEDs). PeSCs and PeLEDs have the same device structures, mainly differing in their operational conditions. These devices utilize a multilayered structure consisting of a hole transport material (HTM), an active layer, and an electron transport material (ETM), all sandwiched between two electrodes, as shown in Fig. (**1**). PeSCs and PeLEDs utilize an active perovskite material with the stoichiometry of ABX_3, where A is a monovalent cation, B is a divalent cation and X is a monovalent anion. The materials which maintain the above-mentioned stoichiometry (ABX$_3$) are termed halide perovskites [2]. These perovskites possess some exceptional properties, which are generally unexpected for solution-processable semiconductors. High defect tolerance [3 - 6], long carrier diffusion lengths [7 - 11], and narrow emission full-width half maxima (FWHM) [3, 12] are the characteristic features of these materials. A high-efficiency solar cell is required to maximize the photon absorption, efficient charge separation, charge transport as well as charge extraction at the electrode interfaces.

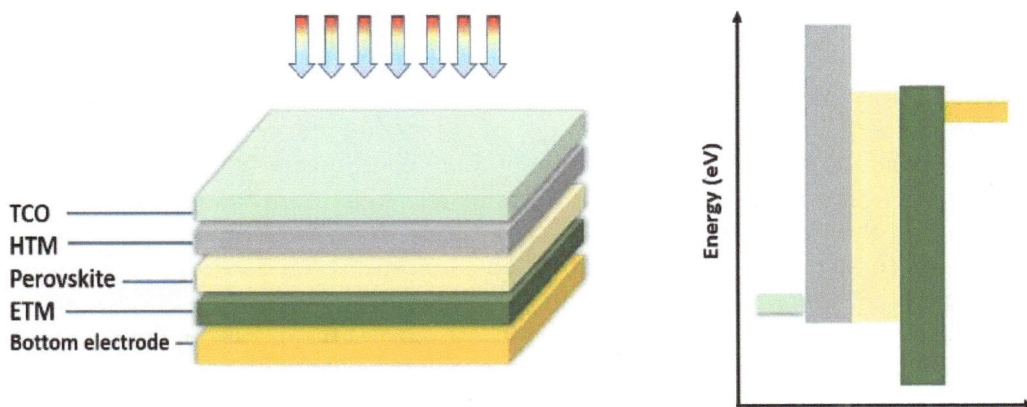

Fig. (1). Schematic diagram of **(a)** perovskite solar cell, **(b)** band diagram of a typical perovskite solar cell. (TCO- transparent conducting oxide; HTM: hole transport material; ETM: electron transport material.)

On the other hand, a light emitting diode is required to have a balanced charge injection, high radiative recombination rate and high light outcoupling efficiency. In recent years, PeLEDs have gained great attention among researchers due to high photoluminescence quantum yield (PLQY) in these materials. On the other hand, perovskite based solar cells have already achieved device efficiencies over 25%, which are comparable to the single crystal-based silicon solar cells. This extraordinary success of these materials forced the scientific community to reconsider the critical need for high temperature processable crystalline materials. This is a remarkable achievement towards large-scale production since these devices can be fabricated on a flexible substrate in a roll-to-roll manner. However, there are still many challenges to improving the operational stability of perovskites [13 - 19]. A number of graphene nanocomposites have been utilized to improve device efficiencies and stability. Large carrier diffusion lengths in perovskites can sometimes be the reason for enhanced non-radiative recombination events in perovskites, since the generated/injected charge carriers can easily diffuse towards the non-radiative sites [20]. These non-radiative sites in perovskite are generally present at the surfaces/interfaces and the grain boundaries due to the presence of uncoordinated atoms or ions at the grain boundaries. Therefore, the most important factor in deciding the performance of perovskite-based devices is to optimize the film thickness, grain size, and film roughness; these are the key parameters that must be tuned carefully by optimization. Superior crystallinity of perovskite films does not only mean achieving large crystallite size with fewer traps per unit volume, but the morphology of perovskite layers must favor the interfacial contact (must be a conformal coating) at the interface of perovskite and charge transport layer. Surface traps are the killing factors for performance in all kinds of devices. Research have demonstrated that the performance of PeSCs and PeLEDs is predominantly determined by the interfacial trap states. These trap states mostly contribute to the non-radiative recombination loss in the absorber layer or *via* minority carrier recombination at the interface [21]. A suitable doping of graphene nanocomposites into the perovskite active layers or a thin graphene nanocomposite-based interlayer can substantially reduce the density of trap states and thus improve device efficiency as well as the stability of the devices. Researchers have adopted various techniques for the passivation of defects at the grain boundaries and also at the perovskite/charge transport layer interface [21]. A thin layer of insulators such as lithium fluoride (LiF) or tri-n-octyl- phosphine (TOPO) has been utilized to successfully passivate surface defect states at the perovskite/charge transport layer interface [22]. However, defects at the grain boundaries in polycrystalline perovskite films cannot be eliminated with the surface passivation alone. The development of solution processable graphene, such as chemically exfoliated graphite into graphene oxide can allow the functionalization of graphene nanocomposites into various layers of organic or perovskite solar cells. Graphene nanocomposite such as pristine graphene quantum dots (GQDs) or doped GQDs can be successfully incorporated into the perovskite matrix itself. Interaction/coordination of GQDs with perovskite could help to reduce under coordinated atoms at the grain boundaries and reduce the density of non-radiative sites. Apart from the defect passivation graphene nanocomposites can help to improve light absorption, reduce charge transfer resistance (improve charge extraction) and improve device stability; some of these applications are discussed below.

GRAPHENE OXIDE IN PEROVSKITE-BASED DEVICES

GO is a layered material with various oxygen containing functional groups. Heterogeneous chemical and electronic structure of GO can lead to reduced electron recombination [23]. Charge transport layers in PeSCs and PeLEDs strongly influence the power conversion efficiency (PCE), stability and hysteresis in perovskite-based devices. Due to appropriate energy levels, high transparency and good electrical conductivity, poly (3,4-ethylenedioxythiophene)-polystyrene sulfonate (PEDOT: PSS) has served as an efficient HTM both in PeSCs and PeLEDs. However, its high acidic and hygroscopic properties can deteriorate perovskite-based devices, reducing operational stability in ambient conditions. To overcome these difficulties, various inorganic HTMs such as nickel oxide (NiO) and Copper(I) thiocyanate (CuSCN) have been utilized, although the device efficiency remains poor as compared to PEDOT:PSS as HTM based devices [24]. Graphene oxide in perovskite solar cells is mostly utilized to improve the electrical properties of charge transport materials. Apart from their use as an additive in the charge transport layer, graphene oxide itself can be utilized as a hole transport material in high efficiency PeSCs. Yang *et al*. utilized GO as a hole transport layer (HTL) to fabricate high-performance perovskite solar cells with enhanced stability [24]. GO based PeSCs retain > 80% of their initial PCE after 2000 hours of stability test. Optical reflectance measurements showed that GO shows lesser reflectance and improved absorbance both in UV (350 – 450 nm) and visible regions (650 – 800 nm), which results in improved short circuit current (J_{sc}) (Fig. **2**). GO can be modified by attaching it with hydrophilic edges and hydrophobic center to make them act as an amphiphilic interface modifier between perovskite and a hydrophobic charge transport layer [25]. Li *et al*. had found that due to mismatched polarity between $CH_3NH_3PbI_3$ (methylammonium lead iodide) and 2,2',7,7'-tetrakis(N,N' -di-pmethoxyphenylamine)-9-9'-spirobifluorene (Spiro-OMeTAD) containing chlorobenzene resulted in contact angels as large as 13.4 degrees, whereas after spin coating a thin layer of GO dispersed in chlorobenzene on top of perovskite, contact angle decreased to almost zero degree [25]. Using a combination of x-ray photoelectron spectroscopy (XPS), Kelvin probe force microscopy (KPFM) and electrochemical impedance spectroscopy (EIS), Li *et al*. showed that GO could interact with the perovskite by forming Pb–O bonds. GO as a buffer layer helps retard the charge recombination in solar cells and shows significantly improved open circuit voltage (V_{oc}) and fill factor (FF). There are several reports on the application of graphene oxide into perovskite solar cells; however their use in PeLED is still understudied. Recently though Zhou *et al*. has shown the effect of PEDOT:PSS:GO composite on the luminance of PeLEDs; GO doping in PEDOT:PSS leads to achieve maximum luminance of 3302 cd/m^2 and maximum current efficiency of 1.92 cd/A, these values were enhanced by 43.3% and 73.0% in comparison with the undoped device respectively. Improvements in PCE in these devices were assigned to band alignment of PEDOT:PSS with perovskite active layer due to doping of GO into the HTM. Better alignment at the highest occupied molecular orbital (HOMO) level of PEDOT:PSS-GO composite and perovskite effectively reduces the energy barrier for hole injection.

Zhou *et al*. found that upon increasing the GO ratio from 0.1 to 0.5, the work function of the GO-doped films increases from 4.92 eV to 5.02 eV, which helps improve hole extraction efficiency and PCE.

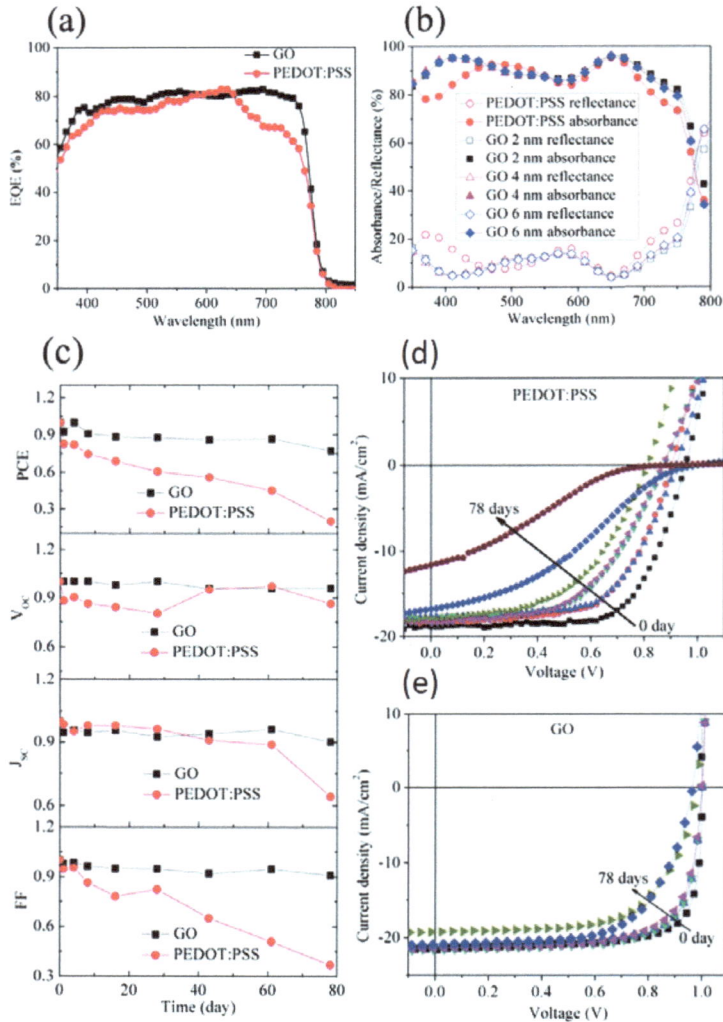

Fig. (2). (a) EQE spectrum measured as a function of wavelength, **(b)** UV-VIS absorption/reflectance spectrum show improved absorption in case of GO both in UV and Visible region, **(c)** light soaking experiments stability test of PEDOT:PSS and GO based perovskite solar cells. GO based PeSCs show much improved stability compared to PEDOT: PSS based devices. J-V characteristics of **(d)** PEDOT: PSS and **(e)** GO based PeSCs. (*Reprinted with permission from Ref. 24 (Copyright (2017) J. Mater. Chem. A*).

Reduced Graphene Oxide in Perovskite-based Devices

Graphene oxides sometimes could be unsuitable for device application due to their low conductivity, which could be significantly improved by the reduction of graphene oxide. Functionalized reduced graphene oxide has been successfully employed to reduce the charge recombination pathways, thereby improving the charge extraction properties in perovskite-based solar cells. Although where Spiro-OMeTAD is the most commonly.

Used hole transport layer in efficient PeSCs, though spiro-OMeTAD shows low intrinsic charge carrier mobility, which needs to be improved by suitable doping of some high mobility materials. We must note that the presence of hygroscopic dopants can accelerate the deterioration of the device at high temperatures; therefore hydrophobic graphene nanocomposites can be utilized to improve the carrier mobility as well as device stability. Cho *et al.* fabricated n-i-p structured PeSCs and investigated the effect of the introduction of rGO nanoflakes in each layer and compared the results with controlled device. Incorporation of rGO drastically varies device power; mixing rGO in mesoporous TiO_2 (m-TiO_2) improves device performance conversion efficiency from 18.8% to 19.5% [26]. However, when rGO was incorporated into spiro-OMeTAD layer, no significant improvements were observed. On the other hand, device efficiency was reduced by incorporating rGO directly into the perovskite matrix. Although these observations were contrary to what was already observed in the literature which the incorporation of rGO always improves the device efficiency [25, 27 - 29]. Cho *et al.* found that the detrimental effect due to rGO was observed because of the presence of disrupted crystal quality and additional shunt pathways when rGO was incorporated into perovskite or spiro-OMeTAD. Cho *et al.* demonstrated that the presence of rGO did quench the PL signal faster compared to the controlled sample, therefore indicating a faster electron transfer in the presence of rGO.

Doped Graphene in Perovskite-based Devices

Despite having good optical transmittance and charge carrier mobility in TiO_2, the requirement of high temperature (~450 °C) sintering remains a challenge for scale-up of perovskite solar cells fabrication. Therefore, scientists and engineers are looking forward to find out some alternative ETMs for low temperature processable PeSCs. Zinc oxide (ZnO) and tin oxide (SnO_2) are most widely used as an alternative to TiO_2; however, ZnO/perovskite interface is found to be unstable due to low thermal stability. ZnO based PeSCs are seen to degrade when heated above 100 °C [30]. On the other hand, electrical and optical properties of SnO_2 heavily depend on the oxidation state of Sn [31 - 33], and therefore, it is critically important to control the number of oxygen vacancies in SnO_2 in order to achieve high PCE. Oxygen vacancies in SnO_2 can increase the conductivity of SnO_2 but also at the gap trap states leading to enhanced non-radiative recombination centers [34]. Graphene quantum dots in perovskite solar cells can work as a superfast electron transport channel. Recently, Gao *et. al* have demonstrated graphene/copper indium disulfide quantum dots (graphene-CuInS2 QDs) composite based high-performance perovskite solar cell. These solar cells display excellent charge transfer at the interface, as well as higher light absorption properties. With the application of graphene-CuInS2 QDs in PeSCs they have achieved 17.1% power conversion efficiency, with improved device stability. Graphene-CuInS2 QDs based perovskite solar cells retained their PCE above 80% for 30 days. Apart from the direct incorporation of graphene nanocomposites into the active layer of PeSCs, modifications of HTLs and ETLs to improve the device efficiencies and stability have also been reported very frequently. Zhou *et al* have developed a facile strategy to improve electron mobility by utilizing quantum dot/SnO_2 blend. SnO_2 as an alternative to TiO_2 has suitable band alignment with $MAPbI_3$, it has excellent optical transmittance in the visible region. Contrary to TiO_2, SnO_2 can be processed at low temperatures, not exceeding 150 °C. This is sufficiently low temperature to achieve stability of Indium tin oxide (ITO) substrates. Low temperature processing of SnO_2 allows us to fabricate PeSCs on a flexible substrate that is compatible with roll-to-roll fabrication method. Graphene quantum dot blended with SnO_2 shows better energy alignment and superior charge transfer properties as compared to pristine SnO_2. Using graphene quantum dot/SnO_2 composites Zhou *et al.* have achieved over 19.6% power conversion efficiency with outstanding reproducibility.

Incorporation of GQDs into Perovskite Active Layer

Upto now, we have mainly discussed the doping of graphene nanocomposites in the charge transport layer. However, these nanocomposites could be directly incorporated into the perovskite matrix. In hybrid perovskite optoelectronic devices, the perovskite serves as an active layer is the heart of photovoltaic and optoelectronic devices, where photogenerated electron−hole pair recombination takes place. Therefore, the most premier concern is to enhance the stability and morphology of the perovskite active layer. People have used a variety of techniques to improve the morphology of perovskite film, such as one step or sequential spin coating, antisolvent dripping method [35], solvent-additive method, gas-assisted crystallization, infrared annealing, and chemical deposition method. Incorporation of doped and undoped GQDs into the perovskite precursor provides high quality, pin-holes free perovskite film morphology, reduced exciton dissociation sites and improved stability of perovskite active layers. In this section, various reports on the use of GQDs to boost perovskite based optoelectronic devices' performance as well as the stability will be discussed. Zhang *et al.* have demonstrated high quality, pinholes free perovskite films with large grain sized active layer morphology using GQDs. Also, GQDs incorporated perovskite active layer exhibited higher absorption and better charge carrier extraction compared to that of pristine perovskite active layer in planner structured PeSCs, owing to the merit of dangling bond passivation capability and excellent conductivity of GQDs. Liu's group has demonstrated the high photoluminescence quantum yield (PLQY=80%) of nitrogen-doped graphene quantum dots (N-GQDs) material [36]. The N-GQDs remained chemically stable in a 50% humidity and a high-temperature due to their stable chemical bonds. Fig. (3) depicts the strong UV absorption spectrum and higher light emission at visible wavelengths window. The fast Fourier transform (FFT) of the N-GQDs exhibits a six-fold symmetry, suggesting a high crystalline structure of N-GQDs, as shown in Fig. (3c). Therefore, the N-GQDs incorporated $CsPbI_3$ perovskite solar cells yielded a remarkable improvement in the PCE up to 16.02% owing to high light harvesting coefficient and charge carrier extraction.

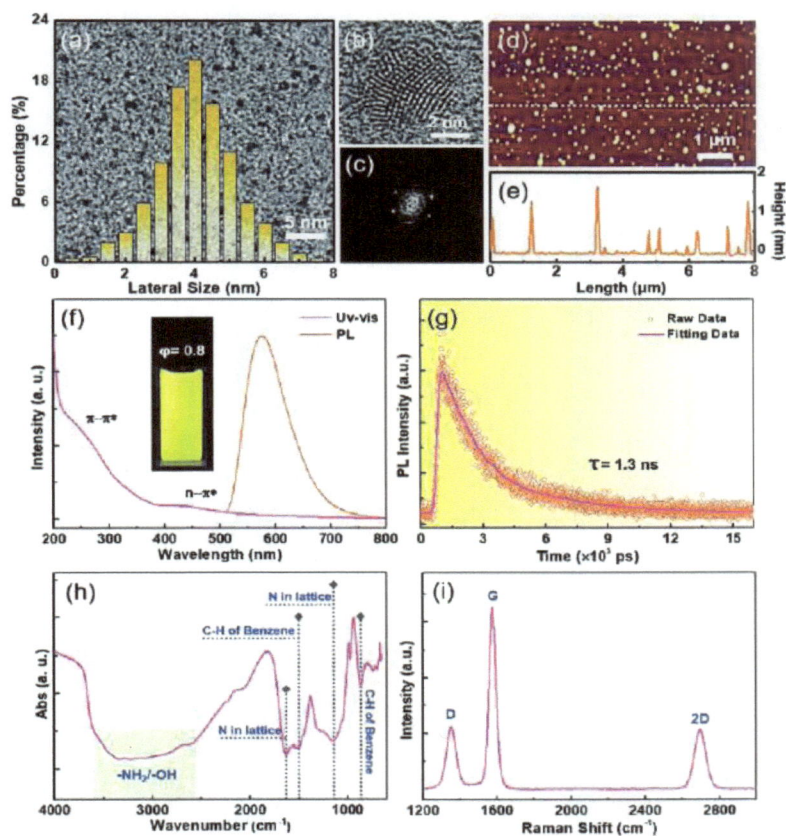

Fig. (3). Characterization of N-GQDs. **(a)** Transmission electron microscopy (TEM) image **(b)** high resolution TEM (HR-TEM) image; **(c)** FFT; **(d)** Atomic force microscope (AFM) image **(e)** AFM height profile; **(f)** normalized UV-vis and PL spectra of aqueous solution. **(g)** PL decay curve; **(h)** FT-IR spectrum; **(i)** Raman spectrum. (*Reprinted with permission from Ref. 36 (Copyright (2019) J. Mater. Chem. A*).

Moreover, Gun *et al.* have reported a high PCE of 19.8% with concomitant reductions in hysteresis of PeSCs using graphite-nitrogen doped GQDs (GN-GQDs) functional semiconductor additive in perovskite photoactive layer (Fig. **4**) [37]. The GN-GQDs additive facilitated the crystallization mechanism of the perovskite active layer with concomitant passivation of trap states by Lewis acid and Lewis base interaction. In addition, GN-GQDs at grain boundaries show matched energy levels with the active layer which in other way helps in better charge transport at grain boundaries.

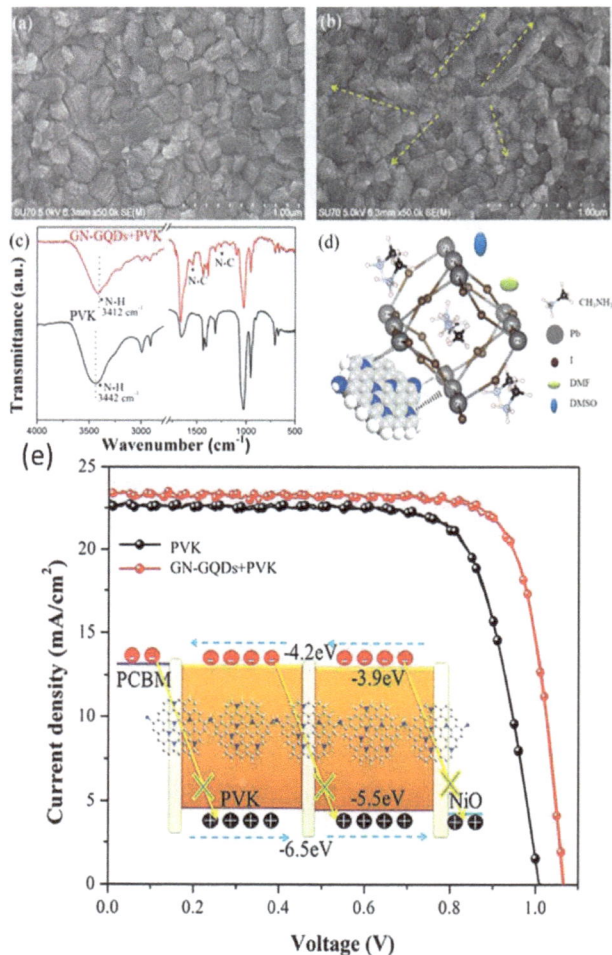

Fig. (4). FE-SEM top-view images of the **(a)** PVK film and **(b)** GQDs + PVK film. **(c)** FT-IR pattern of GN-GQDs + PVK solution and pure PVK solution. **(d)** The scheme of the interaction between GN-GQDs and Pb^{2+}. **(e)** Current density–voltage characteristics of GN-GQDs incorporated PSCs. (*Reprinted with permission from Ref. 37 (Copyright (2019) ACS Appl. Mater. Interfaces*).

CONCLUSION

Graphene nanocomposites have been widely used in perovskite-based devices due to their exceptional optoelectronic and mechanical properties. Solution processable nature of chemically exfoliated graphite into nano-flakes graphene opens up various possibilities to mix these nanocomposites in different layers of perovskite based optoelectronic devices. Perovskites are generally fragile and unstable, and therefore, graphene nanocomposites act as a support system for perovskites. These nanocomposites not only provide the needed mechanical strength (improved stability) but also these graphene nanocomposites can improve charge extraction/injection by suitable doping. In this chapter, we have discussed the application of graphene nanocomposites in various combinations in PeSCs and PeLEDs. It is observed that the graphene nanocomposites can act as a trap passivation agent and work- function modifier when mixed with the perovskites or an interlayer is used between perovskite and charge transport layer. GO is found to improve the device efficiencies as well as stability of the PeSCs. GO utilizing devices are found to retain 80% of their initial PCE after 2000 hours of stability test. PEDOT:PSS-GO composite is found to have better band alignment properties to be used in PeSCs. Incorporation of rGO in mesoporous TiO_2 improves device efficiencies, whereas addition of rGO in perovskite matrix or in spiro-OMETAD could lead to detrimental effects on device PCE. Using graphene quantum dot/SnO_2 composites, Zhou *et. al* have demonstrated a 19.6% PCE and outstanding reproducibility in PeSCs. We have seen that these nanocomposites have been mostly utilized for improving charge extraction properties in perovskite solar cells, whereas their use in perovskite-based LEDs is still limited and holds many more possibilities still to be unveiled. Overall, in this chapter, we have discussed the application of graphene oxide, reduced graphene oxide and doped graphene quantum dots in PeSCs and PeLEDs in order to improve their PCE and stability. Yet one can say that the full potential of graphene-based nanocomposite in high efficiency perovskite solar cells or light emitting diode application is yet to be achieved.

CONSENT FOR PUBLICATION

Not applicable.

CONFLICT OF INTERESTS

The authors declare no conflict of interest, financial or otherwise.

ACKNOWLEDGEMENTS

M.B. acknowledges the Department of Science and Technology, INDIA under award no. DST/INT/SWD/VR/P-13/2019 dated 14/12/2020 for partial support to carry out this work. R.K. and J.K. acknowledge the Ministry of Education, Government of India for PhD fellowship.

REFERENCES

[1]   Guo, H.L.; Wang, X.F.; Qian, Q.Y.; Wang, F.B.; Xia, X.H. A green approach to the synthesis of graphene nanosheets. *ACS Nano,* **2009**, *3*(9), 2653-2659.
        [http://dx.doi.org/10.1021/nn900227d] [PMID: 19691285]

[2]   Kumar, R.; Srivastava, P.; Bag, M. Role of a-site cation and x-site halide interactions in mixed-cation mixed-halide perovskites for determining anomalously high ideality factor and the super-linear power law in ac ionic conductivity at operating temperature. *ACS Appl. Electron. Mater.,* **2020**, *2*(12), 4087-4098.
        [http://dx.doi.org/10.1021/acsaelm.0c00874]

[3]   Steirer, K.X.; Schulz, P.; Teeter, G.; Stevanovic, V.; Yang, M.; Zhu, K.; Berry, J.J. Defect tolerance in methylammonium lead triiodide perovskite. *ACS Energy Lett.,* **2016**, *1*(2), 360-366.
        [http://dx.doi.org/10.1021/acsenergylett.6b00196]

[4]   Pandey, M.; Jacobsen, K.W.; Thygesen, K.S. Band gap tuning and defect tolerance of atomically thin two-dimensional organic-inorganic halide perovskites. *J. Phys. Chem. Lett.,* **2016**, *7*(21), 4346-4352.
        [http://dx.doi.org/10.1021/acs.jpclett.6b01998] [PMID: 27758095]

[5]   Kovalenko, M. V.; Protesescu, L.; Bodnarchuk, M. I. Properties and Potential Optoelectronic Applications of Lead Halide Perovskite Nanocrystals. *Science (80-.).,,* **2017**, *358*(6364)
        [http://dx.doi.org/10.1126/science.aam7093]

[6]   Zakutayev, A.; Caskey, C.M.; Fioretti, A.N.; Ginley, D.S.; Vidal, J.; Stevanovic, V.; Tea, E.; Lany, S. Defect tolerant semiconductors for solar energy conversion. *J. Phys. Chem. Lett.,* **2014**, *5*(7), 1117-1125.
        [http://dx.doi.org/10.1021/jz5001787] [PMID: 26274458]

[7]   Shi, D.; Adinolfi, V.; Comin, R.; Yuan, M.; Alarousu, E.; Buin, A.; Chen, Y.; Hoogland, S.; Rothenberger, A.; Katsiev, K.; Losovyj, Y.; Zhang, X.; Dowben, P. A.; Mohammed, O. F.; Sargent, E. H.; Bakr, O. M. Low trap-state density and long carrier diffusion in organolead trihalide perovskite single crystals. *Science (80-.).,,* **2015**, *347*(6221), 519-522.
        [http://dx.doi.org/10.1126/science.aaa2725]

[8]   Zhang, F.; Yang, B.; Li, Y.; Deng, W.; He, R. Extra long electron-hole diffusion lengths in CH3NH3PbI3-: XClx perovskite single crystals. *J. Mater. Chem. C Mater. Opt. Electron. Devices,* **2017**, *5*(33), 8431-8435.
        [http://dx.doi.org/10.1039/C7TC02802D]

[9]   Fang, Y.; Wei, H.; Dong, Q.; Huang, J. Quantification of re-absorption and re-emission processes to determine photon recycling efficiency in perovskite single crystals. *Nat. Commun.,* **2017**, *8*, 14417.
        [http://dx.doi.org/10.1038/ncomms14417] [PMID: 28220791]

[10]   Stranks, S.D.; Eperon, G.E.; Grancini, G.; Menelaou, C.; Alcocer, M.J.P.; Leijtens, T.; Herz, L.M.; Petrozza, A.; Snaith, H.J. Electron-hole diffusion lengths exceeding 1 micrometer in an organometal trihalide perovskite absorber. *Science,* **2013**, *342*(6156), 341-344.
        [http://dx.doi.org/10.1126/science.1243982] [PMID: 24136964]

[11]   Alcocer, M. J. P.; Leijtens, T.; Herz, L. M.; Petrozza, A.; Snaith, H. J. Electron-hole diffusion lengths exceeding trihalide perovskite absorber. *Science, 2013, 342, 341–344. , 342*, 341-344.

[12] El-Hajje, G.; Momblona, C.; Gil-Escrig, L.; Ávila, J.; Guillemot, T.; Guillemoles, J.F.; Sessolo, M.; Bolink, H.J.; Lombez, L. Quantification of spatial inhomogeneity in perovskite solar cells by hyperspectral luminescence imaging. *Energy Environ. Sci.,* **2016**, *9*(7), 2286-2294.
[http://dx.doi.org/10.1039/C6EE00462H]

[13] Li, Y.; Xu, X.; Wang, C.; Ecker, B.; Yang, J.; Huang, J.; Gao, Y. Light-induced degradation of ch3nh3pbi3 hybrid perovskite thin film. *J. Phys. Chem. C,* **2017**, *121*(7), 3904-3910.
[http://dx.doi.org/10.1021/acs.jpcc.6b11853]

[14] Boyd, C.C.; Cheacharoen, R.; Leijtens, T.; McGehee, M.D. Understanding degradation mechanisms and improving stability of perovskite photovoltaics. *Chem. Rev.,* **2019**, *119*(5), 3418-3451.
[http://dx.doi.org/10.1021/acs.chemrev.8b00336] [PMID: 30444609]

[15] Bisquert, J.; Juarez-Perez, E.J. The causes of degradation of perovskite solar cells. *J. Phys. Chem. Lett.,* **2019**, *10*(19), 5889-5891.
[http://dx.doi.org/10.1021/acs.jpclett.9b00613] [PMID: 31536358]

[16] Surendran, A.; Yu, X.; Begum, R.; Tao, Y.; Wang, Q.J.; Leong, W.L. All inorganic mixed halide perovskite nanocrystal-graphene hybrid photodetector: from ultrahigh gain to photostability. *ACS Appl. Mater. Interfaces,* **2019**, *11*(30), 27064-27072.
[http://dx.doi.org/10.1021/acsami.9b06416] [PMID: 31265238]

[17] Anaya, M.; Galisteo-López, J.F.; Calvo, M.E.; Espinós, J.P.; Míguez, H. Origin of light-induced photophysical effects in organic metal halide perovskites in the presence of oxygen. *J. Phys. Chem. Lett.,* **2018**, *9*(14), 3891-3896.
[http://dx.doi.org/10.1021/acs.jpclett.8b01830] [PMID: 29926730]

[18] Srivastava, P.; Kumar, R.; Bag, M. Discerning the role of an a-site cation and x-site anion for ion conductivity tuning in hybrid perovskites by photoelectrochemical impedance spectroscopy. *J. Phys. Chem. C,* **2021**, *125*(1), 211-222.
[http://dx.doi.org/10.1021/acs.jpcc.0c09443]

[19] Srivastava, P.; Kumar, R.; Bag, M. The curious case of ion migration in solid-state and liquid electrolyte-based perovskite devices: unveiling the role of charge accumulation and extraction at the interfaces. *Phys. Chem. Chem. Phys.,* **2021**, *23*(18), 10936-10945.
[http://dx.doi.org/10.1039/D1CP01214B] [PMID: 33912893]

[20] Kumar, J.; Kumar, R.; Frohna, K.; Moghe, D.; Stranks, S.D.; Bag, M. Unraveling the antisolvent dripping delay effect on the Stranski-Krastanov growth of $CH_3NH_3PbBr_3$ thin films: a facile route for preparing a textured morphology with improved optoelectronic properties. *Phys. Chem. Chem. Phys.,* **2020**, *22*(45), 26592-26604.
[http://dx.doi.org/10.1039/D0CP05467D] [PMID: 33201960]

[21] Stolterfoht, M.; Wolff, C.M.; Márquez, J.A.; Zhang, S.; Hages, C.J.; Rothhardt, D.; Albrecht, S.; Burn, P.L.; Meredith, P.; Unold, T.; Neher, D. Visualization and suppression of interfacial recombination for high-efficiency large-area pin perovskite solar cells. *Nat. Energy,* **2018**, *3*(10), 847-854.
[http://dx.doi.org/10.1038/s41560-018-0219-8]

[22] Dequilettes, D.W.; Koch, S.; Burke, S.; Paranji, R.K.; Shropshire, A.J.; Ziffer, M.E.; Ginger, D.S. Photoluminescence lifetimes exceeding 8 ms and quantum yields exceeding 30% in hybrid perovskite thin films by ligand passivation. *ACS Energy Lett.,* **2016**, *1*(2), 438-444.
[http://dx.doi.org/10.1021/acsenergylett.6b00236]

[23] Zhang, X.; Ji, G.; Xiong, D.; Su, Z.; Zhao, B.; Shen, K.; Yang, Y.; Gao, X. Graphene oxide as an additive to improve perovskite film crystallization and morphology for high-efficiency solar cells. *RSC Advances,* **2018**, *8*(2), 987-993.
[http://dx.doi.org/10.1039/C7RA12049D]

[24] Yang, Q.D.; Li, J.; Cheng, Y.; Li, H.W.; Guan, Z.; Yu, B.; Tsang, S.W. Graphene oxide as an efficient hole-transporting material for high-performance perovskite solar cells with enhanced stability. *J. Mater. Chem. A Mater. Energy Sustain.,* **2017**, *5*(20), 9852-9858.

[http://dx.doi.org/10.1039/C7TA01752A]

[25]   Li, W.; Dong, H.; Guo, X.; Li, N.; Li, J.; Niu, G.; Wang, L. Graphene oxide as dual functional interface modifier for improving wettability and retarding recombination in hybrid perovskite solar cells. *J. Mater. Chem. A Mater. Energy Sustain.*, **2014**, *2*(47), 20105-20111.
[http://dx.doi.org/10.1039/C4TA05196C]

[26]   Cho, K.T.; Grancini, G.; Lee, Y.; Konios, D.; Paek, S.; Kymakis, E.; Nazeeruddin, M.K. Beneficial role of reduced graphene oxide for electron extraction in highly efficient perovskite solar cells. *ChemSusChem*, **2016**, *9*(21), 3040-3044.
[http://dx.doi.org/10.1002/cssc.201601070] [PMID: 27717168]

[27]   Wang, J.T.W.; Ball, J.M.; Barea, E.M.; Abate, A.; Alexander-Webber, J.A.; Huang, J.; Saliba, M.; Mora-Sero, I.; Bisquert, J.; Snaith, H.J.; Nicholas, R.J. Low-temperature processed electron collection layers of graphene/TiO$_2$nanocomposites in thin film perovskite solar cells. *Nano Lett.*, **2014**, *14*(2), 724-730.
[http://dx.doi.org/10.1021/nl403997a] [PMID: 24341922]

[28]   Han, G.S.; Song, Y.H.; Jin, Y.U.; Lee, J.W.; Park, N.G.; Kang, B.K.; Lee, J.K.; Cho, I.S.; Yoon, D.H.; Jung, H.S. Reduced graphene oxide/mesoporous tio$_2$nanocomposite based perovskite solar cells. *ACS Appl. Mater. Interfaces*, **2015**, *7*(42), 23521-23526.
[http://dx.doi.org/10.1021/acsami.5b06171] [PMID: 26445167]

[29]   Acik, M.; Darling, S.B. Graphene in Perovskite Solar Cells: Device Design, Characterization and Implementation. *J. Mater. Chem. A Mater. Energy Sustain.*, **2016**, *4*(17), 6185-6235.
[http://dx.doi.org/10.1039/C5TA09911K]

[30]   Cheng, Y.; Yang, Q.D.; Xiao, J.; Xue, Q.; Li, H.W.; Guan, Z.; Yip, H.L.; Tsang, S.W. Decomposition of organometal halide perovskite films on zinc oxide nanoparticles. *ACS Appl. Mater. Interfaces*, **2015**, *7*(36), 19986-19993.
[http://dx.doi.org/10.1021/acsami.5b04695] [PMID: 26280249]

[31]   Liu, K.; Chen, S.; Wu, J.; Zhang, H.; Qin, M.; Lu, X.; Tu, Y.; Meng, Q.; Zhan, X. Fullerene derivative anchored sno2 for high-performance perovskite solar cells. *Energy Environ. Sci.*, **2018**, *11*(12), 3463-3471.
[http://dx.doi.org/10.1039/C8EE02172D]

[32]   Xie, J.; Huang, K.; Yu, X.; Yang, Z.; Xiao, K.; Qiang, Y.; Zhu, X.; Xu, L.; Wang, P.; Cui, C.; Yang, D. Enhanced electronic properties of SnO$_2$*via* Electron Transfer from Graphene Quantum Dots for Efficient Perovskite Solar Cells. *ACS Nano*, **2017**, *11*(9), 9176-9182.
[http://dx.doi.org/10.1021/acsnano.7b04070] [PMID: 28858471]

[33]   Zhang, F.; Ma, W.; Guo, H.; Zhao, Y.; Shan, X.; Jin, K.; Tian, H.; Zhao, Q.; Yu, D.; Lu, X.; Lu, G.; Meng, S. Interfacial oxygen vacancies as a potential cause of hysteresis in perovskite solar cells. *Chem. Mater.*, **2016**, *28*(3), 802-812.
[http://dx.doi.org/10.1021/acs.chemmater.5b04019]

[34]   Hong, J.A.; Jung, E.D.; Yu, J.C.; Kim, D.W.; Nam, Y.S.; Oh, I.; Lee, E.; Yoo, J.W.; Cho, S.; Song, M.H. Improved efficiency of perovskite solar cells using a nitrogen-doped graphene-oxide-treated tin oxide layer. *ACS Appl. Mater. Interfaces*, **2020**, *12*(2), 2417-2423.
[http://dx.doi.org/10.1021/acsami.9b17705] [PMID: 31856562]

[35]   Kumar, R.; Kumar, J.; Srivastava, P.; Moghe, D.; Kabra, D.; Bag, M. Unveiling the morphology effect on the negative capacitance and large ideality factor in perovskite light-emitting diodes. *ACS Appl. Mater. Interfaces*, **2020**, *12*(30), 34265-34273.
[http://dx.doi.org/10.1021/acsami.0c04489] [PMID: 32608224]

[36]   Bian, H.; Wang, Q.; Yang, S.; Yan, C.; Wang, H.; Liang, L.; Jin, Z.; Wang, G.; Liu, S. Nitrogen-doped graphene quantum dots for 80% photoluminescence quantum yield for inorganic γ-CsPbI3 perovskite solar cells with efficiency beyond 16%. *J. Mater. Chem. A Mater. Energy Sustain.*, **2019**, *7*(10), 5740-5747.

[http://dx.doi.org/10.1039/C8TA12519H]

[37]   Gan, X.; Yang, S.; Zhang, J.; Wang, G.; He, P.; Sun, H.; Yuan, H.; Yu, L.; Ding, G.; Zhu, Y. Graphite-n doped graphene quantum dots as semiconductor additive in perovskite solar cells. *ACS Appl. Mater. Interfaces,* **2019**, *11*(41), 37796-37803.
       [http://dx.doi.org/10.1021/acsami.9b13375] [PMID: 31550130]

CHAPTER 12

Ferroelectric Liquid Crystal Nanocomposite for Optical Memory and Next Generation Display Applications

Harris Varghese[1,2], T.K. Abhilash[1,2] and Achu Chandran[1,2,*]

[1] *Materials Science and Technology Division, CSIR-National Institute for Interdisciplinary Science and Technology (NIIST), Thiruvananthapuram-695019, India*

[2] *Academy of Scientific and Innovative Research (AcSIR), Ghaziabad- 201002, India*

Abstract: The dispersion of nanomaterials in ferroelectric liquid crystals (FLC) has turned out to be a promising method for fabricating optical memory devices and tuneable electro-optical materials. In a nanosuspension between FLC and nanoparticles, the presence of the dopant particles creates a synergic interaction with host FLC, which leads to the improvement of electro-optical properties. Tailoring with nanoparticles of suitable size, concentration, and compatibility results in various fascinating effects and new multifaceted composites for electro-optical devices. Adding nano-sized materials such as metallic, semiconducting, insulating or other functional species into the FLC matrix is a fertile method, giving rise to or increases in memory retention and other electro-optical properties that can replace the current electro-optical devices. These advancements depend on the harmony between the guest and host materials. This chapter gives a comprehensive overview of the present technologies and enhancements that have been acquired in nanoparticle/FLC composite systems, especially for optical memory devices and display applications.

Keywords: Composite Systems, Display Devices, Electro-Optical Properties, Ferroelectric, Ferroelectric Liquid Crystal-Nanocomposites, Ferroelectric Liquid Crystals, Insulating NP, Liquid Crystals, Metallic Nanoparticles, Nanoparticles, Optical Memory, Semiconducting Nanoparticles.

INTRODUCTION

Liquid crystals (LC), or mesogens, are an intermediate phase of matter whose properties and structures are an amalgam of both liquids and crystalline solids. It was first discovered by Austrian Botanist Friedrich Reinitzer and German scientist Otto Lehmann on cholesterol in 1888. Successfully understanding the

---

* Corresponding author Achu Chandran: Material Science and Technology Division, CSIR-National Institute for Interdisciplinary Science and Technology (NIIST), Thiruvananthapuram-695019, India;
E-mail: achuchandran@niist.res.in

Gaurav Manik and Sushanta Kumar Sahoo (Eds.)

fundamentals of this new phase grants Lehmann and Reinitzer the title of grandfathers of LC. The LC can be categorised or classified by their structure, composition and phase. According to the shape, there are rod-shaped (calamitic), disc-shaped, and banana-shaped (bent-core) liquid crystals. Similarly, based on their composition, they are classified as thermotropic, where mesophase is controlled solely by temperature, and lyotropic, where the concentration of solvent plays a pivotal role. Further, the mesogens are sorted based on the symmetry of the phase as chiral (Fig. **1a**) and achiral [1]. The achiral liquid crystals are further divided into nematic (Fig. **1b**), which possesses only orientational order, and smectic (Fig. **1c**), which has a layered structure with both orientational and positional order.

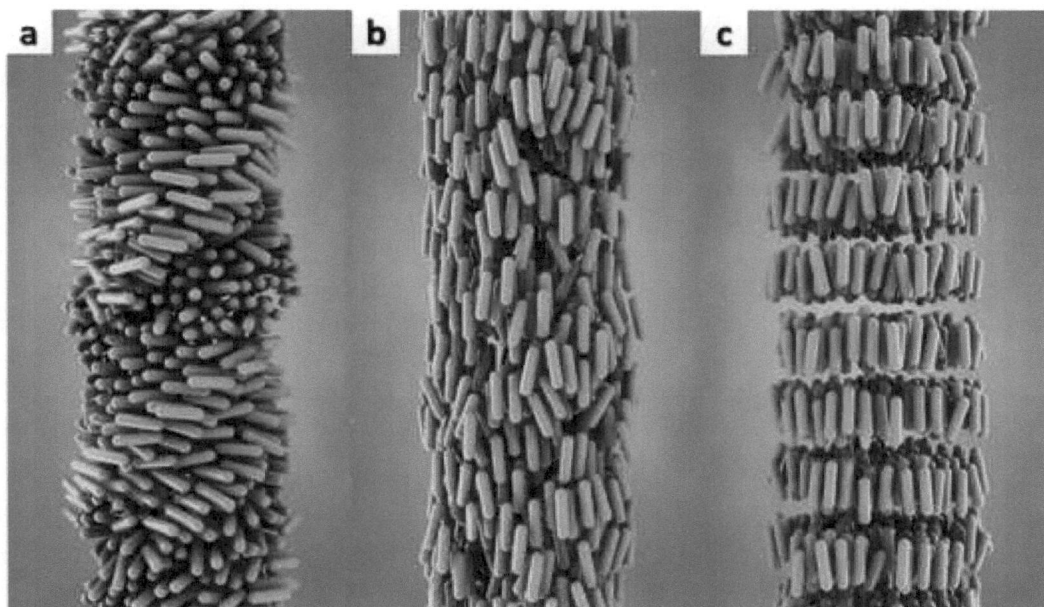

Fig. (1). (a) Chiral Nematic Liquid crystals, (b) Nematic Liquid Crystal and (c) Smectic Liquid Crystal.

Even though non-ferroelectric, ferroelectric and anti-ferroelectric mesophases constitute the chiral branch, at the moment, we are exclusively interested in the ferroelectric liquid crystals (FLC), a remarkable class of LC materials that manifest spontaneous polarisation, consequent to their lower symmetry. Despite theorizing liquid crystalline materials with ferroelectric properties having been done before [2], experimental confirmation was only provided later by Meyer *et al.* [3] by synthesizing the DOBAMBC (2-methylbutyl 4-(--decyloxybenzylideneamino) cinnamate) material, the first FLC material. Nowadays, FLCs are used for multitude of applications such as flat panel displays

[4], spatial light modulators [5], optical antennas [6], and so on. The FLC, Smectic C*, has a helical structure with each molecule having a polarization perpendicular to its long axis, and thus, the net polarisation becomes zero due to the collective ordering of mesogens in each layer. FLCs can be further grouped into surface stabilized ferroelectric liquid crystals (SSFLC), deformed helix ferroelectric liquid crystals (DHFLC), and electroclinic liquid crystals (ELC). In SSFLC, the surface stabilization is obtained by making the cell thickness smaller than the pitch and net polarisation is achieved by unwinding the helix with surface interaction. The DHFLCs are short-pitched FLC where the helix deformation is achieved by application of an electric field while in ELC the electric field is applied in parallel direction to the layers to inhibit the free rotation of molecules and thus produces a net non-zero polarisation. Even though FLCs are superior to nematic with regard to fast response, viewing angle, lack of perfect alignment, domain formation, and other defects are some drawbacks of the FLC for display applications. Doping FLC matrix with materials such as nanoparticles [7], microparticles [8] or colloids [9], having a wide array of properties and dimensions, is a viable and proven method to develop more functionally advanced and stable FLC composite systems without the above-mentioned drawbacks. Within the composite system, the interaction between the molecules and the dopants should be advantageous to our need. The size compatibility, concentration, and dispersion should be proper to have an orderly alignment and avoid any agglomerations or director distortions. Considering all these, nano-dispersion in lower concentrations seems to yield better results. The development and characterization of new nanocomposite paved the way to new regimes, including the development of non-volatile electrooptic memory effect, biosensors, devices for energy storage and conversion, *etc*.

In this chapter, we review the recent advances in FLC by dispersing various nanomaterials into the host LC matrix to enhance its electro-optical properties for memory and next-generation display applications. The nanoparticles (NP) such as metallic, semiconducting, insulating, and others bring about remarkable enrichment in the performance of the FLCs when made into a composite system with FLC. Advanced electro-optical properties, memory retentions, and improved alignments are some of the virtues of these composite FLC-NP systems.

FLC-NANOCOMPOSITES FOR MEMORY APPLICATION

As discussed, the FLC-nanocomposites exhibit faster response time, lower voltage threshold, better contrast, and lasting memory effects than their pure counterpart. Now in this section, we are concentrating more on the soft memory aspects of the FLC. The bistable nature of the FLC [10] with two stable polarisation states facilitate memory applications. The FLC can store the binary 0 and 1 values as these polarization states even after turning off the electric supply, ergo non-volatile memory devices. Bistable switching is when, with sufficient field applied, the director reorient even at the boundaries and this stable state remains as such even after the withdrawal of the applied field, and in FLC these states are generally transient [11]. The SSFLC shows memory effect only depending on the threshold field while for the DHFLC the memory effects are critically related to both potential and its frequency of the applied time delayed square voltage pulse [12]. This is where the NP doping comes; the memory effect exhibited by the mixture depends on the properties, nature, and concentration of both the LC matrix as well as the dopant and the consonance between them [13]. Thus, allowing us to improve upon the soft memory provided by the FLC materials. The subsequent session describes the developments that have been occurred in memory retention of FLC materials by the doping of different nanomaterials onto them.

Metal Nanoparticles in FLC

It is reported that an FLC material, namely Felix 17/100, does not exhibit any memory effect in its pure form but exhibits a non-volatile memory effect when gold (Au) nanoparticles are doped into the LC matrix [14]. The nano-FLC composite showed memory property lasting for up to 30 minutes. This resulted from the higher internal electric field due to the surface plasmon resonance, where the electrons in the conduction band of Au oscillate according to the applied electromagnetic field. Thus, the presence of Au nanoparticles makes the depolarising field, which results from the generation of stable states for charges on the surface, weaker and preserves the memory for long durations. Similarly, gold nanoparticle doped on DHFLC also kept the LC cell in the memory state for a long time (for many days) [15]. Here, the retention of memory is due to charge transfer, induced by the application of electric field, between the nanoparticles and LC molecules together with the stabilization of the helix deformation, which is also the result of the applied field.

Goel *et al.* [16] obtained tailorable optical memory by doping Nickel (Ni) nanoparticles, capped with polyvinylpyrrolidone, into the ferroelectric liquid crystal. The memory effect arises due to the ferromagnetic spin-dependent screening effect at the interface of FLC-Ni (ferroelectric-ferromagnetic). The field generated by the spin-dependent screening counterbalance the depolarising field and the helix of the FLC remains unwind even after the bias removal and, thus, the memory effect. The duration of this memory can be tuned by varying either the nanoparticle concentration or the applied DC field, where a maximum of 25 min of memory was obtained. Additionally, the change in spontaneous polarisation and rotational viscosity resulted due to the fact that doping attributes to a mammoth decrease in the response time. When iron nanopowder is dispersed on commercial FLC, SCE3 [17], the distortion caused by the dopant results in a non-volatile memory lasting up to days. The observed memory resulted from the switch, or rather the reluctance to switch, from a restored initial phase by the application of field to the distorted phase.

Semiconducting Nanoparticles in FLC

A memory effect was observed when cadmium sulphide (CdS) nanorods [18] were doped into KCFLC10R (an FLC material from Kingston Chemicals, UK). For 0.3 wt % of CdS doped FLC, the memory holding was about 15 minutes. Moreover, the memory efficiency was proportional to the increment in dopant concentration. The memory effect only in the doped samples is assumed to be due to the charge transfer from FLC to the nanorods, which creates a higher field around the nanorods and prevents the molecules from recovering even after the removal of the bias. The memory persists till the complete removal of the charges and the duration extends as the number of charge trapping states in the matrix increases. Likewise, cadmium telluride quantum dots (CdTe-QDs) bring about pronounced memory effects in different FLCs materials [19], including namely LAHS19, LAHS18, FLC 6304 and KCFLC 7S (Fig. **2**) in non-surface stabilised configurations. Also, in this case, the quantum dots (QD) are capped with P3HT (Poly-3(hexylthiophene)) polymer to prevent agglomeration, which facilitates charge transfer between the LC and QD and, in turn, memory retention. The duration of memory is determined by the concentration of CdTe-QDs and the nature of the FLC used. LAHS18 with ~4 wt % CdTe-QDs present memory retention of about 10 min. Identically, using CdTe-QDs capped with octadecylamine, Pandey *et al.* [20] obtained good memory property on FLC material, W-327. In the same way, zinc oxide (ZnO)-FLC nanocomposite showed memory retention for 7-15 min [21], resulting from the induced charge transfer between the FLC (W206E) to the nanoparticle and the phenomenon of ion trapping.

Insulating Nanoparticles in FLC

Metal oxide nanomaterials are insulating materials that have a high density of surface states on the edges as well as on surfaces. An FLC material, KCFLC 10S, dispersed with Zirconia nanoparticles (ZNP) exhibited impurity-free memory [22] for about 15 minutes by diminishing the depolarization field and lessening the ionic charges. The nanoparticle addition provided more anchoring and resulted in a more ordered structure for the FLC materials. The impurity charges get trapped around the ZNPs and consequently, the depolarising field decreases. Apart from the non-volatile memory, nano-sized nickel oxide (nNiO) nanorods dispersed into

FLC ZLI 3654 [23] allowed low voltage operation due to low screening and fast alignment of the composite. Here also, the reduction in the depolarizing field by ion adsorption is the cause of the non-volatile memory. Another insulating nanoparticle, Silica (SiO_2), added KCFLC 10R [24] ferroelectric liquid crystal exhibits memory effects, however, the memory holding does not show any considerable enhancement with increase in the concentration of silica.

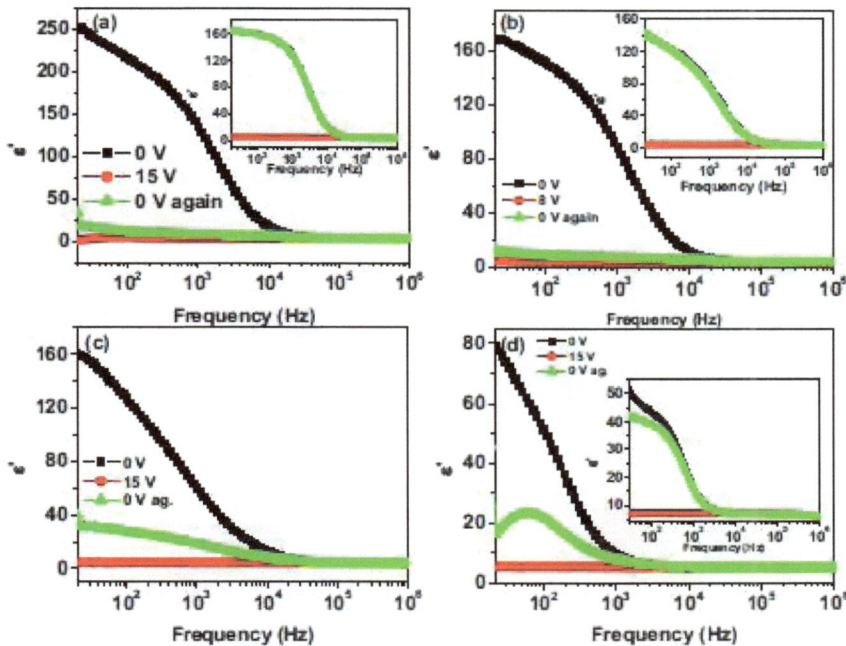

Fig. (2). Dielectric permittivity of (a) LAHS19, (b) LAHS18, (c) FLC 6304 and (d)KCFLC 7S doped with CdTe-QDs at different biases. Insets represent the pure state. The last cell is 10 μm thick while others are 4 μm thick [19]. Reproduced from A. Kumar *et al.*, Appl. Phys. Lett., vol. 97, no. 16, p. 163113, Oct. 2010, with the permission of AIP Publishing.

Other Nanoparticles in FLC

Glycerol is a common and versatile solvent, using which Kumar *et al* [25] obtained memory retention in DHFLC material (FLC 6304, Rolic, Switzerland) and glycerol mixture of more than 12 hrs (Fig. **3**). The glycerol has a major role in stabilizing the helix deformation and depolarization field minimization as well, which results in the memory effect. However, the glycerol-FLC (CS 1026, Chisso, Japan) mixture does not result in any memory effect, which points to the fact that memory effects in both FLC and DHFLC are results of different processes. Using single-wall carbon nanotubes (SWCNTs) layer attached to the deposition of SiO_x on an ITO surface, Petroc *et al* [26] grew smectic C liquid crystals (7OBA) into well oriented distinct local smectic C single crystals (DLSCs) [27]. These exhibited surface memory effects (SME), where the oriented structure in the Sm C phase was memorized and displayed by the LC matrix in the nematic phase. This process emanates owing to the confinement of bulk charges in the LC by the SWCNT. The memory effect is also dependent on the cell thickness, for example, FLC CS-1024 manifested memory [28] on 12 μm thick cells while there was no pronounced memory in 5 μm thick cells. Howbeit, doping the FLC matrix with multi-walled carbon nanotubes (MWCNTs) diminishes memory retention and it is due to the reduction of rotational viscosity as a result of doping.

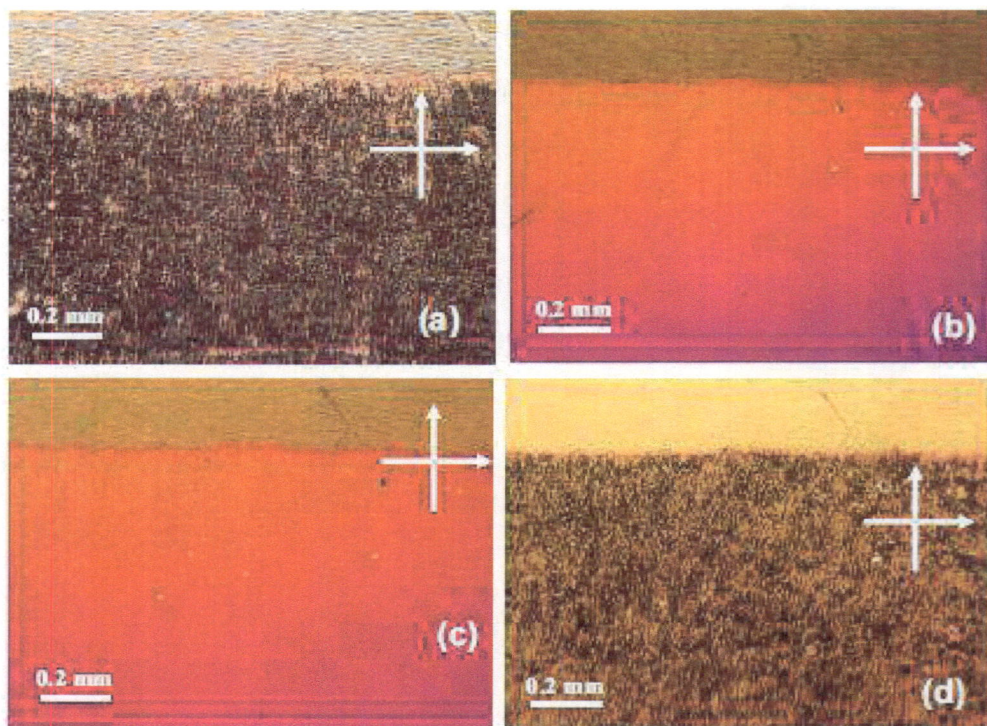

Fig. (3). Optical micrographs of glycerol/DHFLC composite cell of thickness 3 μm at room temperature at (a) 0 V, (b) 15 V bias, (c) after 12 h of removal of bias and (d) again at 0 V [25]. Reproduced from A. Kumar *et al.*, J. Appl. Phys., vol. 105, no. 12, p. 124101, Jun. 2009, with the permission of AIP Publishing.

Likewise, a ferrofluid, cobalt ferrite ($CoFe_2O_4$) generates memory effects in FLCs (SCE13 and Felix017/100) [29]. Here also, the memory originates due to the diminishing of the depolarisation field in the presence of the nanoparticles. It is also found that this memory is temperature-dependent, where it is longer for high temperature, and it is reduced owing to the increase in conductivity of the FF with the temperature. Similarly, a colloidal suspension of barium titanate nanoparticle with SCE13 FLC (SCE113/BT nanocomposite) [30] displays memory effect and higher contrast compared to undoped layer. Under strong fields, the barium titanate ($BaTiO_3$) nanoparticles also *get al*igned to the field direction and interact with the mesogen molecules. It is believed that this interaction results in memory and a high spontaneous polarization value of the dispersion.

FLC-NANOCOMPOSITES FOR NEXT-GENERATION DISPLAY APPLICATIONS

For the next-generation displays and photonic devices, high resolutions, better contrast ratios, faster refresh rates and low power operations are highly desired. For the reduction of motion blur, operating in lower temperatures and also to attain field sequential colour (FSC) a fast response time is required [31]. So, the end goal is to attain higher resolutions, lower electrical energy consumption and extended colour gamut for FLC display, than the liquid crystal display (LCD) in general. As discussed, with the help of nanoparticle doping in FLC, fast electro-optical response, low voltage operation, improved contrast, reduction in impurity ions and improved alignment can be obtained. The thermal and chemical stability of the composite system, desirable concentration, size, shape of dopants favorable to enhance the electro-optical properties are the main challenges. Efforts to improve the FLCs, mainly electro-optical properties, with various NPs for next-generation display devices are summarised in the subsequent session.

Metal Nanoparticles in FLC

The noble metal nanoparticles (MNP), such as gold, silver, platinum, *etc.*, enhances the local electric field in the molecular sites of FLC by plasmon excitation phenomenon and the propagation of surface plasmon polariton (SPP) waves, which results in the improved optical tilt angle of FLCs [14, 31]. The improved intrinsic field inside the FLC/MNP composite lowers the threshold voltage in a considerable manner. Similarly, an increase in photoluminescence (PL) spectral intensity is observed when FLC is doped with metal nanoparticles such as platinum, gold and silver (Fig. **4**). The enhancement in PL intensity is caused by the creation of electromagnetic fields due to the surface plasmon resonance (SPR) when illuminated with light [32]. The improved photoluminescent spectra paved the way for the development of highly bright photoluminescent LCDs.

When gold nanoparticles are doped with FLC materials, a decrease in switching time is observed and it is suitable for faster display applications [33]. Podgornov *et. al* [33] found that the dispersed nanoparticles will trap the ionic impurities and redistribute the applied voltage inside the FLC cell, which reduces response time. However, the rotational viscosity of the pristine material remains unchanged with the doping. The suppression of ionic impurities depends on the thickness of the cell and doping concentration in the host FLCs. A. Kumar *et. al* [34] investigated the effect of Selenium Docosane (SD) capped Palladium nanoparticles (Pd NPs) and found that there is a 50% reduction in operating voltage as compared to the undoped FLC sample. Varying the concentrations of the dopant results in enhanced optical contrast as well as tilt angle and reduction in operating voltage significantly. Kaur *et. al* [35] investigated the doping impact of silver nanoparticles in FLCs, which is in Langmuir-Blodgett thin films configuration, and reported that the photoluminescent properties could be improved by increasing the concentration of the nanoparticles. Neeraj *et. al* [36] studied the influence of Nickel nanoparticles on FLCs and observed an improved transmittance and a fast-switching response suitable for display applications. These improvements are the consequence of strong dipole interactions among the FLC host and doped nanoparticles, which enhances anchoring and ordering of the mesogens. The addition of Nickel nanoparticles leads to the reduction of switching time and enhancement in spontaneous polarization is also reported by Khushboo *et. al* [37]. They found that induced dipole moments by the dopants are responsible for the enhancement in spontaneous polarization.

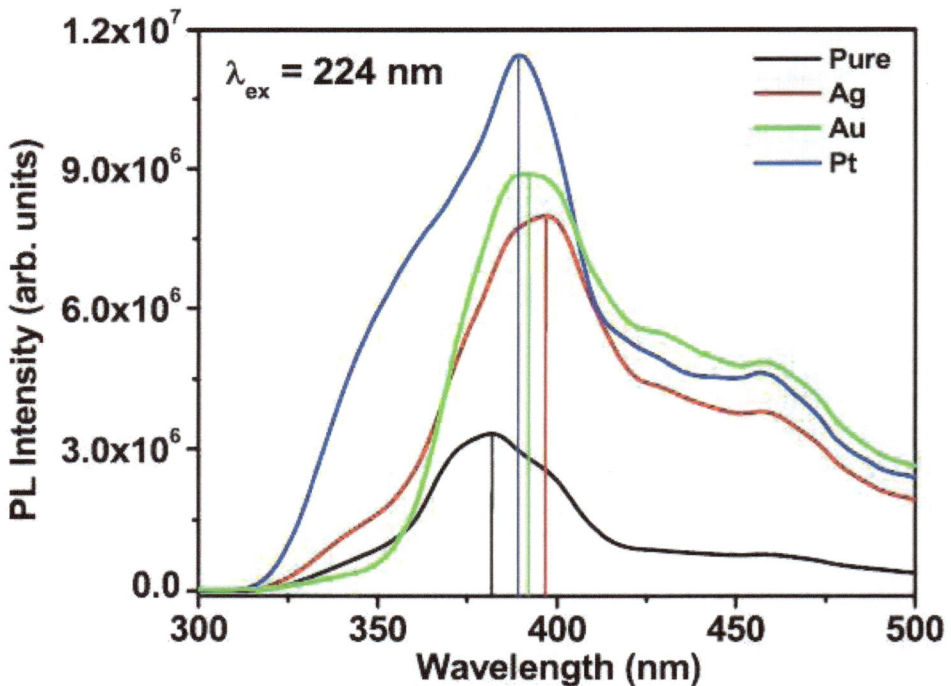

Fig. (4). Enhanced photoluminescent intensity due to metal nanoparticle doping [32]. Reproduced from S. Tripathi *et al.*, Appl. Phys. Lett., vol. 102, no. 6, p. 063115, Feb. 2013, with the permission of AIP Publishing.

Semiconducting Nanoparticles in FLC

Semiconductor Nanoparticles are another promising dopant in FLCs that enhances electro-optical properties suitable for display applications. The inherent dipole moment of these nanoparticles, which on interaction with the dipoles of the host material, improves spontaneous polarization and tilt angle of the FLC nanocomposite [38] [39]. The response time (τs) is related to the surface anchoring energy and is given by;

$$\tau_s = \frac{\eta}{K\pi^2}\left(d^2 + \frac{4dK}{W}\right) \qquad \text{(for strong anchoring)} \qquad (1)$$

$$\tau_s = \frac{4\eta d}{W\pi^2} \qquad \text{(for weak anchoring)} \qquad (2)$$

where η is the Rotational Viscosity, W is the anchoring energy coefficient, d is the cell thickness and K is the bend elastic constant. It is clear that surface anchoring energy is inversely related to the response time of the material. The nano dopants provide strong surface anchoring with the host FLC mesogens, which leads to the fast response time of the composite by the above equations, which is required for the display applications [40]. Moreover, semiconductor nano dopants are capable of reducing the coercive voltage at a suitable concentration [40].

The CdS nanoparticle doping produces a noticeable increment in spontaneous polarization and switching time of the FLC-NP composite. These effects result from the considerable dipole interaction and the high surface anchoring energies among the dopant and the host material. The presence of CdS nanoparticles decreases the dielectric strength and relaxation frequency of the sample [41]. The effect of Co-ZnO/ZnO core/shell quantum dots on FLC was studied by researchers [42] and found that the nanoparticles improved the molecular alignment of the samples. As a consequence of this enhanced molecular ordering, tthe tilt angle, spontaneous polarization and PL intensity and PL intensity got increased significantly. These enhancements make the ZnO QDs be a promising candidate for display applications. Another viable dopant for improved display application is cadmium selenide quantum dots (CdSe QDs). The doping of CdSe QDs produces around a 48% reduction in response time. Also, the doping resulted in the reduction in spontaneous polarization and dielectric strength of the host material [43]. The antiparallel correlation between the mesogens and dopant particles is responsible for these effects.

Insulating Nanoparticles in FLC

Metal oxide nanoparticles form a major group of insulating nanoparticles that enhance dielectric and electro-optical properties of the FLCs on doping. These materials can take on a wide number of structural geometries with an electronic structure which makes them capable of exhibiting metallic, semiconducting, and insulating character. Some metal oxides possess intrinsic dipole moment, and some are capable of absorbing ionic impurities on their surface, making them suitable for display applications in FLCs. Researches show that the addition of metal oxides can improve the FLC alignment and can rigorously change the tilt angle and optical contrast. Recent researches show that the oxygen vacancies on the surface of magnesium oxide (MgO) NPs give them unique chemical properties and help to provide strong surface anchoring energy with the mesogens. The dipole-dipole interaction among the nanoparticles and pure FLC molecules creates a strong intrinsic field inside the composite, which causes the fast response of mesogens against the applied field [44]. These dopants can loosen the helical packing of the FLC molecules, thereby reducing the rotational viscosity, which is reflected in the faster response time of the composite (Fig. 5).

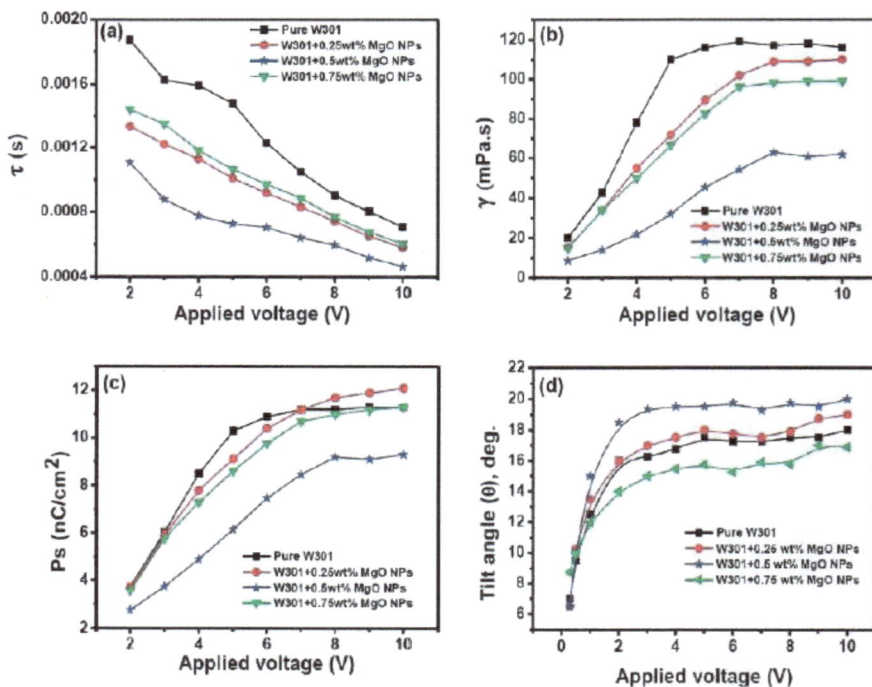

Fig. (5). The variation of switching time, spontaneous polarization, rotational viscosity and tilt angle when MgO nanoparticles are dispersed in FLCs [44]. Reproduced by permission of The Royal Society of Chemistry.

When cupric oxide (CuO) NPs are dispersed in FLC material, the low frequency partially unwound helical mode (p-UHM) present in host FLC (W302) got suppressed [45]. On adding SiO_2 NPs to FLC, it leads to an increase in anchoring energy and dielectric permittivity of host FLC. This increase in anchoring energy is due to a higher intermolecular association between NPs and mesogens. The presence of nanoparticles brought better alignment for molecules and it led to an increase in spontaneous polarization of the composite system. The value of polarization is observed to be increased from layer to layer with an applied electric field and temperature [46]. The switching time can also be reduced by the additional anchoring energy provided by the Silica nanoparticles [40]. Furthermore, on the addition of Alumina NPs, the adsorption of impurity ions by the dopant on its surface sites reduces the impurities in the pristine material and thereby increases the electrical resistance of the compound system [47].

Other nanoparticles in FLC

Carbon nanotubes (CNT) are the materials for the future owing to their potential in a wide range of device applications. Research based on the doping of single-walled and multi-walled carbon nanotubes on FLC materials shows that they can influence the electro-optical and dielectric characteristics of the host material without disturbing its molecular alignment [48]. Carbon nanotubes are capable of trapping the ions on their surface due to delocalized π-electrons, this helps in increasing the localized electric field inside the composite and thereby enabling a faster response to the external electric field. The dipole moment of FLC material is partially screened by the CNTs, which decreases the spontaneous polarization and hence the dielectric strength of the CNT/FLC composite. The variation of rotational viscosity with doping is abrupt sometimes, the CNTs can increase the rotational viscosity and sometimes nanodispersion may lead to the reduction in rotational viscosity [49]. This can be well understood by observing the shift in the Gold Stone mode (GM) relaxation frequency by experimental methods. The effect of carbon nanotubes in DHFLC is studied by Jai Prakash *et al*. [50]. MWCNTs induce short-range ordering in DHFLC molecules. The doping also introduces ionic impurities in the host material. These impurities produce a non-zero spontaneous polarization in the Sm A* phase of the composite.

The effect of nanosized diamond particles is also studied [51]. The addition of diamond NPs is capable of improving the alignment of mesogens. These NPs can absorb the ionic impurities present in the pristine sample. The relative permittivity and conductivity of the host sample are lowered with the dispersion of diamond nanoparticles. The doping of fullerene enhances the localized electric field and leads to a faster switching response of the nanocomposite [52]. Here also, the antiparallel correlation between the mesogens and NPs leads to the reduction in spontaneous polarization of the system. The dielectric strength, rotational viscosity and dielectric constant were reduced while doping with fullerene in FLC. The introduction of graphene oxide (GO) into FLCs opened up the door to low current sensing optoelectronic devices [53]. The ionic charge transport mechanism is observed in GO doped FLC composite. The presence of GO enhances the mobility of the charge carriers and produces a pico-ampere current with a very high breakdown voltage of about 8V. The presence of GO improves the anchoring strength and produces a change in molecular orientation in the composite system. The π-π interaction among the phenyl ring of the FLC molecules and sp^2 domains of GO resulted in an enhancement in the alignment of the mesogenic molecules. Two (photo luminescent) PL emission bands were observed in the FLC-GO composition and, the PL intensity gets increased with increase in temperature due to the electron-phonon coupling [53].

Ferroelectric nanoparticles are characterized by their large permanent dipole moments. They are capable of inducing realignment in neighboring LC molecules, thereby improving the order parameter and reducing threshold voltage. A research team led by Hao-Hsun Liang studied the influence of ferroelectric BaTiO$_3$ NPs, on the FLC CS1024 material [54]. It was reported that the spontaneous polarization (P$_s$) of the pristine material gets enhanced when it is doped with the nanoparticles. The value of P$_s$ is observed to be increased twice that of pure CS1024 material for a doping concentration of 0.1 wt % suspension. Consequently, there is an increased sensitivity of the material to the electric field. The presence of nanoparticles improved the dielectric permittivity of the LCs at a suitable concentration. Moreover, the response time of the doped material changes with concentration and it showed an overall increase with the addition of nanoparticles. These enhanced properties are very suitable for electro-optical applications. It is also observed that the addition of BaTiO$_3$ NPs changes the electro-optical properties without affecting the alignment and tilt angle of the pristine FLC materials [55, 56]. As before, the antiparallel dipole-dipole correlation among the FLC sample and BaTiO$_3$ NPs decreases the P$_s$ of the host material as reported by Mikulko *et al.* [55] and another group led by Rudzki [56].

CONCLUSION

The improved order parameter provided by the nanoparticles changes the properties of the FLCs and generates boosted electro-optical characteristics. Dispersing nanoparticles like metallic, semiconducting, insulating, or other nano-sized materials generated memory retention or extended the existing memory effects to several minutes or days in many FLC systems. In general, the nanoparticles bring about better alignment, charge trappings and consequently reduction in the depolarising field and, ultimately, memory effect in the system.

Similarly, they improve the spontaneous polarisation as well as the tilt angle and reduce both switching time and rotational viscosity, emanating in better display capabilities. These more refined electro-optical performances, such as higher contrast ratio, lesser operation voltages and high refresh rate, are reassuring for the next generation capabilities of the FLC-nanocomposites and for applications in several electro-optical gadgets. So, this vibrant research field has immense potential to drastically upgrade the current memory and display solutions.

CONSENT FOR PUBLICATION

Not applicable.

CONFLICT OF INTEREST

The authors declare no conflict of interest, financial or otherwise.

ACKNOWLEDGEMENTS

The authors sincerely thank Director CSIR-NIIST for his continuous encouragement and support. The authors acknowledge funding support from Science & Engineering Research Board (SERB), DST, India, project no. ECR/2018/001521.

REFERENCES

[1]   Friedel, G. Les États Mésomorphes de La Matière. *Ann. Phys. (Paris)*, **1922**, *9*(18), 273-474.
       [http://dx.doi.org/10.1051/anphys/192209180273]

[2]   McMillan, W.L. Simple Molecular Theory of the Smectic C Phase. *Phys. Rev. A*, **1973**, *8*(4), 1921-1929.
       [http://dx.doi.org/10.1103/PhysRevA.8.1921]

[3]   Meyer, R.B.; Liebert, L.; Strzelecki, L.; Keller, P. Ferroelectric liquid crystals. *J. Phys. Lettres*, **1975**, *36*(3), 69-71.
       [http://dx.doi.org/10.1051/jphyslet:0197500360306900]

[4]   d'Alessandro, A.; Campoli, F.; Beccherelli, R.; Wnek, M.; Galloppa, A.; Galbato, A.; Maltese, P. Performance of a Passive Matrix Ferroelectric Liquid Crystal Display with Analog Grey Levels. *Proceedings of 8th Mediterranean Electrotechnical Conference on Industrial Applications in Power Systems, Computer Science and Telecommunications (MELECON 96)*, *Vol. 3*, pp. 1513-1516.
       [http://dx.doi.org/10.1109/MELCON.1996.551238]

[5]   Cohen, G.B.; Pogreb, R.; Vinokur, K.; Davidov, D. Spatial light modulator based on a deformed-helix ferroelectric liquid crystal and a thin a-Si:H amorphous photoconductor. *Appl. Opt.*, **1997**, *36*(2), 455-459.
       [http://dx.doi.org/10.1364/AO.36.000455] [PMID: 18250693]

[6]   Swenson, C.M.; Steed, C.A.; De La Rue, I.A.; Fugate, R.Q. Low-Power FLC-Based Retromodulator Communications System. Mecherle, G. S., Ed.; 1997; pp 296–310. , **1997**.

[7]   Shukla, R.K.; Liebig, C.M.; Evans, D.R.; Haase, W. Electro-optical behaviour and dielectric dynamics of harvested ferroelectric $LiNbO^3$ nanoparticle-doped ferroelectric liquid crystal nanocolloids. *RSC Advances*, **2014**, *4*(36), 18529.

[http://dx.doi.org/10.1039/c4ra00183d]

[8]   Pratap Singh, D.; Kumar Gupta, S.; Pandey, S.; Singh, K.; Manohar, R. Electro-optical, uv absorbance, and uv photoluminescence analysis of Se 95 In 5 chalcogenide glass microparticle doped ferroelectric liquid crystal. *J. Appl. Phys.,* **2014**, *115*(21), 214103.
[http://dx.doi.org/10.1063/1.4880997]

[9]   Popova, E.V.; Gamzaeva, S.A.; Krivoshey, A.I.; Kryshtal, A.P.; Fedoryako, A.P.; Prodanov, M.F.; Kolosov, M.A.; Vashchenko, V.V. Dielectric properties of magnetic nanoparticles' suspension in a ferroelectric liquid crystal. *Liq. Cryst.,* **2015**, *42*(3), 334-343.
[http://dx.doi.org/10.1080/02678292.2014.988763]

[10]   Clark, N.A.; Lagerwall, S.T. Submicrosecond Bistable Electro□optic Switching in Liquid Crystals. *Appl. Phys. Lett.,* **1980**, *36*(11), 899-901.
[http://dx.doi.org/10.1063/1.91359]

[11]   Jákli, A.; Markscheffel, S.; Saupe, A. Helix Deformation and Bistable Switching of Ferroelectric Liquid Crystals. *J. Appl. Phys.,* **1996**, *79*(4), 1891-1894.
[http://dx.doi.org/10.1063/1.361091]

[12]   Kaur, S.; Thakur, A.K.; Chauhan, R.; Bawa, S.S.; Biradar, A.M. Bistability in deformed helix ferroelectric liquid crystal. *J. Appl. Phys.,* **2004**, *96*(5), 2547-2551.
[http://dx.doi.org/10.1063/1.1775047]

[13]   A.; Kresse, H.; Reshetnyak, V.; Reznikov, Y.; Yaroshchuk, O. Memory effect in filled nematic liquid crystals. *Liq. Cryst.,* **1997**, *23*(2), 241-246.
[http://dx.doi.org/10.1080/026782997208505]

[14]   Kaur, S.; Singh, S.P.; Biradar, A.M.; Choudhary, A.; Sreenivas, K. Enhanced electro-optical properties in gold nanoparticles doped ferroelectric liquid crystals. *Appl. Phys. Lett.,* **2007**, *91*(2), 023120.
[http://dx.doi.org/10.1063/1.2756136]

[15]   Prakash, J.; Choudhary, A.; Kumar, A.; Mehta, D.S.; Biradar, A.M. Nonvolatile memory effect based on gold nanoparticles doped ferroelectric liquid crystal. *Appl. Phys. Lett.,* **2008**, *93*(11), 112904.
[http://dx.doi.org/10.1063/1.2980037]

[16]   Goel, P.; Biradar, A.M. Tunability of optical memory in ferroelectric liquid crystal containing polyvinylpyrrolidone capped ni nanoparticles for low power and faster device operation. *Appl. Phys. Lett.,* **2012**, *101*(7), 074109.
[http://dx.doi.org/10.1063/1.4746766]

[17]   Marino, L.; Scaramuzza, N.; Marino, S. "Non-volatile memory effect in iron nanoparticles dispersed ferroelectric liquid crystal," IJRDO -. *J. Appl. Sci.,* **2019**, *5*(9), 11-24.

[18]   Chaudhary, A.; Malik, P.; Mehra, R.; Raina, K.K. Observation of memory behaviour in cadmium sulphide nanorods doped ferroelectric liquid crystal mixture. *Phase Transit.,* **2013**, *86*(12), 1256-1266.
[http://dx.doi.org/10.1080/01411594.2012.748908]

[19]   Kumar, A.; Prakash, J.; Khan, M.T.; Dhawan, S.K.; Biradar, A.M. Memory effect in cadmium telluride quantum dots doped ferroelectric liquid crystals. *Appl. Phys. Lett.,* **2010**, *97*(16), 163113.
[http://dx.doi.org/10.1063/1.3495780]

[20]   Pandey, S.; Singh, D.P.; Agrahari, K.; Srivastava, A.; Czerwinski, M.; Kumar, S.; Manohar, R. CdTe quantum dot dispersed ferroelectric liquid crystal: transient memory with faster optical response and quenching of photoluminescence. *J. Mol. Liq.,* **2017**, *237*, 71-80.
[http://dx.doi.org/10.1016/j.molliq.2017.04.035]

[21]   Chaudhary, A.; Shukla, R.K.; Malik, P.; Mehra, R.; Raina, K.K. ZnO/FLC Nanocomposites with low driving voltage and non-volatile memory for information storage applications. *Curr. Appl. Phys.,* **2019**, *19*(12), 1374-1378.
[http://dx.doi.org/10.1016/j.cap.2019.08.026]

[22]   Chandran, A.; Prakash, J.; Ganguly, P.; Biradar, A.M. Zirconia nanoparticles/ferroelectric liquid

crystal composites for ionic impurity-free memory applications. *RSC Advances,* **2013**, *3*(38), 17166.
[http://dx.doi.org/10.1039/c3ra41964a]

[23]   Chandran, A.; Prakash, J.; Gangwar, J.; Joshi, T.; Srivastava, A.K.; Haranath, D.; Biradar, A.M. Low-voltage electro-optical memory device based on NiO nanorods dispersed in a ferroelectric liquid crystal. *RSC Advances,* **2016**, *6*(59), 53873-53881.
[http://dx.doi.org/10.1039/C6RA04037C]

[24]   Malik, P.; Chaudhary, A.; Mehra, R.; Raina, K. K. Dielectric studies and memory effect in nanoparticle doped ferroelectric liquid crystal films. *d Mol. Cryst. Liq. Crys,* **2011**, *541*(1)
[http://dx.doi.org/10.1080/15421406.2011.569246]

[25]   Kumar, A.; Prakash, J.; Choudhary, A.; Biradar, A.M. Dielectric and Electro-Optical Studies of Glycerol/Ferroelectric Liquid Crystal Mixture at Room Temperature. *J. Appl. Phys.,* **2009**, *105*(12), 124101.
[http://dx.doi.org/10.1063/1.3149781]

[26]   Petrov, M.; Katranchev, B.; Rafailov, P.M.; Naradikian, H.; Dettlaff-Weglikowska, U.; Keskinova, E.; Smectic, C. Liquid Crystal Growth and Memory Effect through Surface Orientation by Carbon Nanotubes. *J. Mol. Liq.,* **2013**, *180*, 215-220.
[http://dx.doi.org/10.1016/j.molliq.2013.01.015]

[27]   Blinov, L.M.; Chigrinov, V.G. *Electrooptic Effects in Liquid Crystal Materials*; Springer New York: New York, NY, **1994**.
[http://dx.doi.org/10.1007/978-1-4612-2692-5]

[28]   Shukla, R.K.; Chaudhary, A.; Bubnov, A.; Raina, K.K. Multi-Walled Carbon Nanotubes-Ferroelectric Liquid Crystal Nanocomposites: Effect of Cell Thickness and Dopant Concentration on Electro-Optic and Dielectric Behaviour. *Liq. Cryst.,* **2018**, *45*(11), 1672-1681.
[http://dx.doi.org/10.1080/02678292.2018.1469170]

[29]   Goel, P.; Singh, G.; Pant, R.P.; Biradar, A.M. Investigation of Dielectric Behaviour in Ferrofluid–Ferroelectric Liquid Crystal Nanocomposites. *Liq. Cryst.,* **2012**, *39*(8), 927-932.
[http://dx.doi.org/10.1080/02678292.2012.687118]

[30]   Coondoo, I.; Goel, P.; Malik, A.; Biradar, A.M. Dielectric and Polarization Properties of BaTio 3 Nanoparticle/Ferroelectric Liquid Crystal Colloidal Suspension. *Integr. Ferroelectr.,* **2011**, *125*(1), 81-88.
[http://dx.doi.org/10.1080/10584587.2011.574078]

[31]   Guo; Yan; Chigrinov; Zhao; Tribelsky. Ferroelectric Liquid Crystals: Physics and Applications. *Crystals (Basel),* **2019**, *9*(9), 470.
[http://dx.doi.org/10.3390/cryst9090470]

[32]   Tripathi, S.; Ganguly, P.; Haranath, D.; Haase, W.; Biradar, A.M. Optical Response of Ferroelectric Liquid Crystals Doped with Metal Nanoparticles. *Appl. Phys. Lett.,* **2013**, *102*(6), 063115.
[http://dx.doi.org/10.1063/1.4792687]

[33]   Podgornov, F.V.; Gavrilyak, M.; Karaawi, A.; Boronin, V.; Haase, W. Mechanism of Electrooptic Switching Time Enhancement in Ferroelectric Liquid Crystal/Gold Nanoparticles Dispersion. *Liq. Cryst.,* **2018**, *45*(11), 1594-1602.
[http://dx.doi.org/10.1080/02678292.2018.1458256]

[34]   Kumar, A.; Singh, G.; Joshi, T.; Rao, G.K.; Singh, A.K.; Biradar, A.M. Tailoring of Electro-Optical Properties of Ferroelectric Liquid Crystals by Doping Pd Nanoparticles. *Appl. Phys. Lett.,* **2012**, *100*(5), 054102.
[http://dx.doi.org/10.1063/1.3681381]

[35]   Kaur, R.; Bhullar, G.K.; Raina, K.K. Effects of Silver Nanoparticles Doping on Morphology and Luminescence Behaviour of Ferroelectric Liquid Crystals Langmuir–Blodgett Films. *Liq. Cryst.,* **2016**, *43*(12), 1760-1767.
[http://dx.doi.org/10.1080/02678292.2016.1200678]

[36] Neeraj; Raina, K. K. Nickel Nanoparticles Doped Ferroelectric Liquid Crystal Composites. *Opt. Mater. (Amst)*, **2013**, *35*(3), 531-535.
[http://dx.doi.org/10.1016/j.optmat.2012.10.014]

[37] Khushboo; Bhargava, N.; Anand, K.; Malik, P.; Sharma, P.; Jayoti, D.; Raina, K. K. Dielectric and Polarization Switching Studies in Nickel Nanoparticles Dispersed Ferroelectric Liquid Crystal Mixtures. *Integr. Ferroelectr.*, **2017**, *184*(1), 192-198.
[http://dx.doi.org/10.1080/10584587.2017.1368657]

[38] Li, L-S.; Huang, J.Y. Tailoring Switching Properties of Dipolar Species in Ferroelectric Liquid Crystal with ZnO Nanoparticles. *J. Phys. D Appl. Phys.*, **2009**, *42*(12), 125413.
[http://dx.doi.org/10.1088/0022-3727/42/12/125413]

[39] Tripathi, P.K.; Yadav, S.P.; Singh, S. Impact of Silica Nanoparticles Dispersion on the Dielectric and Electro-Optical Properties and Absorption Spectra of Host Ferroelectric Liquid Crystal. *Liq. Cryst.*, **2018**, *45*(7), 953-960.
[http://dx.doi.org/10.1080/02678292.2017.1397784]

[40] Chaudhary, A.; Malik, P.; Mehra, R.; Raina, K.K. Electro-optic and dielectric studies of silica nanoparticle doped ferroelectric liquid crystal in smc* phase. *Phase Transit.*, **2012**, *85*(3), 244-254.
[http://dx.doi.org/10.1080/01411594.2011.624274]

[41] Malik, P.; Chaudhary, A.; Mehra, R.; Raina, K.K. Electrooptic and dielectric studies in cadmium sulphide nanorods/ferroelectric liquid crystal mixtures. *Adv. Condens. Matter Phys.*, **2012**, *2012*, 1-8.
[http://dx.doi.org/10.1155/2012/853160]

[42] Doke, S.; Ganguly, P.; Mahamuni, S. Improvement in molecular alignment of ferroelectric liquid crystal by Co-ZnO/ZnO Core/Shell quantum dots. *Liq. Cryst.*, **2020**, *47*(3), 309-316.
[http://dx.doi.org/10.1080/02678292.2019.1645898]

[43] Shukla, R.K.; Galyametdinov, Y.G.; Shamilov, R.R.; Haase, W. Effect of cdse quantum dots doping on the switching time, localised electric field and dielectric parameters of ferroelectric liquid crystal. *Liq. Cryst.*, **2014**, *41*(12), 1889-1896.
[http://dx.doi.org/10.1080/02678292.2014.959571]

[44] Chandran, A.; Prakash, J.; Naik, K.K.; Srivastava, A.K.; Dąbrowski, R.; Czerwiński, M.; Biradar, A.M. Preparation and characterization of mgo nanoparticles/ferroelectric liquid crystal composites for faster display devices with improved contrast. *J. Mater. Chem. C Mater. Opt. Electron. Devices*, **2014**, *2*(10), 1844.
[http://dx.doi.org/10.1039/c3tc32017k]

[45] Khan, S.; Chauhan, S.; Chandran, A.; Czerwiński, M.; Herman, J.; Biradar, A.M.; Prakash, J. Enhancement of dielectric and electro-optical parameters of a newly prepared ferroelectric liquid crystal mixture by dispersing nano-sized copper oxide. *Liq. Cryst.*, **2020**, *47*(2), 263-272.
[http://dx.doi.org/10.1080/02678292.2019.1643506]

[46] Misra, A.K.; Roy, A.; Pratap Singh, B.; Pandey, K.K.; Shrivas, R.; Saluja, J.K.; Tripathi, P.K.; Manohar, R. Influence of SiO_2 Nanoparticles on the Dielectric Properties and Anchoring Energy Parameters of Pure Ferroelectric Liquid Crystal. *J. Dispers. Sci. Technol.*, **2020**, *41*(14), 2136-2142.
[http://dx.doi.org/10.1080/01932691.2019.1653195]

[47] Joshi, T.; Prakash, J.; Kumar, A.; Gangwar, J.; Srivastava, A.K.; Singh, S.; Biradar, A.M. Alumina nanoparticles find an application to reduce the ionic effects of ferroelectric liquid crystal. *J. Phys. D Appl. Phys.*, **2011**, *44*(31), 315404.
[http://dx.doi.org/10.1088/0022-3727/44/31/315404]

[48] Podgornov, F.V.; Suvorova, A.M.; Lapanik, A.V.; Haase, W. Electrooptic and dielectric properties of ferroelectric liquid crystal/single walled carbon nanotubes dispersions confined in thin cells. *Chem. Phys. Lett.*, **2009**, *479*(4–6), 206-210.
[http://dx.doi.org/10.1016/j.cplett.2009.08.005]

[49] Arora, P.; Mikulko, A.; Podgornov, F.; Haase, W. Dielectric and electro-optic properties of new ferroelectric liquid crystalline mixture doped with carbon nanotubes. *Mol. Cryst. Liq. Cryst. (Phila. Pa.),* **2009**, *502*(1), 1-8.
[http://dx.doi.org/10.1080/15421400902813592]

[50] Prakash, J.; Kumar, A.; Joshi, T.; Mehta, D. S.; Biradar, A. M.; Haase, W. *Spontaneous polarization in smectic a phase of carbon nanotubes doped deformed helix ferroelectric liquid crystal.,* **2011**.
[http://dx.doi.org/10.1080/15421406.2011.570216]

[51] Agrahari, K.; Pathak, G.; Vimal, T.; Kurp, K.; Srivastava, A.; Manohar, R. Dielectric and spectroscopic study of nano-sized diamond dispersed ferroelectric liquid crystal. *J. Mol. Liq.,* **2018**, *264*, 510-514.
[http://dx.doi.org/10.1016/j.molliq.2018.05.097]

[52] Shukla, R.K.; Raina, K.K.; Haase, W. Fast switching response and dielectric behaviour of fullerene/ferroelectric liquid crystal nanocolloids. *Liq. Cryst.,* **2014**, *41*(12), 1726-1732.
[http://dx.doi.org/10.1080/02678292.2014.949889]

[53] Singh, D.P.; Duponchel, B.; Boussoualem, Y.; Agrahari, K.; Manohar, R.; Kumar, V.; Pasricha, R.; Pujar, G.H.; Inamdar, S.R.; Douali, R.; Daoudi, A. Dual photoluminescence and charge transport in an alkoxy biphenyl benzoate ferroelectric liquid crystalline–graphene oxide composite. *New J. Chem.,* **2018**, *42*(20), 16682-16693.
[http://dx.doi.org/10.1039/C8NJ02985G]

[54] Liang, H-H.; Xiao, Y-Z.; Hsh, F-J.; Wu, C-C.; Lee, J-Y. Enhancing the electro-optical properties of ferroelectric liquid crystals by doping ferroelectric nanoparticles. *Liq. Cryst.,* **2010**, *37*(3), 255-261.
[http://dx.doi.org/10.1080/02678290903564403]

[55] Mikułko, A.; Arora, P.; Glushchenko, A.; Lapanik, A.; Haase, W. Complementary Studies of BaTiO3 Nanoparticles Suspended in a Ferroelectric Liquid-Crystalline Mixture. EPL (Europhysics Lett.), 2009, 87 (2), 27009.

[56] Rudzki, A.; Evans, D.R.; Cook, G.; Haase, W. Size dependence of harvested BaTiO$_3$ nanoparticles on the electro-optic and dielectric properties of ferroelectric liquid crystal nanocolloids. *Appl. Opt.,* **2013**, *52*(22), E6-E14.
[http://dx.doi.org/10.1364/AO.52.0000E6] [PMID: 23913089]

CHAPTER 13

Next-Generation Energy Storage and Optoelectronic Nanodevices

Debabrata Panda[1] and **Krunal M. Gangawane**[1,*]

[1] *Department of Chemical Engineering, National Institute of Technology Rourkela, Rourkela-769008, Odisha, India*

Abstract: Among the variety of nanostructures that have been explored as a favorable material for the application of higher energy storage devices as supercapacitors, catalysts in high-performance batteries, proton exchange membranes in fuel cells, optoelectronic devices, and so on, 2D & 3D nanostructure of graphene-based derivatives, metal oxides and dichalcogenides have received the most potential attention for building high-performance nano-devices due to their extraordinary properties. Over the past decade, several efforts have been implemented to design, develop, and evaluate electrodes' structures for enhanced energy storage devices. A significant modification has achieved the remarkable performance of these synthesized devices in terms of energy storage capacity, conversion efficiency, and the reliability of the devices to meet practical applications' demands. Light-emitting diode (LED) in quantum well or quantum dots is considered an important aspect for an enhanced optoelectronic device. This current study outlines different 3D nanostructures for next-generation energy storage devices. It provides a systematic summary of the advantages of 3D nanostructures in perspective to next-generation energy storage devices, photocatalytic devices, solar cells, a counter electrode for metal-ion batteries, and supercapacitors, optoelectronic nano-devices.

Keywords: 2D & 3D nanostructure, Counter electrode, Fuel cell, Graphene derivatives, LED, Li-ion Battery, Metal-ion battery, Nanodevices, Optoelectronic, Photocatalytic device, Quantum dots, Supercapacitor, Solar cells.

INTRODUCTION

Over the current decade, due to the large fluctuations in energy demand and supply, energy storage has become an essential area of discussion as more energy is being obtained from intermittent sources.

A conventional capacitor generally possesses a comparatively high power density

---

* **Corresponding author Krunal M. Gangawane:** Department of Chemical Engineering, National Institute of Technology, Rourkela, Odisha, India; E-mail: gangawanek@nitrkl.ac.in

Gaurav Manik and Sushanta Kumar Sahoo (Eds.)

in comparison with batteries and can rapidly discharge the power. On the other end, batteries will be able to accumulate a large amount of energy in comparison with conventional capacitors. However, batteries cannot meet the demands, such as high dynamic charge and electrode polarization resulting from energy conversion. Electrochemical capacitors (EC) are primarily used to improvise the performance of batteries and fuel cells inside a hybrid vehicle for quick acceleration, electricity storage devices developed from renewable energy like solar wind energy, and allow additional power for the recovery of energy by storing energy within a solid/electrolyte interface. Those layers' critical thickness exhibits a concentration of electrolyte with adequate size in the range of 5-10 A^0 and capacitance of 10-20μF/cm^2 for a smooth electrode [1]. The main reason behind ECs attracting customers for an extensive range of applications is illustrated in Fig. (**1**) by a Ragone plot.

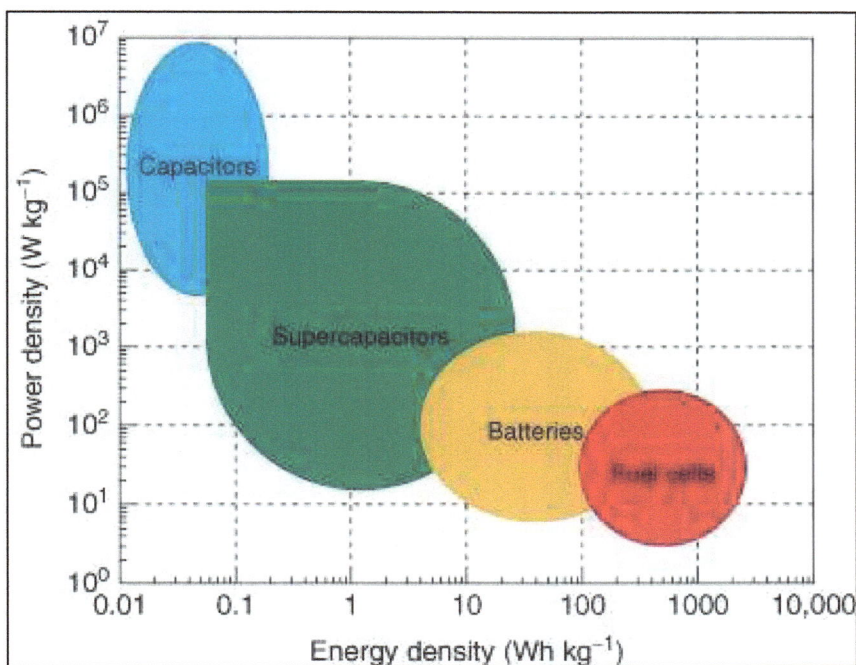

Fig. (1). Ragone plot of various energy storage systems [1].

Electrochemical double-layer Supercapacitors (EDLCs) are generally high-performance energy storage micro-devices that provide a capacitance value higher than a capacitor but with low voltage limits that channel the gap between electrochemical devices and rechargeable batteries. It consists of an electrochemical cell containing an anode, cathode, electrolyte, and a separator. The electrode composition is usually Nickel (Ni), cobalt (Co), molybdenum (M), tungsten (W), *etc.,* which is separated by an electrolyte of basic, acidic, or neutral. The different types of supercapacitors made from different sources and for different applications are represented in Fig. (**2**). A rapid growth in the field of supercapacitors (2D and 3D nanostructures) was used in miniaturized microelectronics (*e.g.*, portable, wearable) and macro devices (*e.g.*, hybrid electric vehicle, sensitive automation, computer chips, automobiles, constructions equipment) due to its enhanced characteristics (high cycle life, charging and discharging characteristics, and specific power) [2]. A developed supercapacitor (SCs) of rectangular tube polyaniline structure bridges the gap between the rechargeable battery and conventional capacitor for enhanced power storage for quick charge/discharge [3], with an enhanced power density of 1091 Wkg^{-1} and a specific capacitance of 4007 F g^{-1}, which shows extreme flexibility of about 96% beneficial for energy recovery [4]. However, there is an increase in demand to develop sodium (Na) based energy storage as an alternative to large-scale energy storage [5]. A critical comparison between capacitors, SCs, and batteries for their characteristics is summarized in Table **1**.

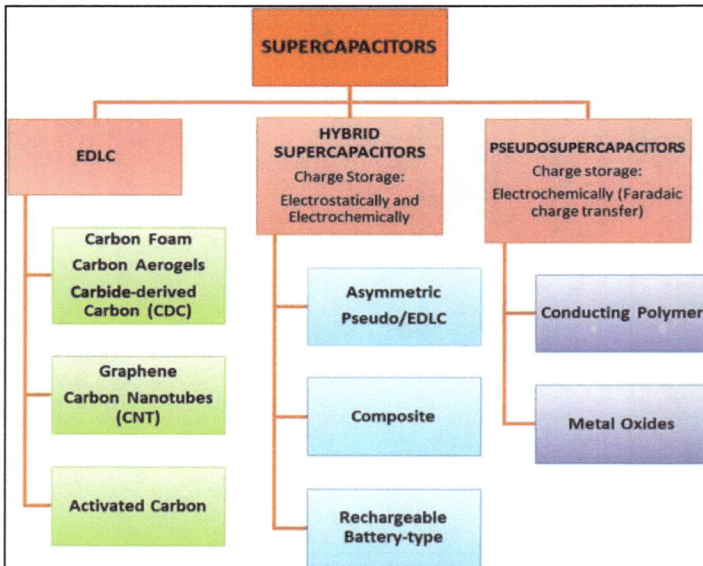

Fig. (2). Types of supercapacitors for technical applications.

Table 1. Assessment of essential characteristics between various energy storage systems.

Sl No	Characteristic	Conventional Capacitors	Supercapacitor	Rechargable Battery	References
1	Specific energy (Wh/kg)	0.01-0.1	1-1000	10-100	[4, 5]
2	Specific power (W/kg)	>10000	500-10000	< 1000	[4]
3	Discharging time	10^{-6}-10^{-3} Sec	1 minute	0.025-2.5 h	[6]
4	Charging time	10^{-6}-10^{-3} Sec	1 minute	6 h	[6]
5	Coulomb efficiency	~100%	~95%	70-85%	[6]
6	Cycle life	Maximum	>500000	~1000	[6]
7	Temperature range (^{0}C)	-40-125^{0}C	-40-70^{0}C	-20-70^{0}C	[6]
8	Maximum charge (Volt)	4-630 V	1.2-3.8 V	2.5-4.2 V	[6]
9	Capacitance (F)	< 2.7 F	0.1-3500 F	-	[3]

On the other end, novel materials like graphene or graphene oxide (GO) derivatives, metal or transition metal oxide dichalcogenides (TMDs), two-dimensional (2D) or three-dimensional nanostructured materials, perovskites, and other nanostructures (quantum dots) have shown tremendous results over the technical fields [7 - 9]. With enhanced physical, thermal, and optoelectrical properties of two-dimensional atomic structure, interlayer distance, functionalization, bandgap modification, these functionalized nano-materials had been explicitly used in the development of innovative optoelectronic applications, power electronics, spintronic, and bio-sensing devices [8, 10, 11]. Most optoelectronic devices include LEDs, dye synthesized solar cells (DSCs), photodetectors, transistors, piezo electronic devices, and bio-sensors in medical applications (Fig. **3**).

Feng. *et al.* [13], presented a combined study on surface plasmon and quantum dots by incorporating silver nanoparticles inside the hole of a P-GaN layer of photonic crystal green LEDs (Fig. **3a**) and a 3D FDM method is applied to delineate the impact of light. Yan *et al.* [14] assessed the performance characteristics of CdSe green and red quantum dot LEDs at a distinct frequency by analyzing their photoluminescence of individual thin films. In this study, the excitation wavelength of 365, 385, 405, and 455 nm were compared, out of which green and red quantum dot LEDs exhibited similar kinds of patterns. The energy conversion efficiency of the 455nm was 14.1% higher than 365nm LED with an incident radiant flux of 20Mw (Fig. **3b**). Stylianakis *et al.* [15] compared the field emitter performance of polymeric and fullerene derivatives towards the development of rGO based on various ratios of a cold cathode (Fig. **3c**). The prepared composites exhibit excellent stability of a high field enhancement factor.

The combination of two individual sheets was analyzed by Zhao *et al.* [16] for a Ga-N-based Ultraviolet (UV) LED to increase the light extraction efficiency, and minimize ohmic contact resistance with p-GaN (Fig. **3d**). In this context, a graphene-based material with favourable energy levels on active layers of the synthesized solar cell was prepared, which exhibited a performance improvement of 13%, and the improved power conversion efficiency was attributed to the transport of electrons from active layers to the bottom of the electrode [17].

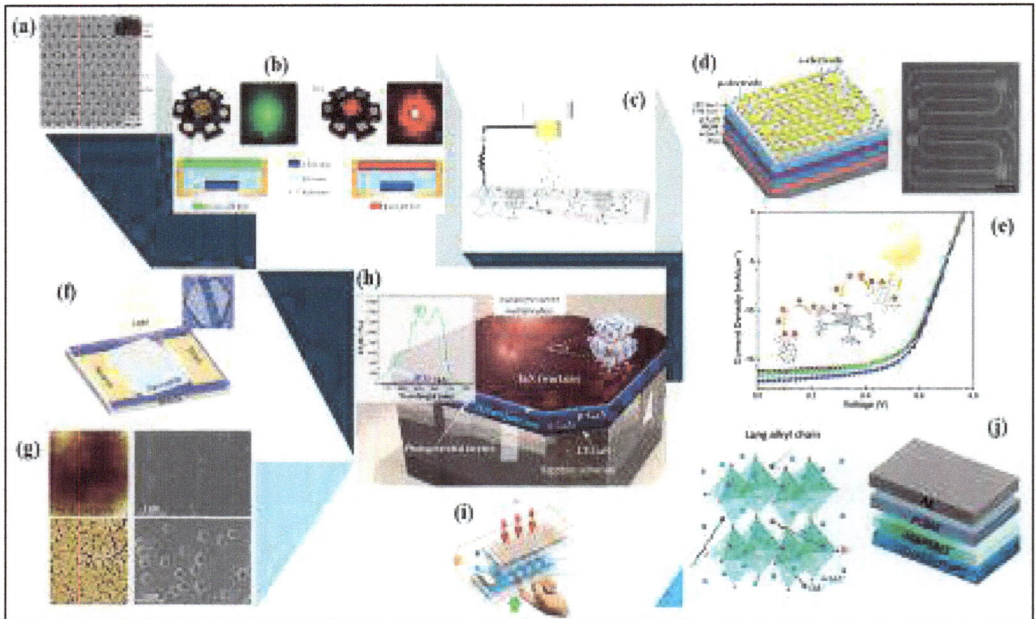

Fig. (3). Optoelectronic application of Nanodevices [12].

Xue *et al.* [18] explored a dual approach regarding a 2D perovskite platelet phototransistor, which is compatible with conventional capacitors with a melting temperature of over 100°C (Fig. **3f**). Besides, the role of a V pit entrenched with InGaN/GaN superlattices along with green LEDs enactment was analyzed to expand the quantum efficiency up to 30%. An SEM micro-image of cathodo-luminescence was used to authenticate the light emission characteristics of quantum wells which acts as an obstacle for carriers' inclusion into the non-recombination centers [19]. Svrcek *et al.* [20] covenants in the development of InN thin film over a metalorganic gaseous stage and an optimal condition of InN/p-GaN electron extraction efficiency was analyzed. On the other hand, Lee *et al.* [21] reported an ultrahigh sensible CdTe micro-dots-based photodetector over a Bi coated Indium titanate oxide, which exhibited excellent stability, better effectiveness under variable stress environments due to the piezo-phototronic effect (Fig. **3i**). However, Chen *et al.* [22] organized a highly fluorescent $MaPbBr_3$ thin film with octyl ammonium bromide as an additive within the pervskite precursors, which showed higher current efficiency values due to its higher excited nanocrystals grain size with efficient binding energy (Fig. **3j**).

LATEST TRENDS IN SUPERCAPACITORS

Most of the miniaturized power electronic nanodevices (portable, wearable, implantable) have been incorporated into day-to-day life in the form of control units in electronics, wireless receivers, signal sensors, energy harvesting units, and miniaturized energy storage units. In comparison with conventional capacitors, a supercapacitor is able to store 10-100 times more energy per unit volume and delivers energy much faster than batteries. A critical discussion on varieties of electrochemical supercapacitors such as EDLC, PCs, 3D nanoporous SCs, coated SCs, fiber and yarn type SCs, paper structured SCs, 3D printed and battery type SCs.

Electrochemical Double-layer Capacitors (EDLC)

In EDLCs, energy exhibits an enhanced form in comparison with conventional capacitors and rechargeable batteries because it utilizes an interface between electrodes and an electrolyte through its enhanced active surface areas and minimum separation distances. Since the charge separation is across a small distance, the interphase can be set up between electrodes and adjacent electrolytes (KOH, H_2SO_4, or Na_2CO_3). Moreover, a higher amount of energy can be kept on the electrode with large surface areas, provided by a large number of pores. The energy storage mechanism can respond because of a simple movement of ions between electrode and electrolyte and, the interaction between them and the surface wettability [3, 4, 23]. Usually, a specific grade of abundant and inexpensive carbon is widely used in EDLCs, which exhibits a specific capacitance of 100-140 F/g in an organic electrolyte. For further improvement in capacitance, sophisticated carbon structures (SWCNT, MWCNT, graphene) were used.

Pseudo Capacitors (PCs)

A relatively high capacitance under a non-electrostatic base is found in pseudocapacitors because of the increased number of active sites and a reversible faradic type charge transfer between 3D porous structures. PCs are based upon faradic charge processes (electron exchange across the electrode interface) that can be reversible or irreversible. It is renowned for its redox chemical reaction with the help of transition metal oxides (titanium dioxide(TiO_2), Zinc oxides(ZnO), *etc.*), and conducting polymers. It exhibits 10-100 times more capacitance than that of EDLCs.

3D Porous Supercapacitor

An effective method to improve the specific capacitance of a supercapacitor is to increase active areas and micro-porous structures in a 3D form of the electrode. An electrode employing Nickel hydroxide ($Ni(OH)_2$) along with a nanoporous gold nanostructure provides a specific capacitance of 3168 Fg^{-1} and volumetric capacitance of 2223 Fcm^{-3} with a current density of 5 A/g because of its flexible 3D structure [24]. The fabricated electrode also shows an excellent energy density of 98Wh/kg and a power density of 50 kW/kg. Placing a thin layer of $Ni(OH)_2$ over a carbon-coated 3DCu electrode can provide a higher electrical conductivity by introducing some active areas on 3D-C/Cu. It can also provide a reduction of a pathway for electron transport between electrodes and provides a specific capacitance of 1860 F/g^{-1} respectively [25].

Paper-like Supercapacitor

A paper-like SC can easily be modified into other desirable shapes (2D / 3D) or structures according to requirements. It can be used as wearable electronics for health monitoring, electronic textiles, and artificial skins due to its customized architecture (pyramid, honeycomb, living-hinge) and stretchability in a random direction. For instance, an ultra-long MnO_2 nanowire composite SC of honeycomb structure exhibits a specific capacitance of 220-228 $mF\ cm^{-2}$ and a stretchability of more than 500% without any degradation electrochemical properties. The composite also shows 98% of its initial capacitance after 10,000 cycles under 400% of tensile strain [26]. Some of the advanced architectures for paper-like supercapacitors are illustrated in Table **2**.

Table 2. Summary of architectures for supercapacitors technology.

Sl No	Material	Structure	Electrolyte	Measurement	Capacitance	Refs
1	Mn_3O_4/rGO	Paper	KOH	0.5 Ag^{-1}	410 F/g	[27]
2	CNF/BDD	Battery	H_2SO_4	10mVs^{-1}	118 mF/cm^2	[28]
3	CNF/PEDOT	Thinfilm	$HClO_4$	-	235 F/g	[29]
4	NG/MnO_2	3D Flower	Na_2SO_4	0.5 Ag^{-1}	225 F/g	[30]

(Table 2) cont.....

Sl No	Material	Structure	Electrolyte	Measurement	Capacitance	Refs
5	BPC/MnO_2	Anchored MnO_2 on Biomass	Na_2SO_4	0.5 Ag^{-1}	385 F/g	[31]
6	BPC/Fe_2O_3	Porous carbon with Fe_2O_3	KOH	1 Ag^{-1}	990 F/g	[32]
7	3D NiCo LDH/NiNw	3D cotton	KOH	0.125 Ag^{-1}	465 F/g	[33]
8	NiCo hydroxide/NF	3D Flower	KOH	10 $mAcm^{-2}$	1450 $\mu Ah/cm^2$	[34]
9	MnO_2/CNT	Ultra-thin	$LiClO_4$	5mvs^{-1}	135 mF/cm^2	[35]
10	HC/ rGO	laminated	KOH	1 Ag^{-1}	1665 F/g	[36]
11	N_2 doped graphene	Bubble	KOH	1 Ag^{-1}	480 F/g	[37]
12	$CoMoO_4$@$Co_{1.5}Ni_{1.5}S_4$	Rambutan	KOH	1 Ag^{-1}	1410 F/g	[38]
13	Ni-Co-LDH@NiOH	Battery	KOH	1 Ag^{-1}	2630 F/g	[39]
14	NiCo-S	Nanosheet	KOH	0.5 Ag^{-1}	2555 F/g	[40]
15	N_2-doped hierarchical carbon	Fiber web	H_2SO_4	1 Ag^{-1}	298 F/g	[41]
16	$Ni_{0.4}Co_{0.6}(OH)_2$	Peony	KOH	1 Ag^{-1}	1818 F/g	[42]
17	3D/PGLS	Meso-porous	BF_4	0.2 Ag^{-1}	91-92 F/g	[43]

Fiber and Yarn-like Supercapacitor

Several methods, other than coating a conducting polymeric layer on existing fabric, require using woven or knitted materials into various stretchable circuitries to prepare an electronic device for future wearable electronics such as to analyze healthcare data, physiological signals, and motion-sensing inside the human bodies. Fiber-like SCs are specially developed for energy storage devices with outstanding electrochemical performance.

3D Printed Supercapacitor

Additive manufacturing always offers a distinctive stage of rapid prototyping for various applications with enhanced physical and electrochemical properties. These beneficial properties made it suitable for preparing higher energy generation/storage devices. Foster *et al.* [44] prepared a graphene oxide-based polylactic acid (GO/PLA) and a graphene oxide-based acrylonitrile butadiene styrene (GO/ABS) conductive filament with a graphene loading of 5.6% with the help of fused deposition modeling. The 3D printed electrodes were characterized both electrochemically and physio-chemically and proven as an excellent freestanding anode for Li-ion battery and supercapacitor as shown in Figs. (**4**) and (**5**). Aqueous inks composed of 1-3 nm of 2D $Ti_3C_2T_x$ with an enhanced lateral pore size of 8μm possess an enhanced specific surface area of active sites with a high specific capacitance of 2-2.5 F cm^{-2} and a gravimetric capacitance of 240-242.5 F/g with retention of above 90% over 10000 cycles. It also exhibits a higher energy density of 0.02-0.024 mWh/cm^2 and a power density of 0.6-0.64 mW/$cm^{2,}$ respectively, making it suitable for functional materials for integrated devices.

Battery type Supercapacitor

A battery-type supercapacitor refers to those ECs with high power density, energy density, ultra-high capacitance, high retention cycle to meet future demands of power electronics, hybrid electric vehicle (HEV), and industrial automation. A vertically aligned carbon nanofiber (CNF) of boron-doped diamond films was employed earlier as a capacitor electrode with enhanced surface area, higher electrical and thermal stability, and exhibited a specific capacitance of 30-48 mF/cm^2 at 10 mV/s, a power density of 25.3-27.3 Kw/Kg, and energy density of 22.9-44.1 Wh/kg suitable for EDLC devices [28].

Sodium (Na) based miniaturized device, as an alternative to Lithium-ion battery, can be prepared with the help of highly porous Na/S mesoporous carbon as a cathode and non-flammable liquid as an electrolyte operating at 100% columbic efficiency with retention of over 3000 cycles. The fabricated composite exhibits an energy density of 716 mA hg^{-1} with a voltage of 3.8 V. At maximum operating conditions, these devices achieved a power density of 1463 W. kg^{-1} at 50°C, respectively [5].

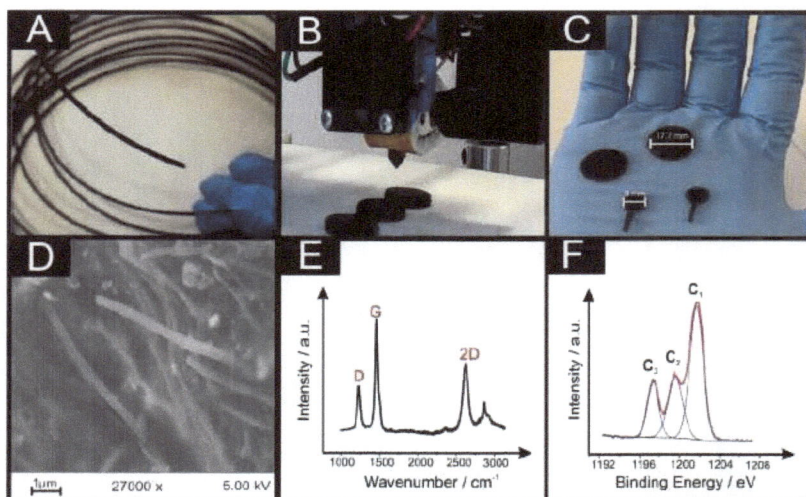

Fig. (4). (a) Graphical image of graphene/PLA electrode (b) FDM process (c) Printed 3DEs (d) SEM microimages of printed electrode (e) Raman spectroscopy of electrode (f) XPS analysis of electrodes [44].

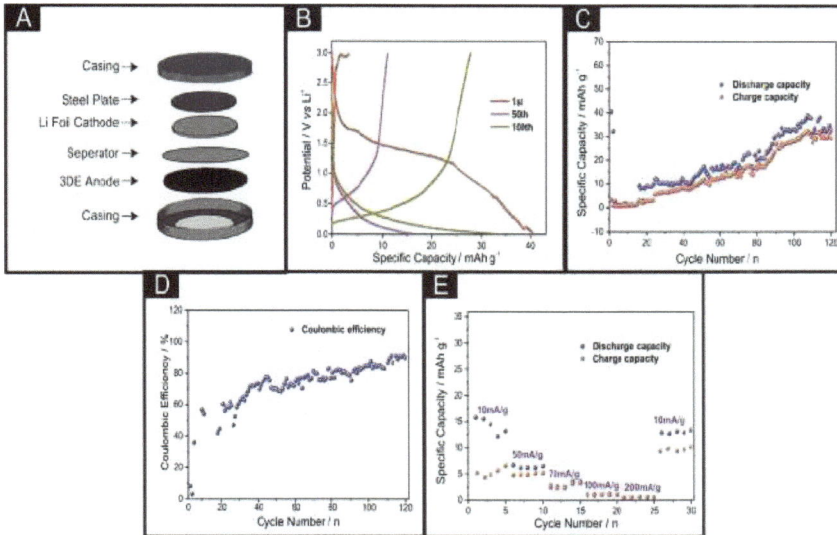

Fig. (5). (a) Schematic representation of coin cell fabrication **(b)** Charge and discharge profile **(c)** Cycling stability **(d)** Columbic efficiency **(e)** Rate of capability [44].

BATTERIES

Lithium-Ion Batteries

A Li-ion battery is an advanced rechargeable battery used in most portable power electronics because of its higher energy density per unit mass in comparison with other energy storage devices. It also has higher energy efficiency, excellent temperature resistance, and a low self-discharge rate. During charging conditions, Li^+ ions are extracted from the cathode and inserted into the lattice of an anode, whereas it is reversed during discharge, as shown in Fig. (6). Most of the PHEVs and EVs nowadays use Li-ion batteries due to their high efficiency, safety, and low cost. Some of the disadvantages like high cost, medium life period, and safety concerns during overheating made it a point of interest to study deeply. With a change in the anode and cathode-based materials as (LMO, LFP, LNMC, LTO, Li-s, *etc.*), the disadvantages can be somehow overcome. A comparison between the characteristics of various common Li-ion batteries is illustrated in Table **3**.

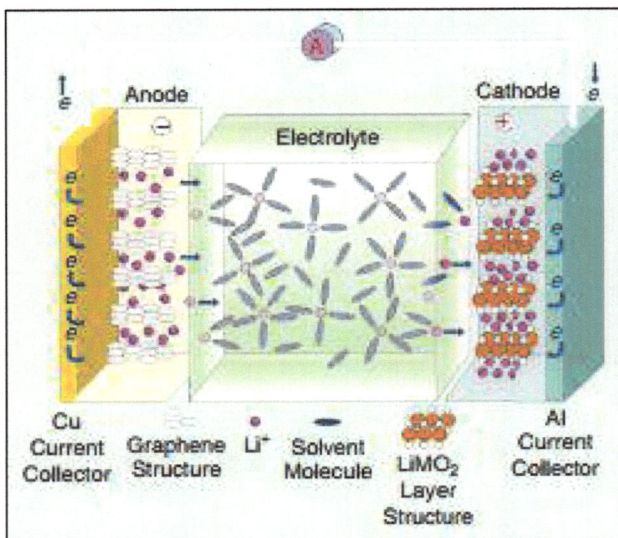

Fig. (6). Schematic representation of Li-Ion battery.

Nickel-Metal Hydride (NiMH) Batteries

A Ni-MH battery primarily comprises a metal-based positive electrode (anode) (Mostly Nickel), metallic cadmium (Cd) based negative electrode, and an alkaline electrolyte (KOH, NaOH *etc.*). In replacement to Ni-Cd, the Ni-MH battery has an enhanced power density, environmental friendliness and is less prone to memory effect. However, it fails in self-discharging rate, low service life, and a very low columbic efficiency (60-65%). During fast charging of these batteries, a huge amount of heat is exhausted from the cell, which can cause cell rupture capacity decay. However, NiMH batteries provide a much higher life cycle in comparison with lead-acid batteries. These batteries have been popularly used in EVs and HEVs since the 2000s [45].

Table 3. Comparison of energy characteristics for various kinds of Li-ion battery [46].

Characteristics	LMO	LFP	LNMC	LTO	Li-S
Energy density (Wh/Kg)	150-170	110-125	190-200	60-80	550
Power density (W/kg)	180-200	180-200	180-200	1000	-
Cycle Life	>2000	>2500	>2000	>10000	>100
Cost	Medium	Medium	Medium	High	-
Safety	High	High	High	High	High

Lead-Acid Batteries

These are intended to provide high power density, inexpensiveness, safety, and reliability. Conversely, the low specific energy, poor temperature profile, and a concise cycle life obstructs or limits their use (EVs for high loads).

Na-S battery

A Na-S battery consists of a molten anode of sulfur, a molten cathode of sodium, and solid alumina ceramics of electrolyte. The charging and discharging profiles have to be maintained at a temperature of 300^0C to keep both the materials in a molten state. These batteries' intriguing potential can deliver a higher energy density of 250 Wh/Kg, cycle efficiency of 85%, and a long retention life cycle of 4000 cycles. A comparison between various attributes of rechargeable batteries is illustrated in Table **4**.

Table 4. Comparison of characteristics between various batteries [45].

Characteristics	Lead-acid	Ni-MH	Li-ion	NaS	VRB
Energy density (Wh/kg)	30-55	65-130	80-200	160-250	15-35
Power density (W/Kg)	80-320	255-1000	550-2000	160-240	70-200
Cycle life (%)	Upto 1000	Upto 2000	Upto 10000	Upto 4000	>12500
Round- trip efficiency	~70	~68	~98	~90	75-90
Self-discharging rate	Low	High	Medium	-	Negligible

OPTOELECTRONIC NANO-DEVICES

Optoelectronics is the communication between optics and electronics, including studying, designing, and manufacturing a device that converts electrical energy into light energy with the help of semiconductors. These are crystalline materials, which are lighter than metals and heavier in comparison with insulators. The optoelectronic nano-devices can be found in space and military services, telecommunications, navigation control by a photodiode, DSCs, LEDs, optical Fiber, and laser diodes. The interaction of light energy sources with structures is much smaller than its wavelength but can be enhanced by the excitation of plasmons, which acquire energy between photons and electrons. As the size of the metal nanostructure and optoelectronic nanodevices approaches atomic scale, the quantization effect of electronics and plasmon structure increases the relevance of light scattering. A 3D hybrid halide perovskite fabricated by electron beam lithography features a nanoscale pattern electrode and superior photodetection of

5.34 mA/W and photo-electron sensitivity of $1.07*10^{13}$ cm Hz/W. Such an orthogonal form of processing and patterning methods enabled high resolution, and creation of complex perovskite-based electronics for the future. A single crystalline n-type CdS nanosheets synthesized by CVD of 40-100nm thick, 10-300 μm has a high on-off ratio of 1.7×10^9, and a peak trans-conductance of 14.1 μS [46 - 48].

CONCLUSION

The subsequential development of next-generation energy storage devices such as electrochemical supercapacitors, rechargeable batteries like Li-ion, Na-S, NiMH batteries will play an important role in the recent future of sustainable energy because of their widespread use in portable electronics, electric/hybrid vehicles, stationary power systems, *etc.* to meet the ever-growing demand of high-performance energy and power density. The extraordinary properties of nanomaterials like high specific surface area, maximum pore size distribution, low thermal conductivity, and high sorption capacity, *etc.,* will make them attractive for researchers and industrialists working in the applications of energy storages, LEDs, image pickup devices, optical communication systems, remote sensing systems, optical fibers and photodiodes. To benchmark the state-of-the-art needed in the development, comprehensive research into the energy storage area would require physical and chemical modifications employing a multidisciplinary approach involving the development of functional nanomaterials for implementation in high-performance electrochemical systems. This shall help meet the widening requirements of exciting applications in the real energy industry.

CONSENT FOR PUBLICATION

Not applicable.

CONFLICT OF INTEREST

The authors declare no conflict of interest, financial or otherwise.

ACKNOWLEDGEMENTS

Declared none.

REFERENCES

[1]   Kim, B.K.; Sy, S.; Yu, A.; Zhang, J. Electrochemical supercapacitors for energy storage and conversion. Handb. Clean Energy Syst, **2015**; pp. 1-25.
[http://dx.doi.org/10.1002/9781118991978.hces112]

[2]   Ke, Q.; Wang, J. Graphene-Based Materials for Supercapacitor Electrodes – A Review. *J. Mater.,* **2016**, *2*(1), 37-54.

[3]   Huang, J.; Sumpter, B.G.; Meunier, V. A universal model for nanoporous carbon supercapacitors applicable to diverse pore regimes, carbon materials, and electrolytes. *Chemistry,* **2008**, *14*(22), 6614-6626.
[http://dx.doi.org/10.1002/chem.200800639] [PMID: 18576455]

[4]   Hashemi, M.; Rahmanifar, M.S.; El-Kady, M.F.; Noori, A.; Mousavi, M.F.; Kaner, R.B. The use of an electrocatalytic redox electrolyte for pushing the energy density boundary of a flexible polyaniline electrode to a new limit. *Nano Energy,* **2017**, *2018*(44), 489-498.

[5]   Mendes, T.C.; Zhou, F.; Barlow, A.J.; Forsyth, M.; Howlett, P.C.; MacFarlane, D.R. An ionic liquid based sodium metal-hybrid supercapacitor-battery. *Sustain. Energy Fuels,* **2018**, *2*(4), 763-771.
[http://dx.doi.org/10.1039/C7SE00547D]

[6]   Pandolfo, A.G.; Hollenkamp, A.F. Carbon properties and their role in supercapacitors. *J. Power Sources,* **2006**, *157*(1), 11-27.
[http://dx.doi.org/10.1016/j.jpowsour.2006.02.065]

[7]   Cheng, J.; Wang, C.; Zou, X.; Liao, L. Recent advances in optoelectronic devices based on 2D materials and their heterostructures. *Adv. Opt. Mater.,* **2019**, *7*(1), 1-15.
[http://dx.doi.org/10.1002/adom.201800441]

[8]   Stylianakis, M.M.; Maksudov, T.; Panagiotopoulos, A.; Kakavelakis, G.; Petridis, K. Inorganic and hybrid perovskite based laser devices: A Review. *Materials (Basel),* **2019**, *12*(6), 1-28.
[http://dx.doi.org/10.3390/ma12060859] [PMID: 30875786]

[9]   Viskadouros, G.; Zak, A.; Stylianakis, M.; Kymakis, E.; Tenne, R.; Stratakis, E. Enhanced field emission of WS$_2$ nanotubes. *Small,* **2014**, *10*(12), 2398-2403.
[http://dx.doi.org/10.1002/smll.201303340] [PMID: 24610733]

[10]   Noori, K.; Konios, D.; Stylianakis, M. M.; Kymakis, E.; Giustino, F. Energy-Level Alignment and Open-Circuit Voltage at Graphene/Polymer Interfaces: Theory and Experiment. 2D Mater., 2016, 3(1), 15003.

[11]   Bhushan, B. *Encyclopedia of Nanotechnology*; Encycl. Nanotechnol, **2012**.
[http://dx.doi.org/10.1007/978-90-481-9751-4]

[12]   Stylianakis, M.M. Optoelectronic Nanodevices. *Nanomaterials (Basel),* **2020**, *10*(3), E520.
[http://dx.doi.org/10.3390/nano10030520] [PMID: 32183135]

[13]   Feng, Y.; Chen, Z.; Jiang, S.; Li, C.; Chen, Y.; Zhan, J.; Chen, Y.; Nie, J.; Jiao, F.; Kang, X.; Li, S.; Yu, T.; Zhang, G.; Shen, B. Study on the coupling mechanism of the orthogonal dipoles with surface plasmon in green LED by cathodoluminescence. *Nanomaterials (Basel),* **2018**, *8*(4), E244.
[http://dx.doi.org/10.3390/nano8040244] [PMID: 29659499]

[14]   Yan, C.; Du, X.; Li, J.; Ding, X.; Li, Z.; Tang, Y. Effect of excitation wavelength on optical performances of quantum-dot-converted light-emitting diode. *Nanomaterials (Basel),* **2019**, *9*(8), E1100.
[http://dx.doi.org/10.3390/nano9081100] [PMID: 31374836]

[15]   Stylianakis, M.M.; Viskadouros, G.; Polyzoidis, C.; Veisakis, G.; Kenanakis, G.; Kornilios, N.; Petridis, K.; Kymakis, E. Updating the role of reduced graphene oxide ink on field emission devices in synergy with charge transfer materials. *Nanomaterials (Basel),* **2019**, *9*(2), 1-15.
[http://dx.doi.org/10.3390/nano9020137] [PMID: 30678208]

[16]   Zhao, J.; Ding, X.; Miao, J.; Hu, J.; Wan, H.; Zhou, S. Impng. *Nanomaterials (Basel),* **2019**, *9*(2), E203.
[http://dx.doi.org/10.3390/nano9020203] [PMID: 30720748]

[17]   Stylianakis, M.M.; Kosmidis, D.M.; Anagnostou, K.; Polyzoidis, C.; Krassas, M.; Kenanakis, G.;

Viskadouros, G.; Kornilios, N.; Petridis, K.; Kymakis, E. Emphasizing the operational role of a novel graphene-based ink into high performance ternary organic solar cells. *Nanomaterials (Basel),* **2020,** *10*(1), 1-13.
[http://dx.doi.org/10.3390/nano10010089] [PMID: 31906494]

[18]   Xue, Y.; Yuan, J.; Liu, J.; Li, S. Controllable synthesis of 2D perovskite on different substrates and its application as photodetector. *Nanomaterials (Basel),* **2018,** *8*(8), 1-10.
[http://dx.doi.org/10.3390/nano8080591] [PMID: 30081503]

[19]   Liu, M.; Zhao, J.; Zhou, S.; Gao, Y.; Hu, J.; Liu, X.; Ding, X. An InGaN/GaN Superlattice to Enhance the Performance of Green LEDs: Exploring the Role of V-Pits. *Nanomaterials (Basel),* **2018,** *8*(7), E450.
[http://dx.doi.org/10.3390/nano8070450] [PMID: 29933543]

[20]   Svrcek, V.; Kolenda, M.; Kadys, A.; Reklaitis, I.; Dobrovolskas, D.; Malinauskas, T.; Lozach, M.; Mariotti, D.; Strassburg, M.; Tomašiūnas, R. Significant carrier extraction enhancement at the interface of an InN/p-GaN heterojunction under reverse bias voltage. *Nanomaterials (Basel),* **2018,** *8*(12), E1039.
[http://dx.doi.org/10.3390/nano8121039] [PMID: 30545138]

[21]   Lee, D.J.; Mohan Kumar, G.; Ilanchezhiyan, P.; Xiao, F.; Yuldashev, S.U.; Woo, Y.D.; Kim, D.Y.; Kang, T.W. Arrayed CdTeMicrodots and their enhanced photodetectivity *via* piezo-phototronic effect. *Nanomaterials (Basel),* **2019,** *9*(2), 1-13.
[http://dx.doi.org/10.3390/nano9020178] [PMID: 30717115]

[22]   Chen, L.C.; Tseng, Z.L.; Lin, D.W.; Lin, Y.S.; Chen, S.H. Improved performance of perovskite light-emitting diodes by quantum confinement effect in perovskite nanocrystals. *Nanomaterials (Basel),* **2018,** *8*(7), 1-10.
[http://dx.doi.org/10.3390/nano8070459] [PMID: 29941783]

[23]   Arvind, D.; Hegde, G. Activated carbon nanospheres derived from bio-waste materials for supercapacitor applications - a review. *RSC Advances,* **2015,** *5*(107), 88339-88352.
[http://dx.doi.org/10.1039/C5RA19392C]

[24]   Kim, S.I.; Kim, S.W.; Jung, K.; Kim, J.B.; Jang, J.H. Ideal nanoporous gold based supercapacitors with theoretical capacitance and high energy/power density. *Nano Energy,* **2016,** *24*, 17-24.
[http://dx.doi.org/10.1016/j.nanoen.2016.03.027]

[25]   Kang, K.N.; Kim, I.H.; Ramadoss, A.; Kim, S.I.; Yoon, J.C.; Jang, J.H. Ultrathin nickel hydroxide on carbon coated 3D-porous copper structures for high performance supercapacitors. *Phys. Chem. Chem. Phys.,* **2018,** *20*(2), 719-727.
[http://dx.doi.org/10.1039/C7CP07473E] [PMID: 29231217]

[26]   Lv, Z.; Luo, Y.; Tang, Y.; Wei, J.; Zhu, Z.; Zhou, X.; Li, W.; Zeng, Y.; Zhang, W.; Zhang, Y.; Qi, D.; Pan, S.; Loh, X.J.; Chen, X. Editable supercapacitors with customizable stretchability based on mechanically strengthened ultralong MnO_2 nanowire composite. *Adv. Mater.,* **2018,** *30*(2), 1-9.
[http://dx.doi.org/10.1002/adma.201704531] [PMID: 29134702]

[27]   Wang, W.; Zhang, Y.; Zhang, L.; Shi, Y.; Jia, L.; Zhang, Q.; Xu, X. Flexible Mn_3O_4 Nanosheet/Reduced Graphene Oxide Nanosheet Paper-like Electrodes for Electrochemical Energy Storage and Three-Dimensional Multilayers Printing. *Mater. Lett.,* **2018,** *213*, 100-103.
[http://dx.doi.org/10.1016/j.matlet.2017.11.025]

[28]   Yu, S.; Yang, N.; Vogel, M.; Mandal, S.; Williams, O.A.; Jiang, S.; Schönherr, H.; Yang, B.; Jiang, X. Battery-like supercapacitors from vertically aligned carbon nanofiber coated diamond: design and demonstrator. *Adv. Energy Mater.,* **2018,** *8*(12), 1-10.
[http://dx.doi.org/10.1002/aenm.201702947]

[29]   Edberg, J.; Inganäs, O.; Engquist, I.; Berggren, M. Boosting the capacity of all-organic paper supercapacitors using wood derivatives. *J. Mater. Chem. A Mater. Energy Sustain.,* **2017,** *6*(1), 145-152.

[http://dx.doi.org/10.1039/C7TA06810G]

[30]   Dong, J.; Lu, G.; Wu, F.; Xu, C.; Kang, X.; Cheng, Z. Facile synthesis of a nitrogen-doped graphene flower-like MnO_2 Nanocomposite and Its Application in Supercapacitors. *Appl. Surf. Sci.,* **2018**, *427*, 986-993.
       [http://dx.doi.org/10.1016/j.apsusc.2017.07.291]

[31]   Chen, Q.; Chen, J.; Zhou, Y.; Song, C.; Tian, Q.; Xu, J.; Wong, C.P. Enhancing pseudocapacitive kinetics of nanostructured MnO_2 through anchoring onto biomass-derived porous carbon. *Appl. Surf. Sci.,* **2018**, *440*, 1027-1036.
       [http://dx.doi.org/10.1016/j.apsusc.2018.01.224]

[32]   Fang, K.; Chen, J.; Zhou, X.; Mei, C.; Tian, Q.; Xu, J.; Wong, C.P. Decorating Biomass-Derived Porous Carbon with Fe_2O_3 Ultrathin Film for High-Performance Supercapacitors. *Electrochim. Acta,* **2018**, *261*, 198-205.
       [http://dx.doi.org/10.1016/j.electacta.2017.12.140]

[33]   Zhao, C.; Wang, C.; Gorkin, R., III; Beirne, S.; Shu, K.; Wallace, G.G. Three Dimensional (3D) Printed Electrodes for Interdigitated Supercapacitors. *Electrochem. Commun.,* **2014**, *41*, 20-23.
       [http://dx.doi.org/10.1016/j.elecom.2014.01.013]

[34]   Gou, J.; Xie, S.; Liu, C. Flower-like Ni-Co Hydroxides on Ni Foam for High-Performance Supercapacitor Applications. *New J. Chem.,* **2018**, *42*(6), 4175-4181.
       [http://dx.doi.org/10.1039/C7NJ04663D]

[35]   Patil, B.; Ahn, S.; Park, C.; Song, H.; Jeong, Y.; Ahn, H. Simple and Novel Strategy to Fabricate Ultra-Thin, Lightweight, Stackable Solid-State Supercapacitors Based on MnO2-Incorporated CNT-Web Paper. *Energy,* **2018**, *142*, 608-616.
       [http://dx.doi.org/10.1016/j.energy.2017.10.041]

[36]   Xi, C.; Zhu, G.; Liu, Y.; Shen, X.; Zhu, W.; Ji, Z.; Kong, L. Belt-like Nickel Hydroxide Carbonate/Reduced Graphene Oxide Hybrids: Synthesis and Performance as Supercapacitor Electrodes. *Colloids Surf. A Physicochem. Eng. Asp.,* **2017**, *2018*(538), 748-756.

[37]   Zhang, S.; Sui, L.; Kang, H.; Dong, H.; Dong, L.; Yu, L. High Performance of N-Doped Graphene with Bubble-like Textures for Supercapacitors. *Small,* **2018**, *14*(5), 1-11.
       [http://dx.doi.org/10.1002/smll.201702570] [PMID: 29251420]

[38]   Liang, H.; Lin, J.; Jia, H.; Chen, S.; Qi, J.; Cao, J.; Lin, T.; Fei, W.; Feng, J. Hierarchical NiCo-LDH@NiOOH Core-Shell Heterostructure on Carbon Fiber Cloth as Battery-like Electrode for Supercapacitor. *J. Power Sources,* **2017**, *2018*(378), 248-254.

[39]   Zha, D.; Fu, Y.; Zhang, L.; Zhu, J.; Wang, X. Design and Fabrication of Highly Open Nickel Cobalt Sulfide Nanosheets on Ni Foam for Asymmetric Supercapacitors with High Energy Density and Long Cycle-Life. *J. Power Sources,* **2017**, *2018*(378), 31-39.

[40]   Liu, C.; Liu, J.; Wang, J.; Li, J.; Luo, R.; Shen, J.; Sun, X.; Han, W.; Wang, L. Electrospun mulberry-like hierarchical carbon fiber web for high-performance supercapacitors. *J. Colloid Interface Sci.,* **2018**, *512*, 713-721.
       [http://dx.doi.org/10.1016/j.jcis.2017.10.093] [PMID: 29107922]

[41]   Wu, X.; Meng, L.; Wang, Q.; Zhang, W.; Wang, Y. A Novel and Facile Step-by-Step Hydrothermal Fabrication of Peony-like Ni0.4Co0.6(OH)2 Supported on Carbon Fiber Cloth as Flexible Electrodes for Advanced Electrochemical Energy Storage. *Sol. Energy Mater. Sol. Cells,* **2017**, *2018*(174), 325-332.

[42]   Xia, J.; Zhang, N.; Chong, S.; Li, D.; Chen, Y.; Sun, C. Three-Dimensional Porous Graphene-like Sheets Synthesized from Biocarbon *via* Low-Temperature Graphitization for a Supercapacitor. *Green Chem.,* **2018**, *20*(3), 694-700.
       [http://dx.doi.org/10.1039/C7GC03426A]

[43]   Zhou, W.J.; Xu, M.W.; Zhao, D.D.; Xu, C.L.; Li, H.L. Electrodeposition and Characterization of

Ordered Mesoporous Cobalt Hydroxide Films on Different Substrates for Supercapacitors. *Microporous Mesoporous Mater.,* **2009**, *117*(1–2), 55-60. [http://dx.doi.org/10.1016/j.micromeso.2008.06.004]

[44]   Foster, C.W.; Down, M.P.; Zhang, Y.; Ji, X.; Rowley-Neale, S.J.; Smith, G.C.; Kelly, P.J.; Banks, C.E. 3D Printed Graphene Based Energy Storage Devices. *Sci. Rep.,* **2017**, *7*(January), 42233. [http://dx.doi.org/10.1038/srep42233] [PMID: 28256602]

[45]   Hu, X.; Zou, C.; Zhang, C.; Li, Y. Technological Developments in Batteries: A Survey of Principal Roles, Types, and Management Needs. *IEEE Power Energy Mag.,* **2017**, *15*(5), 20-31. [http://dx.doi.org/10.1109/MPE.2017.2708812]

[46]   Li, X.H.C.Z.C.Z.Y. Technological Developments in Batteries. *IEEE Power Energy Mag.,* **2017**, (July), 42-44.

[47]   Lin, C.H.; Cheng, B.; Li, T.Y.; Retamal, J.R.D.; Wei, T.C.; Fu, H.C.; Fang, X.; He, J.H. Orthogonal Lithography for Halide Perovskite Optoelectronic Nanodevices. *ACS Nano,* **2019**, *13*(2), 1168-1176. [PMID: 30588789]

[48]   Ye, Y.; Yu, B.; Gao, Z.; Meng, H.; Zhang, H.; Dai, L.; Qin, G. Two-dimensional CdS nanosheet-based TFT and LED nanodevices. *Nanotechnology,* **2012**, *23*(19), 194004. [http://dx.doi.org/10.1088/0957-4484/23/19/194004] [PMID: 22538931]

CHAPTER 14

Nanomaterials' Synthesis Approaches for Energy Storage and Electronics Applications

Ravi Verma[1,*], Shanky Jha[2], D. Harimurugan[2], H. N. Nagendra[3,4], Srinivasan Kasthurirengan[3], N. C. Shivaprakash[4] and **Upendra Behera[3]**

[1] *Department of Control and Instrumentation Engineering, Dr. B. R. Ambedkar National Institute of Technology, Jalandhar-140011, India*

[2] *Department of Electrical Engineering, Dr. B. R. Ambedkar National Institute of Technology, Jalandhar-140011, India*

[3] *Centre for Cryogenic Technology, Indian Institute of Science, Bangalore560012, India*

[4] *Instrumentation and Applied Physics, Indian Institute of Science, Bangalore560012, India*

Abstract: Nanomaterials are materials with cross-sectional dimensions varying from one to hundreds of nanometers and lengths ranging from hundreds of nanometers to millimeters. Nanomaterials either occur naturally or can be produced purposefully by performing a specialized function. Until recently, most nanomaterials have been made from carbon (carbon nanotubes), transition metals, and metal oxides such as titanium dioxide and zinc oxide. In a few cases, nanoparticles may exist in the form of nanocrystals comprising a number of compounds, including but not limited to silicon and metals. The discovery of nanomaterials has played a vital role in the emerging field of research and technology. Recently, a large amount of research efforts has been dedicated to developing nanomaterials and their applications, ranging from space to electronics applications. In this chapter, we describe the role of nanoparticles in electronics and energy storage applications, with examples including chips, displays, enhanced batteries, and thermoelectric, gas sensing, lead-free soldering, humidity sensing, and super capacitor devices. The chapter also attempts to provide an exhaustive description of the developed advanced nanomaterials and different conventional and advanced techniques adopted by researchers to synthesize the nanoparticles *via* bottom-up techniques (pyrolysis, chemical vapor deposition, sol-gel, and biosynthesis) and top-bottom approaches (mechanical milling, nanolithography, laser ablation, and thermal decomposition).

Keywords: Bottom-up Technique, Bio-synthesis, Carbon Nanotubes, Chemical Vapour Deposition, Electronics Applications, Energy Storage, Graphene, Humidity Sensor, Laser Ablation, Mechanical Milling, Nanolithography,

---

* **Corresponding author Ravi Verma:** Department of Control and Instrumentation Engineering Dr. B. R. Ambedkar National Institute of Technology, Jalandhar-140011, India; Tel: +91-1332-285650; Fax: +91-1332-273560; E-mail: vermaravi@nitj.ac.in

Nanomaterials, Pyrolysis, Sole Gel, Spinning, Sputtering, Super Capacitor, Thermal Deposition, Thermo-electric, Top-down Technique.

INTRODUCTION

The production, storage, and utilization of energy in an optimal way are some of the important issues that are lately confronting the researchers of the world. The production of energy is not a major achievement nowadays. However, the storage and delivery of energy on demand has increasingly become even more important. Further, portable electronic devices and transportation systems are such applications where we do need to store energy appropriately. The use of energy storage devices is thus increasingly gathering importance in the field of science and technology. There are three different but important ways, such as electrical, electro-chemical, and chemical, in which energy can be stored in energy storage devices. There are large numbers of suitable materials which can be used for energy storage. However, considering the cost and energy per weight ratio of the materials, the options of suitable material for energy storage reduced drastically. Nanomaterials, such as carbon nanotubes and graphene, are suitable options for use as energy storage devices.

The high energy capacity and surface area are important parameters for the fabrication of the energy storage device. Carbon is one such lightest element used in various forms, such as carbon nanotubes (CNT's) for the fabrication of energy storage devices [1 - 5]. The single walled carbon nanotubes (SWCNTs) are not only light weight but they also attribute some other advantages, such as a large surface area of approximately 1315 m^2 g^{-1} when compared to graphite which offers it in a poor range of 10-20 m^2 g^{-1}.

Nanomaterials offer drastically improved electronic conductivity in comparison with conventional super capacitor materials and conventional batteries. They also have faster ion diffusion and higher specific capacities. All these features make nanomaterials offer an encouraging solution for high power and high energy devices. After extensive research and development in the area, a library of nanomaterials with various morphologies has been explored. Some of them are zero dimensional nanoparticles such as quantum dots; one dimensional such as nanotubes, nano-belts and nanowires; two dimensional nano-sheets and nano-flakes and three dimensional porous nano-networks. These nanoscale building blocks have been combined with lithium ions for creating energy storage technology which is not achievable using conventional materials. The atoms and molecules in a nanomaterial act differently and disclose unmatched chemical, physical and electronics properties. Nanomaterials show physical properties such as reflection, absorption and light dispersion. They also show extraordinary chemical properties such as oxidation, reduction, sensitivity, and stability towards humidity, atmosphere, heat, and light, etc. When nanomaterials are layered on a surface in the form of a solution, their reflection and adsorption properties make them a seamless choice in different applications. These properties make them ideal candidates for use in electronic, drug delivery, optical, mechanics, catalysis, bio-encapsulation and wastewater treatment, especially adsorption applications [6 - 15].

Along with the carbon nanotubes, graphene, fullerenes, carbon nano-fibres and activated carbons are also used as alternative energy storage materials [16 - 18]. Most of them are chemically inert, cheaper and light weight. They have special electrochemical, electrical, optical and mechanical properties. For example, graphene's surface area is approximately 2630 m^2 g^{-1}. The large surface area of graphene makes it the most suitable choice for energy storage applications. Fig. (**1**) shows the classification of different types of nanomaterials.

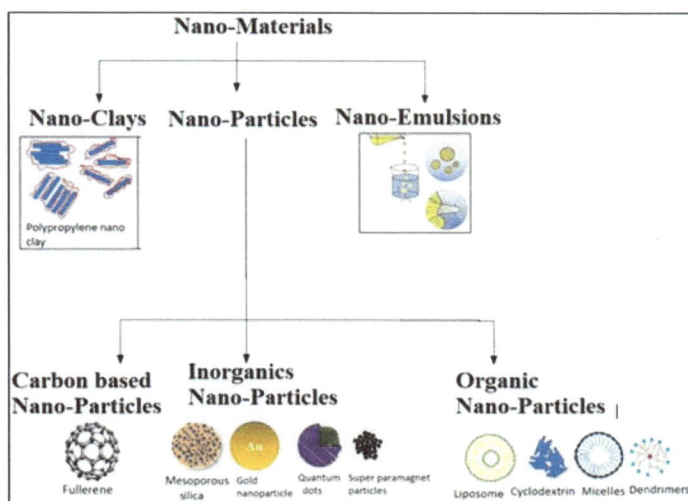

Fig. (1). Classification of nanomaterials.

The growing efforts towards clean and renewable energy drive breakthroughs in energy storage technology. Among different energy storage systems, electrochemical systems are used in portable electronic devices. Higher volumetric and gravimetric energy density is required for portable electronics devices. For stationary applications such as grid-scale energy storage, the lower capital cost is the key. The basic difference between batteries and electrochemical is in the charge storage mechanism. A battery store charge via faradic reactions, whereas electro-chemical stores charge near the surface. Hence, electrochemical systems have more specific power compared to batteries, which have a more specific energy. The battery is useful when a constant supply of energy is required, whereas electrochemical finds applications where high frequency charge or discharge is needed, such as elevator operation. In the last few years, nanocomposite materials have been applied in electro-chemical energy storage devices. In this chapter, we cover the different applications of nanomaterials towards energy storage applications. Fig. (**2**) shows the application of nanoparticles in different fields.

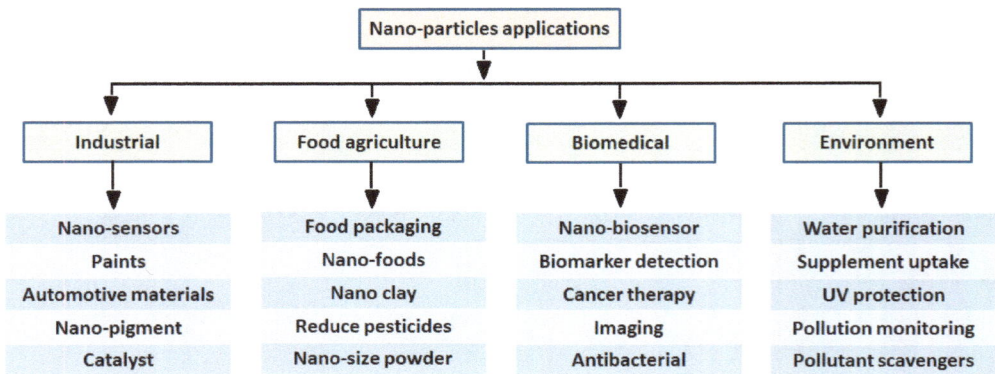

Nano-particles applications			
Industrial	**Food agriculture**	**Biomedical**	**Environment**
Nano-sensors	Food packaging	Nano-biosensor	Water purification
Paints	Nano-foods	Biomarker detection	Supplement uptake
Automotive materials	Nano clay	Cancer therapy	UV protection
Nano-pigment	Reduce pesticides	Imaging	Pollution monitoring
Catalyst	Nano-size powder	Antibacterial	Pollutant scavengers

Fig. (2). Illustration of applications of nanoparticles in different fields.

BASIC APPROACHES TO SYNTHESIZE NANOMATERIALS

The method of preparation of the nanoparticles carries immense importance as these may decide the final properties exhibited by them. Therefore, there is a need to emphasize two commonly used techniques employed for the synthesis of nanoparticles [19]. These are a) Bottom-up technique and b) Top-down technique, which are detailed below-

Bottom-up Technique

In this method, the material component is minimized till it reaches its atomic level. At this stage, the self-assembly results in the preparation of the nanoparticles. The different methods used under bottom up techniques are chemical vapour deposition (CVD), sol gel, bio-synthesis, and pyrolysis. Fig. (**3**) shows the schematic diagram of bottom-up and top down techniques.

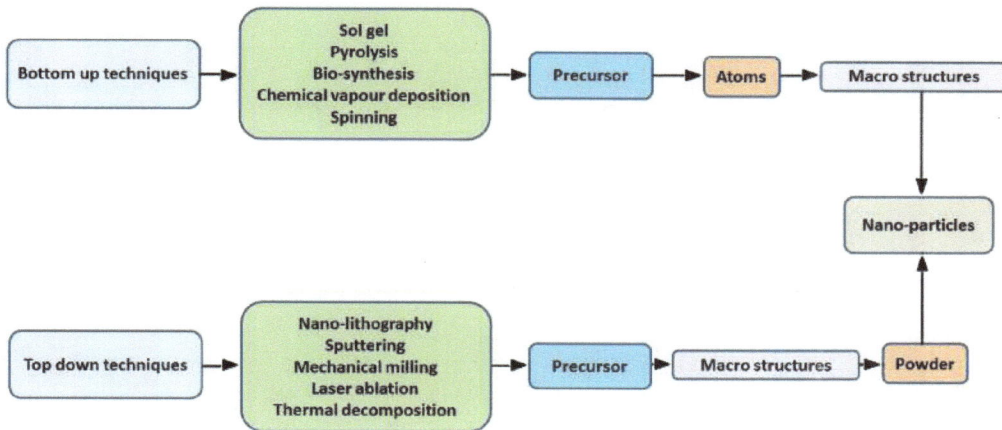

Fig. (3). A schematic illustration of bottom-up and top-down techniques.

Chemical Vapour Deposition (CVD)

In the CVD method, a thin film of gaseous reactants is coated over a substrate. The deposition takes place when the amalgamation of gas molecules occurs in the reaction chamber at room temperature over a catalyst substrate. The heated substrate comes under the vicinity of the combined gas and the formation of the thin film takes place on the surface of the substrate due to the chemical reaction. Various catalysts such as mono dispersed Fe-Mo nanoparticles and gold nanoparticles have been employed in the past for synthesizing CNTs [20, 21]. The formed thin film is reusable and can be used for various applications. The substrate's temperature is an influencing factor in the formation of nanoparticles. The nanoparticles formed by the chemical vapour deposition method are of high purity, uniform in size, are highly pure, and have high and strong mechanical stability. The disadvantages of CVD method are the production of highly toxic gases and the need for a specially designed equipment. Fig. (4) shows the photographic view of a typical CVD method used for the production of CNTs using an iron catalyst in a furnace heated to 1300K.

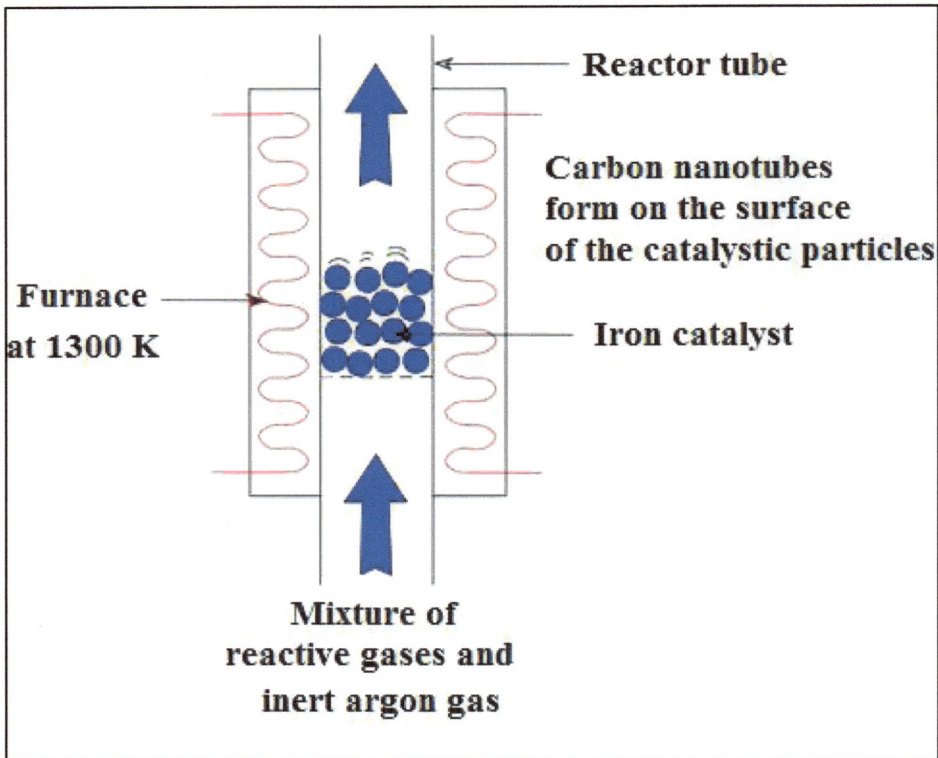

Fig. (4). A schematic representation of a chemical vapour deposition process.

Sol Gel

The simplest and most commonly used method for the synthesis of nanoparticles is sol gel. Sol gel is a chemical method where metal chlorides and metal oxides are commonly used as precursor. In this method, the start of the reaction and the system's pH value are controlled with the help of a catalyst [22 - 24]. Different methods to fabricate nanoparticles are sedimentation, filtration and centrifugation, which separate the particles from the liquid phase. The main advantages of this process are uniform nanostructure at low temperatures, purity and controlled composition. The main disadvantages of this process are the different drying steps involved in it, which makes it difficult to scale up.

Bio-synthesis

This process is one of the inexpensive, safe, green, environment friendly, and decomposable methods. This method that uses plant extracts, fungi and bacterium in combination with precursors for bio-reduction instead of conventionally used methods has been reviewed extensively in the past [25, 26]. The method, due to its unique and distinct properties, finds distinct applications in the medical field.

Pyrolysis

This method is used for the large scale production of nanoparticles in industries. The liquid or vapour can be a precursor in this method. Precursors are added to a furnace and burnt in the absence of air or oxygen. By-product gases produced in the furnace collect the nanoparticles [27, 28]. This is a simple method to prepare the nanoparticles and also has high sensitivity and product yield.

Top-Down Techniques

The top down techniques include thermal decomposition, mechanical milling, laser ablation, sputtering and nanolithography. It starts with a configuration produced on higher scale and then compressed to nanoscale. This technique is contradictory to the previous approach. The details of the different types of top-down techniques are as provided below.

Mechanical Milling

Among various top down techniques, this one is the most widely used one to produce nanoparticles and has been extensively reviewed in the literature [29 - 31]. In this technique, different elements are milled in the vicinity of an inert atmosphere. The particles are annealed after the milling process.

Nanolithography

This technique is used for fabricating nanoparticles of sizes ranging from 1-100 nm and employs various nanolithographic forms like electron-pillar, optical etc. The main advantage of nanolithography is to produce a bunch of required shapes and sizes from a single nanoparticle [32 - 34]. While this process requires sophisticated equipment, it is quite cost-effective.

Laser-ablation

In this method, a solution containing various solvents is used for the preparation of the nanoparticles. In this process, due to irradiation by a laser beam, the chemical reduction of metals (immersed in a solution) occurs, thereby producing metal-based inorganic nanoparticles [35 - 37]. This process is a green process. Fig. (**5**) shows the schematic diagram of a typical laser ablation setup.

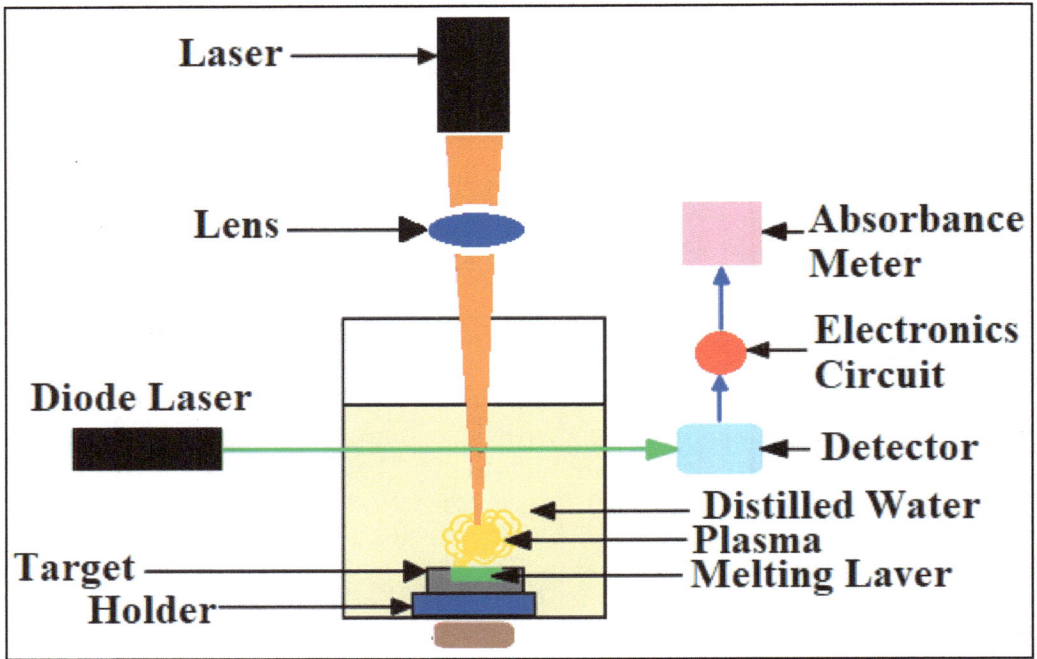

Fig. (5). Schematic diagram of a laser ablation setup.

Thermal Decomposition

This process is an endothermic chemical process. In this process, the chemical bonds of the compound are decomposed by applying heat. Nanoparticles are produced by rotting a metal at a temperature known as decomposition temperature. Due to this decomposition, secondary products are also formed [38]. This method is quite suitable for the synthesis of carbon-based and metal oxide-based nanoparticles.

NANOMATERIAL IN ELECTRONICS: LAB-ON-A-CHIP TECHNOLOGY

Lab-on-a-Chip platform is an important tool for sample analysis and cells studies because of its low cost and ease of portability. The platform has the advantages of a wide range of nanomaterials such as quantum dots, nanowires, nanowires and other forms. They along with nanomaterials offer enormous improvement in properties for different applications like detector sensitivity enhancement along with other applications. The application of nanomaterials for the development of sensors is one of the key areas of research reviewed in detail earlier [39]. The unique electrical and optical properties of nanomaterials make them basic building blocks for designing different types of higher-performance sensing devices. Nanomaterials have been one of the best options for many years for ensuring the improved performance of the transducer. Transducer is a very important part of the sensing system that is integrated on Lab-on-a-Chip. The transducers improved using nanomaterials can enhance the electrochemical measurements like potentiometric, optical measurements like absorbance and other mixed signals like electro-chemiluminescence. The nanomaterials support for the Lab-on-a-chip platform may involve the following:

Electrochemical detection

Different researchers have reported nanomaterials based electrochemical detector which is integrated with Lab-on-a-chip system. Among all these, the carbon nanotubes were studied in detail by the researchers [40, 41]. This is perhaps due to the fact that they exhibit extraordinary thermal conductivity and electrical conductivity along with excellent mechanical strength.

Carbon nanotubes can be deposited easily on different surfaces using a simple technique of either Chemical Vapour Deposition or Plasma Enhanced Chemical Vapour Deposition. The transducer modified with the help of carbon nanotubes has a higher surface area. This increases the current density, which in turn improves the current stability, and hence, the sensitivity of the detector.

Optical Detection

The performance of opto-fluidic detectors, based upon Surface Raman Spectroscopy, can be improved with nanomaterials. The Raman scattering by the molecules adsorbed on the surface can be improved by Surface Raman Spectroscopy, which is a surface sensitive technique. The use of nanomaterials like gold nanowires not only provides many possibilities for sensitive detection but also provides a large increase in a Raman cross-section [42]. The combination of Surface Raman Spectroscopy and nanowires is good for microchip integration. It also increases the sensitivity of the detectors due to the higher surface-t--volume ratio of the advanced nanomaterials. Mixing silver nanoparticles with eluent stream in laminar flow offers higher Surface Raman Spectroscopy reproducibility.

Other Detection Methods

Other detection methods like electro-chemiluminescence are also applied in the lab-on-a-chip technology. It is a type of luminescence in which a material emits light under the effect of an electric field. It was investigated experimentally that the emission efficiency is a function of the properties of gold nanoparticles. The efficiency increases as the size of gold nanoparticles decreases as reported earlier

[43]. Generally, the gold nanoparticles have sizes in the range of 10-15 nm, although gold nanoparticles with size of 5 nm produce stronger ECL.

Nanomaterials as Performance Transparent Displays

The transparent displays based on the nanoparticles are one of the best ways to realize transparent monitors. The aperiodic arrangement of the array of nanoparticles plays a very vital role in this case. The different morphologies of nanoparticles with varied arrays like deterministic aperiodic array and periodic array have been investigated by the different researchers for high performance transparent displays [44]. The different morphologies of silicon and silicon di-oxide nanoparticles (Si-SiO$_2$) are compared to obtain the maximum scattering cross-section using Red, Yellow and Blue wavelength. The figure of merit of transparent displays depends on the optical properties such as scattering, extinction, wavelength, and absorption cross-section. It also depends upon the ratio of the scattered to absorbed radiation by the nanomaterials. The optical properties (on the absorption and scattering cross section) of different classes of nanoparticles such as cube, disk, sphere, prolate ellipse, oblate ellipse and pyramid with different arrays such as the Thue-mose and Fibonacci have been investigated by the researchers. The extinction cross-section is acquired by scattering cross-section plus absorption cross-section. The plane-wave field irradiates the incident light onto the array of nanoparticles to determine the absorption and scattering cross-section. Researchers used Finite-Difference Time-Domain numerical method to simulate and calculate the periodic and aperiodic arrays of nanoparticles. It has also been investigated and found that the structure's size affects the specific amount of wavelength. It was found that coated oblate ellipse with the Fibonacci structure has a sharper and narrower scattering cross-section.

NANOMATERIALS IN ENERGY STORAGE

Battery Storage Applications

There are many methods and forms in which energy can be stored. Some of the forms are grid energy, pumped water, compressed air, thermal energy, flywheel, superconducting magnetic energy, hydrogen fuel cells and battery.

Battery stores electrical energy in the form of chemical energy on cathode and anode during charging and generates electrical energy as an output during discharge. The battery must have high specific energy, high power density, low cost and long cycle life. Lithium-ion based batteries have fetched more attention due to their superior performance [45]. These batteries also attract the attention of different researchers across the globe because of much higher energy density. The performance of the batteries has lately improved significantly with the development of novel materials. The research for the anode and cathode material of lithium-ion batteries has focused on the use of advanced nanomaterials. Nanometer thick coatings are used to improve ionic and electronic conduction pathways and stop irreversible side reactions.

In Lithium-sulphur battery, the cathodes are made up of either sulphur-polymer or sulphur-carbon composites. The use of nanocomposites helps to significantly improve electronic thermal conductivity. The sulphur-carbon composite supports electro-chemical reaction inside the carbon due to strong adsorption, and therefore, researchers have achieved approximately 500 cycles with a capacity of 800 mAh. The carbon structures with sizes in the range of approximately 2 to 50 nm can provide better electrical contact and larger capacities. Suphur-graphene oxide nanocomposites also have shown promising results. Graphene oxide is highly conductive and cycle life of approximately 1500 cycles can be achieved using these systems.

Thermo-Electric Material Applications

Thermoelectric materials are of utmost importance for the conversion of renewable energy into usable forms. They can prove to be the solution to energy crisis across the globe. They are suitable for the space technologies like space craft and missiles. The performance of thermo-electric devices depends mainly on the type of the materials and their thermal and electrical properties. It also depends on their Seebeck coefficient and thermal stability. It has been found in recent research that high performance thermoelectric materials such as carbon nanomaterials are suitable for power generation in the upcoming time [46].

The many allotropes of carbon show high thermal and electrical conductivities but controlling these parameters for thermoelectric applications is a very difficult task. The thermal conductivity of carbon materials can be improved drastically using the nanostructure of carbon such as carbon nanotubes, carbon nanowires, and graphene. However, from among all mentioned carbon nanomaterials, the carbon nanotubes have attained more significance due to their high thermal stability and electrical conductivity and large surface area. Carbon nanotubes are one dimensional carbon nanomaterials with single and multiple walls. They are considered as possible materials for thermo-electric devices. The operating temperature of multi-walled carbon nanotubes for thermoelectric application varies from 87 °C to 567 °C. The use of graphene as a thermo-electric material is matter of discussion due to the following two drawbacks: its Seebeck coefficient is very small due to zero band gap, and its j factor is high. However, the researchers have still reported the thermo-electric power of approximately 80

$\mu V.K^{-1}$ at room temperature. Hence, graphene can also be used as thermo-electric material compared to elemental semiconductors.

Nanomaterials in Gas Sensing

A sensor produces a response when exposed to a stimulus. The response may be in one or more forms of sensor properties such as electrical conductivity, mass, etc. Hence, the sensors allow us to observe the environment surrounding and use this data for different applications [47]. The use of nanomaterials enables us to produce effective sensors for various applications. The special properties of nanomaterials like modulation of the electrical signal, high adsorptive capacity owing to the large surface to volume ratio, and tuning electrical properties by controlling the composition of nanomaterial makes them a suitable choice for sensor applications. The electronic properties of carbon nanotubes are found highly sensitive to chemical environment. The high chemical sensitivity makes them a desirable choice for the design of the chemical sensors. Researchers have investigated sensor conductance in the environment of nitrogen dioxide and ammonia vapours. In the environment of ammonia, the valence band of the nanotube shifts away from the fermi level. It will result in hole depletion and decreased conductance. With exposure of nitrogen dioxide gas, the nanotube fermi level moves closer to the valence band. This results in the large hole carrier's concentration and increased sample concentration.

Nanomaterials in Lead-free Soldering

There is a significant advancement in science and technology with the fabrication of nanomaterials such as nanoparticles, nanowires and carbon nanotubes. Many methods have been developed to assemble nanoparticles in ordered 1-D, 2-D and 3-D structures. The interconnect formation and joining between these self-assembled structures is still lesser developed and fundamentally less understandable. However, it plays a vital role in the device and electronic integration. There needs to be a significant improvement in the field of nanoscale manufacturing. There has been some continuous progress towards the joining of the nano-scale components. The two commonly known techniques that are used to join nano-scale components are: annealing and Focused Ion-beam or focused E-beam irradiation [48]. The process of annealing is used at normal high temperatures approximately in the temperature range of 400 °C to 500 °C. This is an effective way to reduce the contact resistance between the components, mainly between metals to metals. However, the use of annealing at a higher temperatures may damage many components, including polymers and organics involved. Focused Ion-beam or focused E-beam soldering have been successfully used to adhere nanotubes to nanotubes carbon nanotubes to substrate but the contamination due to partial precursor decomposition is the main drawback of this technique.

With the evolution and emergence of nanotechnology, there is a high demand for developing nanoscale lead-free solders for upcoming nano-electronics industries. Nano-solders are not only having improved properties but also have lower melting points. This will greatly reduce the operating temperature and save energy. The researchers have also fabricated tin (Sn)-based and indium (In)-based solders. These soldiers have novel platforms for nanocomponent assembly and integration.

Nanomaterials in Humidity Sensing

The humidity measurement is essential for electronics industries. It has been found by various researchers that due to the use of nanoparticles and polymer nanocomposites as humidity sensing materials, the performance of the sensor is greatly enhanced. For nanocomposites materials, the surface effect is governing due to their high surface to volume ratio. This property is thus quite desirable to manufacture the humidity sensors. Researchers have used the electrostatic spray deposition technique to fabricate different functional nanomaterial thin films. Compared to the other film fabrications techniques, the electrostatic spray deposition technique is a low-cost and simple process having control of the morphology of deposited layers. Some researchers have used the electrostatic spray deposition technique to prepare silver nanoparticles thin film and investigated the structural, optical, and surface morphology for humidity sensor applications. Some researchers have reported ZnO/SnO_2 composite nanorods and graphene based materials as the potential materials for use as humidity sensors [49]. The use of ZnO/SnO_2 composite nanorods increases the sensitivity of the sensor. The recovery time and response time of ZnO/SnO_2 composite nanorods are faster compared to the ZnO nanoparticles and SnO_2 nanorods.

Nanomaterials as High Performance Super Capacitor

The energy consumption along with the emission of carbon dioxide is increasing at a very faster rate because of the rapid increase in the world population and increasing human dependence on energy consuming equipments. The rapid increase in energy consumption across the globe evidently has an adverse effect on human health. This initiated the development of clean and sustainable energy conversion and storage devices such as batteries and super capacitors for electric vehicles with lowest possible exhaust emission. Super capacitors are popularly known as ultra-capacitor. Several Carbon nanotube- and graphene-based nanomaterials have been reviewed in the past for potential supercapacitor applications [50]. They are electro-chemical devices with the abilities of high energy storage capabilities like conventional batteries and high power delivery ability like conventional capacitors. As charging and discharging occurs on the surface (surface phenomenon), it does not induce drastic structural changes. They store charge on the electrode surface compared to the conventional batteries in the form of energy. They can obtain high power and longer cycle life compared to the conventional capacitors and batteries, and therefore, may be used for a variety of applications ranging from industrial electric utilities and hybrid electric vehicles. Super capacitor works as a high performance energy source due to their high power and energy density. Super capacitors are of two types: a) electrochemical double layer capacitor and b) Pseudo-capacitor. In an electrochemical double layer capacitor, energy is stored electrostatically at electrode-electrolyte interface, whereas in a pseudo-capacitor, the charge is stored on the electrode surface via redox reactions. A super capacitor consists of mainly three important components: a) electrode, b) electrolyte and c) separator. The electrode is one of the important components for efficient storage and delivery of charge. The common materials for the super capacitor are metal oxide, conducting polymer and carbon-based material. Carbon-based materials such as mesoporous carbon, carbon fibres, activated carbon, graphene are used as electro-active materials in electrochemical double layer capacitor. Electrochemical double layer capacitor performance depends on the surface area of the carbon based material to store charge and, therefore, shows better cyclic ability and output power. The pseudo-capacitor used conducting polymers such as polythiophene or metal oxide like MnO_2 etc. Now-a-days, carbon nanotubes and graphene are used as an electrode in super capacitor because of their outstanding mechanical and electrical properties and high specific surface area.

CONCLUDING REMARKS

This chapter gives detailed information about the type of nanomaterials, developments in the area, and their applications in the field of electronics. This chapter also explains the different techniques to prepare the nanomaterials, such as bottom-up and top-down techniques. Nanomaterials have a great future in the field of electronics and energy storage due to their immense surface properties and performance capabilities. This chapter also gives an in-depth idea about the effect of nanomaterials in the area of electronics.

CONSENT FOR PUBLICATION

Not applicable.

CONFLICT OF INTEREST

The authors declare no conflict of interest, financial or otherwise.

ACKNOWLEDGEMENTS

The first author would like to thank Dr. Lalit Bansal, Senior Assistant Professor, VIT Vellore, for his valuable support towards the writing of this chapter.

REFERENCES

[1] Law, M.; Greene, L.E.; Johnson, J.C.; Saykally, R.; Yang, P. Nanowire dye-sensitized solar cells. *Nat. Mater.,* **2005**, *4*(6), 455-459.
[http://dx.doi.org/10.1038/nmat1387] [PMID: 15895100]

[2] Bao, W.; Wan, J.; Han, X.; Cai, X.; Zhu, H.; Kim, D.; Ma, D.; Xu, Y.; Munday, J.N.; Drew, H.D.; Fuhrer, M.S.; Hu, L. Approaching the limits of transparency and conductivity in graphitic materials through lithium intercalation. *Nat. Commun.,* **2014**, *5*, 4224.
[http://dx.doi.org/10.1038/ncomms5224] [PMID: 24981857]

[3] Wu, H.; Chan, G.; Choi, J.W.; Ryu, I.; Yao, Y.; McDowell, M.T.; Lee, S.W.; Jackson, A.; Yang, Y.; Hu, L.; Cui, Y. Stable cycling of double-walled silicon nanotube battery anodes through solid-electrolyte interphase control. *Nat. Nanotechnol.,* **2012**, *7*(5), 310-315.
[http://dx.doi.org/10.1038/nnano.2012.35] [PMID: 22447161]

[4] Wan, J.; Kaplan, A.F.; Zheng, J.; Han, X.; Chen, Y.; Weadock, N.J.; Hu, L. Two dimensional silicon nanowalls for lithium ion batteries. *J. Mater. Chem. A Mater. Energy Sustain.,* **2014**, *2*(17), 6051-6057.
[http://dx.doi.org/10.1039/C3TA13546B]

[5] Simon, P.; Gogotsi, Y. *Materials for electrochemical capacitors.,* **2010**.
[http://dx.doi.org/10.1142/9789814317665_0021]

[6] Ranjit, K.; Baquee, A.A. Nanoparticle: An overview of preparation, characterization and application. *Int Res J Pharm,* **2013**, *4*, 47-57.

[7] Mageswari, A.; Srinivasan, R.; Subramanian, P.; Ramesh, N.; Gothandam, K.M. Nanomaterials: classification, biological synthesis and characterization.*Nanoscience in Food and Agriculture 3*; Springer: Cham, **2016**, pp. 31-71.
[http://dx.doi.org/10.1007/978-3-319-48009-1_2]

[8] Ealia, S.A.M.; Saravanakumar, M.P. A review on the classification, characterisation, synthesis of nanoparticles and their application. *IOP Conf. Series Mater. Sci. Eng.,* **2017**, *263*(3), 032019. [IOP Publishing.].
[http://dx.doi.org/10.1088/1757-899X/263/3/032019]

[9] Esakkimuthu, T.; Sivakumar, D.; Akila, S. Application of nanoparticles in wastewater treatment. *Pollut. Res.,* **2014**, *33*, 567-571.

[10] Blanco-López, M.C.; Rivas, M. Nanoparticles for bioanalysis. *Anal. Bioanal. Chem.,* **2019**, *411*(9), 1789-1790.
[http://dx.doi.org/10.1007/s00216-019-01680-x] [PMID: 30828758]

[11] Martin, C.R. Nanomaterials: a membrane-based synthetic approach. *Science,* **1994**, *266*(5193), 1961-1966.
[http://dx.doi.org/10.1126/science.266.5193.1961] [PMID: 17836514]

[12] Horri, B.A.; Abdullah, A.Z.; Tan, K.B.; Vakili, M.; Salamatinia, B.; Poh, P.E. Adsorption of dyes by nanomaterials: Recent developments and adsorption mechanisms. *Separ. Purif. Tech.,* **2015**, *150*, 229-242.
[http://dx.doi.org/10.1016/j.seppur.2015.07.009]

[13] Tenne, R. Fullerene-like materials and Nanotubes from inorganic compounds with a layered (2-D) structure. *Colloids Surf. A Physicochem. Eng. Asp.,* **2002**, *208*, 83-92.
[http://dx.doi.org/10.1016/S0927-7757(02)00104-8]

[14]   Zaman, M.; Ahmad, E.; Qadeer, A.; Rabbani, G.; Khan, R.H. Nanoparticles in relation to peptide and protein aggregation. *Int. J. Nanomedicine,* **2014**, *9*, 899-912.
[PMID: 24611007]

[15]   Fawole, O.G.; Cai, X.M.; MacKenzie, A.R. Gas flaring and resultant air pollution: A review focusing on black carbon. *Environ. Pollut.,* **2016**, *216*, 182-197.
[http://dx.doi.org/10.1016/j.envpol.2016.05.075] [PMID: 27262132]

[16]   Gwinn, M.R.; Vallyathan, V. Nanoparticles: health effects--pros and cons. *Environ. Health Perspect.,* **2006**, *114*(12), 1818-1825.
[http://dx.doi.org/10.1289/ehp.8871] [PMID: 17185269]

[17]   Brook, R.D.; Franklin, B.; Cascio, W.; Hong, Y.; Howard, G.; Lipsett, M. Air Pollution and cardio vascular disease: A statement for healthcare professionals from the expert panel on population and prevention. *Science of the American Heart Association,* **2004**, *109*, 2655-2671.
[PMID: 15173049]

[18]   Bharali, D.J.; Klejbor, I.; Stachowiak, E.K.; Dutta, P.; Roy, I.; Kaur, N. Organically modified silica nanoparticles: A non-viral vector for in vivo gene delivery and expression in the brain. *Nat. Methods,* **2005**, *2*, 639.

[19]   Pimpin, A.; Srituravanich, W. **2012**.

[20]   Bhaviripudi, S.; Mile, E.; Steiner, S.A.; Zare, A.T.; Dresselhaus, M.S.; Belcher, A.M. Chemical Vapour Deposition synthesis of single-walled carbon Nanotubes from gold nanoparticle catalysts. *J. Am. Chem. Soc.,* **2007**, *129*, 1516-1517.
[http://dx.doi.org/10.1021/ja0673332] [PMID: 17283991]

[21]   Li, Y.; Liu, J.; Wang, Y.; Wang, Z.L. Preparation of mono dispersed Fe-Mo nanoparticles as the catalyst for Chemical Vapour Deposition synthesis of carbon Nanotubes. *Chem. Mater.,* **2001**, *13*, 1008-1101.
[http://dx.doi.org/10.1021/cm000787s]

[22]   Ramesh, S. *Sol-gel synthesis and characterization of nanoparticles.,* **2013**.
[http://dx.doi.org/10.1155/2013/929321]

[23]   Mann, S.; Burkett, S.L.; Davis, S.A.; Fowler, C.E.; Mendelson, N.H.; Sims, S.D. Sol-gel synthesis of organized matter. *Chem. Mater.,* **1997**, *9*, 2300-2310.
[http://dx.doi.org/10.1021/cm970274u]

[24]   Lu, C.H.; Jagannathan, R. Cerium- ion-doped yttrium aluminum garnet nanophosphors prepared through sol-gel pyrolysis for luminescent lighting. *Appl. Phys. Lett.,* **2002**, *80*, 3608-3636.
[http://dx.doi.org/10.1063/1.1475772]

[25]   Kulkarni, N.; Muddapur, U. Biosynthesis of metal nanoparticles: A review. *J. Nanotechnol.,* **2014**, •••, 1-8.
[http://dx.doi.org/10.1155/2014/510246]

[26]   Shah, M.; Fawcett, D.; Sharma, S.; Tripathy, S.K.; Poinern, G.E.J. Green synthesis of metallic nanoparticles via biological entities. *Materials (Basel),* **2015**, *8*(11), 7278-7308.
[http://dx.doi.org/10.3390/ma8115377] [PMID: 28793638]

[27]   D'Amato, R.; Falconieri, M.; Gagliardi, S.; Popovici, E.; Serra, E.; Terranova, G. Synthesis of ceramic nanoparticles by laser pyrolysis: From research to applications. *J. Anal. Appl. Pyrolysis,* **2013**, *104*, 461-469.
[http://dx.doi.org/10.1016/j.jaap.2013.05.026]

[28]   Johannessen, T.; Jensen, J.R.; Mosleh, M.; Johansen, J.; Quaade, U.; Livbjerg, H. Applications in catalysis and product/process Engineering. *Chem. Eng. Res. Des.,* **2004**, *82*, 1444-1452.
[http://dx.doi.org/10.1205/cerd.82.11.1444.52025]

[29]   Prasad Yadav, T.; Manohar Yadav, R.; Pratap Singh, D. Mechanical milling: A top down approach for

the synthesis of nanomaterials and nanocomposites. *Nanoscience and Nanotechnology,* **2012**, *2*, 22-48.
[http://dx.doi.org/10.5923/j.nn.20120203.01]

[30]   Catalent. Mechanical milling. *Catalent.,* **1996**, *19*, 1-2.

[31]   Mucsi, G. A review on mechanical activation and mechanical alloying in stirred media mill. *Chem. Eng. Res. Des.,* **2019**, *148*, 460-474.
[http://dx.doi.org/10.1016/j.cherd.2019.06.029]

[32]   Seisyan, R.P. Nanolithography in microelectronics: A review. *Tech. Phys.,* **2011**, *56*, 1061-1073.
[http://dx.doi.org/10.1134/S1063784211080214]

[33]   Wang, L.; Major, D.; Paga, P.; Zhang, D.; Norton, M.G.; McIlroy, D.N. High yield synthesis and lithography of silica-based nano spring mats. *Nanotechnology,* **2006**, •••, 17.

[34]   Hassani, S.S.; Sobat, Z. Studying of various nano litho graphy methods by using scanning probe microscope. *Int. J. Nanodimens.,* **2011**, *1*, 159-175.

[35]   Sadrolhosseini, A.R.; Mahdi, M.A.; Alizadeh, F.; Rashid, S.A. *Laser Technology and Its Applications*; Intech Open, **2018**, pp. 117-121.

[36]   Amendola, V.; Meneghetti, M. Laser ablation synthesis in solution and size manipulation of noble metal nanoparticles. *Phys. Chem. Chem. Phys.,* **2009**, *11*(20), 3805-3821.
[http://dx.doi.org/10.1039/b900654k] [PMID: 19440607]

[37]   Kabashin, A.V.; Meunier, M. Synthesis of colloidal nanoparticles during femto second laser ablation of gold in water. *J. Appl. Phys.,* **2003**, *94*, 7941.
[http://dx.doi.org/10.1063/1.1626793]

[38]   Salavati-Niasari, M.; Davar, F.; Mir, N. Synthesis and characterization of metallic copper nanoparticles via thermal decomposition. *Polyhedron,* **2008**, *27*, 3514-3518.
[http://dx.doi.org/10.1016/j.poly.2008.08.020]

[39]   Medina-Sánchez, M.; Miserere, S.; Merkoçi, A. Nanomaterials and lab-on-a-chip technologies. *Lab Chip,* **2012**, *12*(11), 1932-1943.
[http://dx.doi.org/10.1039/c2lc40063d] [PMID: 22517169]

[40]   Joshi, A.; Kim, K.H. Recent advances in nanomaterial-based electrochemical detection of antibiotics: Challenges and future perspectives. *Biosens. Bioelectron.,* **2020**, *153*, 112046.
[http://dx.doi.org/10.1016/j.bios.2020.112046] [PMID: 32056661]

[41]   Wang, J. Nanomaterial-based electrochemical biosensors. *Analyst (Lond.),* **2005**, *130*(4), 421-426.
[http://dx.doi.org/10.1039/b414248a] [PMID: 15846872]

[42]   Guo, Y.; Oo, M.K.; Reddy, K.; Fan, X. Ultrasensitive optofluidic surface-enhanced Raman scattering detection with flow-through multihole capillaries. *ACS Nano,* **2012**, *6*(1), 381-388.
[http://dx.doi.org/10.1021/nn203733t] [PMID: 22176766]

[43]   Richter, M.M. , **2008**.

[44]   Seyyedi, M.; Rostami, A.; Matloub, S. Effect of morphology of nanoparticles on performance of transparent display. *Opt. Quantum Electron.,* **2020**, *52*, 1-27.
[http://dx.doi.org/10.1007/s11082-020-02417-2]

[45]   Banerjee, J.; Dutta, K.; Rana, D. Carbon nanomaterials in renewable energy production and storage applications.*Emerging nanostructured materials for energy and environmental science*; Springer: Cham, **2019**, pp. 51-104.
[http://dx.doi.org/10.1007/978-3-030-04474-9_2]

[46]   Chen, Z.G.; Han, G.; Yang, L.; Cheng, L.; Zou, J. Nanostructured thermoelectric materials: Current research and future challenge. *Prog. Nat. Sci.,* **2012**, *22*, 535-549.
[http://dx.doi.org/10.1016/j.pnsc.2012.11.011]

[47]   Yang, W.; Gan, L.; Li, H.; Zhai, T. Two-dimensional layered nanomaterials for gas-sensing

applications. *Inorg. Chem. Front.,* **2016**, *3*, 433-451.
[http://dx.doi.org/10.1039/C5QI00251F]

[48]   Atalay, F.E.; Avsar, D.; Kaya, H.; Yagmur, V.; Atalay, S.; Seckin, T. *Nanowires of Lead-Free Solder Alloy SnCuAg.,* **2011**.
[http://dx.doi.org/10.1155/2011/919853]

[49]   Lv, C.; Hu, C.; Luo, J.; Liu, S.; Qiao, Y.; Zhang, Z.; Song, J.; Shi, Y.; Cai, J.; Watanabe, A. Recent advances in graphene-based humidity sensors. *Nanomaterials (Basel),* **2019**, *9*(3), 422.
[http://dx.doi.org/10.3390/nano9030422] [PMID: 30871077]

[50]   Yang, Z.; Tian, J.; Yin, Z.; Cui, C.; Qian, W.; Wei, F. Carbon nanotube-and graphene-based nanomaterials and applications in high-voltage supercapacitor: a review. *Carbon,* **2019**, *141*, 467-480.
[http://dx.doi.org/10.1016/j.carbon.2018.10.010]

CHAPTER 15

Nanomaterials for Flexible Photovoltaic Fabrics

Sudheer Kumar[1,*] and **Sukhila Krishnan**[2]

[1] *School for Advanced Research in Petrochemicals (SARP), Laboratory for Advanced Research in Polymeric Materials (LARPM), Central Institute of Petrochemicals Engineering & Technology (CIPET), B/25, CNI Complex, Patia, Bhubaneswar 751024, Odisha, India*

[2] *Sahrdaya College of Engineering and Technology, Department of Applied Science and Humanities, Kodakara, Thrissur-680684, Kerala, India*

Abstract: The development of extremely flexible photovoltaic (PV) devices for energy harvesting and storage applications is currently receiving more attention by the researchers from industries. The presently available energy storage devices are too rigid and extensive and also not suitable for next-generation flexible electronics such as silicon-based solar cells. Thus, the researchers have developed high-performance, lightweight, conformable, bendable, thin, and flexible dependable devices. On the other hand, these energy storage devices require to be functional under different mechanical deformations, for example, bending, twisting, and even stretching. The nanomaterial (TiO_2, ZnO, Ag, *etc.*) coated fabrics also play a vital role in improving the efficiency of the solar cell (devices) to a great extent. The current chapter provides information about the development of nanomaterials-based flexible photovoltaic solar cell devices for wearable textile industry applications. The fabricated carbon ink printed fabrics such as polyester, cotton woven and nonwoven, and polyethylene terephthalate nonwoven can be used as cathode and heating sources of PV devices. The organic and flexible conductive substrate printed with carbon ink can be utilized as heating source fabrics for wearable electronics devices. The flexible substrate-based photovoltaics (PV) device is mostly used in the textile industries due to its flexibility, environmental friendliness, low cost as well as easy processability. The flexible-wearable photovoltaic devices pave the way to be used for enormous applications in various fields.

Keywords: Energy storage and harvesting, Efficiency, Flexible photovoltaic fabrics, Flexible electronics, High-performance, Lightweight, Low cost, Nanomaterials, Photovoltaic fibers, Photovoltaic fiber, Solar cell, Wearable textile industry applications.

---

* **Corresponding author Sudheer Kumar:** School for Advanced Research in Petrochemicals (SARP), Laboratory for Advanced Research in Polymeric Materials (LARPM), Central Institute of Petrochemicals Engineering & Technology (CIPET), B/25, CNI Complex, Patia, Bhubaneswar 751024, Odisha, India; E-mail: sudheerkumar.211@gmail.com

INTRODUCTION

Recently, flexible energy storage materials have been in huge demand by the industries. Day by day, this has increased the utlilization of nanomaterials coated flexible systems which bestows higher flexibility, good electrical performance, excellent mechanical properties as well as environmental stability [1 - 3]. Nanomaterials play a significant role as active layers on the fabric substrate and nanomaterials-based flexible substrate nanodevices have been extensively exploited in different applications in the last decade compared to the available traditional planar technology [4]. On the other hand, the development of the organic transistor was fabricated on a flexible plastic substrate or thin glass substrate, mostly for electronic devices. However, they are demonstrated with poor electrical characteristics as compared to the high-performance inorganic semiconductor materials and were found to be inappropriate for other applications [5, 6]. The materials needed for the flexible electronics are depicted in Fig. (**1**).

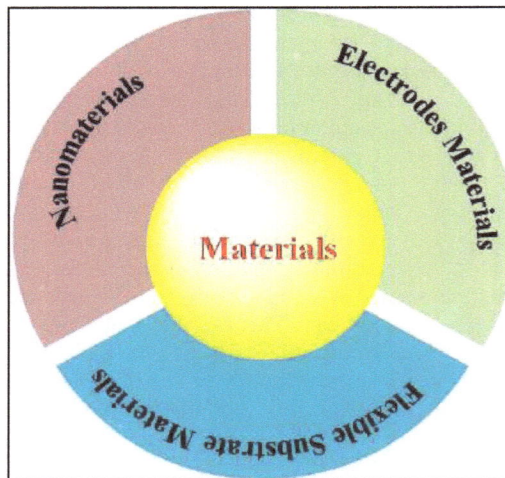

Fig. (1). Materials for flexible electronics applications.

Moreover, the use of elastomeric substrate is very limited for the deposition of the active layers on polymer substrates due to the low processing temperature and reduced performance of the fabricated electronic device. Conversely, the development of novel materials like conducting polymers, carbon-based structures, for example, metallic and ceramic composites; the nanomaterials-based electronic device can be formed on the elastomeric flexible substrate, while exploring a new electronic device field like flexible electronics. The flexible electronics focused on fabricating compatible flexible systems reveal good carrier mobility and better electrical behaviour. The nanomaterials can be utilized as an active layer or substrate to address the restrictions of the inert system and will be immobile with the defining mechanics due to their heterogeneous low-dimensional structure.

Currently, flexible solar cells play a significant role in photovoltaics because they are weightless, tolerable to complicated deformation, can be incorporated into curve surfaces, consistent with reel-to-reel processing, proper storage and transportation [7 - 9]. These are utilized in many applications like portable or wearable electronics, power generated textiles, *etc* [10, 11]. Amorphous silicon solar cells based on flexible photovoltaic devices are available, but they are too expensive with a low bending radius. For that reason, researcher-developed novel solar cells, for example, dye-sensitized solar cells (DSCs), organic solar cells (OSCs), and perovskite solar cells (PSCs), which were found to be of great interest, especially for the low-temperature fabrication, low thickness, and color regulation. In the case of the flexible solar cell, entire functional layers are converted to flexible; particularly the electrode works as a substrate for active layers. Moreover, the interaction among the flexible electrodes and active layers by low-temperature processing is important to alter the morphology and boundary of entire functional layers while tuning the properties of the photovoltaic devices. Table **1** demonstrated the three main types of flexible substrate as a polymer (plastic), metal foil, and thin glass. Flexible photovoltaic fabrics, after the addition of MWCNT were used as a flexible wearable heating source or cathode for DSSCs.

Table 1. Advantages and disadvantages of the different flexible substrates (Koh *et al.*, 2018 [12].

Substrate	Advantages	Disadvantages
Metal foil	• Rough and conformable • Low H_2O and O_2 permeation • High process temperature • Good dimension stability	• Weak dimension stability • Less process temperature and chemical • Resistance • Good H_2O and O_2 permeation
Thin glass	• Conformable • Transparent • Low H_2O and O_2 permeation	• Poor mechanical stability • Less process temperature
Polymer (*e.g.* **Plastic**)	• Rough • Ductile • Transparent	• Weak dimension stability • Less process temperature and chemical resistance • High H_2O and O_2 permeation

This chapter deliberates the present-day research advancement on the fabrication of the nanomaterial-based flexible photovoltaic fabric cells for energy harvesting and storage applications inside the textile industry.

TYPES OF PHOTOVOLTAIC FIBERS

Recently, flexible photovoltaic devices have drawn attention due to their potential applications for wearable electronics and the main intention to reduce the weight of substrate, formation of the novel structure of power-harvesting and stimulate the fabrication technology. Common photovoltaic cell-based solar cells usually are constructed on fiber and wire-like substrate. Fiber electrode is based on the thin film of active PV fabrics on the surface applied layer by layer 3D structure. In contrast, it harvests the energy using light from 3D dimensional space owing to the identical photoanode. In particular, 3D dimensional power-harvesting characteristics tendered chances for high-end power applications. Tuning the length and diameter ratio of the fiber based highly flexible PV devices is feasible for wearable solar energy harvesting and adaptation [13, 14]. Photovoltaics is categorised into different types such as organic, inorganic, dye sensitize, and perovskite PV fibers.

Inorganic Photovoltaic Fibers

Inorganic PV fibers like silicon solar cells based on electron-hole pairs, are categorized through the p-n joints. Initially, polysilicon solar cell based on flexible fiber is synthesized by the vapour deposition method, which prepares long-length fiber, which is highly cost-effective. Afterwards, flexible coaxial p-n photodiode fiber was prepared by chemical vapour deposition at high pressure and a fiber solar device was developed that demonstrated the efficiency of (0.5% up to 10 meters) and performed as a quick-response photodetector for visible laser light [15]. In addition to this, the fiber synthesized by the normal fiber optic method is suitable for the preparation of p-n joint and microfibre solar cells, exhibiting an efficiency of 3.6% [16, 17]. The optimization of the highest quantum efficiency and light-harvesting is based on the width and doping amount in the p and n doped layers. As a result, the fiber polysilicon solar cell displayed 10.5% of efficiency, although details regarding the size of the device were not supplied. However, the amorphous Si solar cell on glass fiber fabrics in textile was revealed *via* electron beam evaporation and achieved 1.4%efficiencies. Further, inorganic materials are very difficult to fabricate on fabric substrates. Because of electron deposition on the Mo wire, a flexible fibre-shaped CuInSe2 solar cell was synthesized and demonstrated an efficiency of 2.3%. Before applying the usual PV layers, Plentz *et al.* [18] prepared low-efficiency amorphous silicon PV cells on glass fibre fabrics with dip-coating of polymer, even though the intense problem of cracking was found. Copper indium gallium diselenide (CIGS) cells exhibit excellent efficiencies when deposited onto glass fibre fabric pre-coated with particle-filled resin, which can withstand the very high processing temperatures of this form of cell [19]. For applying semiconductor, both these inorganic PV cells employ the vacuum method, thermal evaporation of elements for CIGS and plasma-enhanced chemical vapour deposition (CVD) for a-Si:H. Both for metallic and conducting transparent oxide, sputtering was done in a partial vacuum to apply electrical contacts.

The inorganic PV fibers is a better and efficient fabrication materials, which persist good stability for a solid devices. While various types of inorganic photovoltaic materials are available and more attempts have been taken for novel improvements. Along with this, the p-n joint on fiber is used in different wearable electronics [20].

Organic Photovoltaic Fibers

The use of the organic and polymer material for the organic photovoltaic *via* solution procedure provides the functionality for structure investigation. Lee *et al.* [21] fabricated a twisted fiber device coated with two different materials such as poly(2,3-dihydrothieno-1,4-dioxin)-poly(styrene sulfonate) (PEDOT:PSS)/poly (3-hexylthiophene-2,5-diyl) (P3HT): phenyl-C_{61}-butyric acid methyl ester and silver respectively and formed the flexible organic PV fiber that exhibited 3.9% efficiency. The developed flexible and lightweight fiber solar cell is used in photovoltaic textile as well as wearable electronics. Michael *et al.* [22] also fabricated organic fiber devices on cellulose fiber by solution process and large-scale production was widely used to manufacture smart textile industry. The organic materials are simple to synthesize although, inadequate PV active materials were deposited for fiber based devices and organic PV fiber is less efficient than other kinds . there is more opportunity for enhancement in higher efficiency of 10%-13%.

Liu *et al.* [23] prepared a DSSC on a woven polyester cotton blended fabric *via* screen printing technology with an interface layer on the fabric before applying an Ag electrode, leading to DSSCs with a PCE of 2.78%. Arumugam *et al.* [24] also had used the same method for preparing the full spray-coated textile-based OPVs. With rigid platinum-coated FTO glass for a single electrode, this device is inflexible. Yun *et al.* [25] prepared a textile DSSC based on the 3-D structure in which an electrolyte-introduced woven glass-fibre fabric was inserted into the active materials coated stainless steel fabric electrodes. The resultant structure produced a PCE of 1.7%, which was enclosed inside a polyester film small bag with a filling. Cyclic bending tests showed a decreased efficiency (40%) following 1000 twisting cycles, with less drop seen after 2000 bend cycles. After one day with non-volatile electrolyte, the DSSC saw a tiny rise in particular power that was stable for the next six days. Song *et al.* [26] presented a flexible DSSC PV textile, with a quasi-solid electrolyte, exhibiting a PCE of 5.08% which was attained by inserting a DSSC film onto fabric by epoxy polymer. The work illustrated counter-electrode formations which improved the mechanical performance and can be bent several times, resulting in an ultimate flexible DSSC textile.

Dye-sensitized Photovoltaic Fibers

A lot of researchers have designed the DSSC as a tube-like structure to enhance the light-harvesting and decrease the problems in sealing plus increasing the stability, although the device was too rigid. Zeng *et al.* [27] prepared dye-sensitized twisted PV fiber with the help of wire-shaped photoanode and Pt counter electrode with high-temperature resistance, which offers great potential for an employed flexible wearable solar power source. Hence, the developed design avoids conductive transparent glass and reduces the sealing region as well as evaporation of organic solvents in liquid electrolytes. DSSC is simple to fabricate with air tolerance to certain circumstances therefore they are better models for different examinations. Recently, electrodes have been fabricated from the non-metal, lightweight substrate containing carbon nanotube yards and carbon fiber, *etc* [28]. and used in electrode fabrication, photoanodes with various nanostructures, like TiO_2 nanoparticles [29], TiO_2 nanowires, TiO_2 nanotubes [30], TiO_2 nanorods [31], ZnO nanowires [32], ZnO nanosheet [33].

Further, developed the Ti/TiO_2 nanotube arranged fiber photoanode and MoS_2 nanofilm/TiO_2/Carbon fiber counter electrode bestowed for 9.5% efficiency, which is the most suitable method to prepare device of 20-30 cm length. Gratzel *et al.* [34] also described DSSC with an efficiency of 28% in ambient lighting and open extensive areas of applications for self-governing power operation. The dye and quantum dots sensitized PV fibers act in the main part of the indoor environment, and it is the most applied deliberation for textile electronics.

Perovskite Photovoltaic Fibers

The perovskite materials are easily fabricated by the vapour deposition, solution coating, and solid-state preparation techniques at a lower temperature and get deposited into the solar cells, photodetectors, light-emitting devices, memristors, *etc*. PVSC perovskite solar cells (PSC) is an active-active topic currently. Peng *et al.* [20] prepared the perovskite modified steel wire and carbon nanotube (MWCNT) sheet as a cell top electrode-based fibre-shape perovskite solar exhibiting the efficiency of 3.3%-3.8%. Lee *et al.* [35] also prepared a perovskite layer and silver nanowire modified double twisted CNT fiber electrode that demonstrated better flexibility as well as efficiency of 3.0%. Hu *et al.* [36] described semi-transparent fiber perovskite solar cell and top electrode based on the Au electrode and stable photoelectron production and charge transfer, Ti wire helped devices to obtain 5.3% efficiency. Particular layer control by layer coating and electrochemical deposition process bestowed for excess exposure as well as consistency on a curve surface with the efficiency of 7.1-7.5% in the device. The carbon nanotube-aligned sheet-based polyethylene naphthalate/indium tin oxide strip based huge perovskite crystals provides 9.49 efficiency. The latest planar perovskite solar cells report 22.1% efficiency, and day-by-day performance gets improved, which is used in the light-harvesting, film crystalline and charge carrier in the curved surface. Nevertheless, the perovskite layers can be fabricated on lengthy fiber for that reason, the current device length is very small than other types. On the other hand, the fabrication of a huge size device with good power output requires more investigation. Inorganic dye-sensitized perovskite solar fibers have attained a 10% efficiency. However, the earlier two have longer lengths in comparison to devices of both longer length and better efficiency. Organic photovoltaic fibers are comparatively less explored in terms of device efficiency as well as size owing to the complex information of capable organic heterojunction on the curved surface. This is attributed to the fewer materials that were absorbed. The progress of photovoltaics extra penetration of photovoltaic fibers can be anticipated.

FABRICATION OF NANOMATERIAL-BASED PHOTOVOLTAICS FABRICS

Miao *et al.* [37] deposited the porous and amorphous TiO$_2$ thin film on flexible polyester textile and cotton fabrics substrate by magnetron sputtering techniques. Earlier to deposition, the fabrics were properly cleaned *via* ultrasonic technnique with final washing by acetone solution. Subsequent drying treatment in the presence of nitrogen atmosphere than after fabrics cut into 6 cm × 6 cm pieces as per ASTM D1776-08 standard prior sputtering and characterization. Further to enhance the homogeneity of the multilayer films, a specimen holder was employed at 100 rpm rotation speed. It was observed that the woven structure of blank polyester demonstrated a clear and even surface. While the base textile structure of the coated fabrics is not altered following deposition. The porous TiO$_2$ film can examine the surface of the fiber and the usual size of the holes diminishes with the improvement in sputtering time. Finally, the result confirmed that, by adjusting the deposition time the desired hole size can be achieved. Conversely, the deposition technique required for low temperature is appropriate for the flexible textile substrate exposed to high-temperature demolition. So, the flexible amorphous TiO$_2$ thin film was coated on the fabric substrates by suitable techniques.

Arbab *et al.* [38] prepared the layers of the printable composite containing woven/nonwoven fabrics in which MWCNT used as a flexible wearable heating source or cathode for DSSCs. The MWCNT based printable carbon ink was prepared by employing globular protein serum bovine albumin (BSA) by the inclusion of distribution (*e.g.* amino agent) that increased the distribution of MWCNTs and fabricated tubular porous carbon-based matrix. These types of fabrics are electrically conductive and play a crucial role in the electron transport pathway and increase the electrocatalytic movement, improve the heat dissipation and endow a perfect heating source. Moreover, heat production and conductivity of prepared flexible fabric are extremely based on its kinds and conductive ink uptake. The real fabric structure materials revealed less electrical resistance of 15-20 Ω, low charge resistance of 2.69 Ω and power conversion efficiency of about 8% in most of the circumstances. The fabricated carbon ink printed fabrics as cathode and heating source of PV devices.

The organic, flexible and conductive substrate printed with carbon ink can be employed as heating source fabrics for wearable electronics devices. Various carbon ink printed fabrics such as polyester, cotton woven and nonwoven cotton, PET sheet get heated with varying voltage starting 5 to 20 V . The outcome temperature investigated by the thermal sensor and thermal expert camera demonstrated that enhancement in the applied DC voltage could significantly increase the produced heat as well as dissipation of energy in the form of heat. For that reason, woven and nonwoven cotton fabrics exhibited a quick response to the applied voltage as resulted in a heat production range from 110-140°C. However, the PET and polyester obtained fabrics revealed hysteresis occurrence with lower temperatures in the range of 70–80 °C owing to their thermoplastic condition.

On the other hand, carbon ink deposited organic fabrics and cotton woven demonstrated the maximum temperature of 140 °C at 20V. In the case of carbon ink deposited fabrics, the mobility of charge carriers increases with the applied voltage. As a result the produced heat creates the collision of charge carriers and phonons for the reason of quantum and vibration energy of the carbon crystal lattice. This further enhances the charge phonon collision owing to the rise in voltage, the produced temperature also extensively gets improved. The cotton-based fabrics exhibited the maximum energy dissipation and thus better heat production compared to PET sheets degraded quickly *via* a good enhancement in the voltage to larger than 15V, which is attributed to the thermoplastic conditions and less structural density. Therefore, the altering of active layers can extensively influence the conductivity of carbon ink coated organic flexible textiles.

APPLICATION OF NANOSTRUCTURED BASED FLEXIBLE FABRICS FOR ENERGY HARVESTING AND STORAGE

Nanomaterials on the flexible substrate have fascinated many people due to their tremendous physical, electrical and chemical properties. On the other hand, nanomaterial-based flexible substrates and nanodevices have been extensively employed in different applications. The PV textile materials can be employed to generate power in wearable fabrics, cell phone and static electronics devices to transfer light, cool and heat, *etc.*, *via* converting light into electrical energy. Furthermore, PV substances can be incorporated into textile substrate mainly on clothes; conversely, the better favorable results from a well-organized PV fiber are yet to come with different smart textile and its related products.

Photovoltaic Textile (PV)

The flexible substrate-based photovoltaics (PV) device are used in the textile industries due to its characteristics such as more flexibility, environmentally friendly, low cost as well as easy processability [39, 40]. PV gets accumulated in the textile substrate and tunes the PV characteristics within the products . The effect of the produced PV layers on the traditional textile fibers, metallic fibers [41], tapes, and combining in the fabric form through joining or weaving and depositing straight onto fabrics [42]. The development techniques of PVs instead of silicon-based techniques are more complex processes and require elevated temperature, where traditional textile materials cannot stand. Recently, photovoltaic DSSC and perovskite solar cells (PSCs) have been generally investigated for PV techniques due to the high flexibility, easy formation, and compatibility of textile-based fabrication procedures. PV textile is divided into three types based on substrate types, *i.e.* photovoltaic tapes, fibre-structured photovoltaics, and fabric substrate [2].

Wearable Electronics

At this time, wearable electronics devices are the latest topics in academia and industry for the investigation of multifunctional electronics in clothing. The main part of the wearable electronic device *e.g.* the power resource is important for forthcoming formation and its durability. Although, the traditional heavy battery powering wearable electronics devices is deficient in flexibility, ease, weightlessness and maintenance-free. Consequently, researchers developed an alternate solution for a textile-compatible power resource for constant power wearable electronics devices. Amongst energy generating devices, solar cells employ boundless sunlight to create electricity as one of the ecofriendly, utmost appropriate, and potential substitute devices. Thus, lots of attempts have been made to develop photovoltaic textiles, as depicted in Fig. (**2**).

Fig. (2). Application of the flexible-wearable photovoltaic devices in various fields. Reprinted with copyright permission from Hashemi *et al.* Energy Environ. Sci., **2020** [43].

Further, energy-producing textiles employing PV, triboelectric, piezoelectric, and thermoelectric energy harvesting systems have drawn huge interest in fabrication of wearable electronics techniques. Based on the recent research, energy-producing fabric materials are found to resolve the current and future problems. The developing techniques such as the internet of things, wearable and flexible electronics, are based on energy for that reason and energy harvesting devices. Improvements in fabrics and methods of PV techniques will permit the PV textile to be efficient products.

CONCLUSIONS

This review has explored many methods to convert flexible PV devices for various energy harvesting and storage applications. By seeing the level of maturity of these innovations in contrast to other methods in the flexible PV area, it is foreseen that the tendency of employing nanomaterials for flexible PV fabrics will remain as an emerging area, whereas the exhilarating technologies defined in this review will progress more in the coming years.

The development of nanomaterials based flexible photovoltaic fabric solar cell is a versatile material used in various applications due to their higher flexibility, lightweight, easy processability as well as good electrical performance, excellent mechanical properties and cost-effective as compared to the exciting silicon-based

PV solar cell. On the other hand, polymer-based flexible substrate PV solar cells exhibited str*etch*able, twistable, bendable and required low temperature for fabrication. The developed new PV Solar cell reveals the great potential to explore new facilities for flexible electronics.

The future development and, eventually, the usage of flexible PV fabrics will need the textile properties to be fully enumerated and the strength of the materials for use as power generated textiles or wearable electronics to be tackled. It will be thought-provoking due to the unawareness of standards now accessible for the testing of PV fabrics (presently, only BS EN 16812:2016 is known). Numerous materials, in their unmodified form, do not exhibit the required performance or applications, and hence, further advancement is desirable; although it is very tough to convincingly remark on it unless data is tested. The stability of many nanomaterials-based PV fabrics is essential to be enhanced for the long-term use need of fabric. The methods to prepare these PV fabrics at a commercial scale will also be required. In conclusion, power management also has to be taken into account. The output from a nanomaterials-based PV fabric will not be a constant process owing to the wearer's movement and Sun's location. The next obstacle to the implementation of nanomaterials-based PV fabrics is the formation of power management electronics which has to be individually combined within a textile.

CONSENT FOR PUBLICATION

Not applicable.

CONFLICT OF INTERESTS

The authors declare no conflict of interest, financial or otherwise.

ACKNOWLEDGEMENTS

The first author would like to thank Dr R. Ananthakumar, Jr. Scientist, SARP-LARPM, Central Institute of Petrochemicals Engineering & Technology (CIPET), for his valuable support in while writing this chapter.

REFERENCES

[1]   Singh, M.K. *Flexible photovoltaic textiles for smart applications*; Sol. Cells New Asp. Solut, **2011**, pp. 43-68.

[2]   Unsal, Ö.F.; Hiçyilmaz, A.S.; Yilmaz, A.N.Y.; Altin, Y.; Borazan, İ.; Bedeloglu, A.C. Energygenerating textiles.*Advances in Functional and Protective Textiles*; Woodhead Publishing, **2020**, pp. 415-455.
[http://dx.doi.org/10.1016/B978-0-12-820257-9.00017-5]

[3]   Wong, W.S.; Salleo, A., Eds. *Flexible electronics: materials and applications*; Springer Science & Business Media, **2009**, Vol. 11, .
[http://dx.doi.org/10.1007/978-0-387-74363-9_6]

[4]   Zhang, H. *Nanomaterials and Devices on Flexible Substrates*; Advanced Nano Deposition Methods, **2016**.
[http://dx.doi.org/10.1002/9783527696406.ch15]

[5]   Lin, Q.; Huang, H.; Jing, Y.; Fu, H.; Chang, P.; Li, D.; Fan, Z. Flexible photovoltaic technologies. *J. Mater. Chem. C Mater. Opt. Electron. Devices,* **2020**, *2*(7), 1233-1247.
[http://dx.doi.org/10.1039/c3tc32197e]

[6]   Lin, Q.; Huang, H.; Jing, Y.; Fu, H.; Chang, P.; Li, D.; Yao, Y.; Fan, Z. Flexible photovoltaic technologies. *J. Mater. Chem. C Mater. Opt. Electron. Devices,* **2014**, *2*(7), 1233-1247.
[http://dx.doi.org/10.1039/c3tc32197e]

[7]   Li, Y.; Meng, L.; Yang, Y.M.; Xu, G.; Hong, Z.; Chen, Q.; You, J.; Li, G.; Yang, Y.; Li, Y. Highefficiency robust perovskite solar cells on ultrathin flexible substrates. *Nat. Commun.,* **2016**, *7*(1), 1-0.

[8]   Chen, J.; Huang, Y.; Zhang, N.; Zou, H.; Liu, R.; Tao, C.; Fan, X.; Wang, Z.L. Micro-cable structured textile for simultaneously harvesting solar and mechanical energy. *Nat. Energy,* **2016**, *1*(10), 1-8.
[http://dx.doi.org/10.1038/nenergy.2016.138]

[9]   Jung, S.; Lee, J.; Seo, J.; Kim, U.; Choi, Y.; Park, H. Development of annealing-free, solutionprocessable inverted organic solar cells with n-doped graphene electrodes using zinc oxide nanoparticles. *Nano Lett.,* **2018**, *18*(2), 1337-1343.
[http://dx.doi.org/10.1021/acs.nanolett.7b05026] [PMID: 29364692]

[10]   Kaltenbrunner, M.; Adam, G.; Głowacki, E.D.; Drack, M.; Schwödiauer, R.; Leonat, L.; Apaydin, D.H.; Groiss, H.; Scharber, M.C.; White, M.S.; Sariciftci, N.S.; Bauer, S. Flexible high power-pe--weight perovskite solar cells with chromium oxide-metal contacts for improved stability in air. *Nat. Mater.,* **2015**, *14*(10), 1032-1039.
[http://dx.doi.org/10.1038/nmat4388] [PMID: 26301766]

[11]   Zhang, Y.; Wu, Z.; Li, P.; Ono, L.K.; Qi, Y.; Zhou, J.; Shen, H.; Surya, C.; Zheng, Z. Fully solution-processed TCO-free semitransparent perovskite solar cells for tandem and flexible applications. *Adv. Energy Mater.,* **2018**, *8*(1), 1701569.
[http://dx.doi.org/10.1002/aenm.201701569]

[12]   Koh, W.S.; Lee, K.M.; Toh, P.Y.; Yeap, S.P. Nano-graphene and its derivatives for fabrication of flexible electronic devices: A quick review. *Energy conversion,* **2018**, *40*, 43.

[13]   Fu, Y.; Wu, H.; Ye, S.; Cai, X.; Yu, X.; Hou, S.; Kafafy, H.; Zou, D. Integrated power fiber for energy conversion and storage. *Energy Environ. Sci.,* **2013**, *6*(3), 805-812.
[http://dx.doi.org/10.1039/c3ee23970e]

[14]   Peng, M.; Yan, K.; Hu, H.; Shen, D.; Song, W.; Zou, D. Efficient fiber shaped zinc bromide batteries and dye sensitized solar cells for flexible power sources. *J. Mater. Chem. C Mater. Opt. Electron. Devices,* **2015**, *3*(10), 2157-2165.
[http://dx.doi.org/10.1039/C4TC02997F]

[15]   He, R.; Day, T.D.; Krishnamurthi, M.; Sparks, J.R.; Sazio, P.J.; Gopalan, V.; Badding, J.V. Silicon p-i-n junction fibers. *Adv. Mater.,* **2013**, *25*(10), 1461-1467.
[http://dx.doi.org/10.1002/adma.201203879] [PMID: 23212830]

[16]   Homa, D.; Cito, A.; Pickrell, G.; Hill, C.; Scott, B. Silicon fiber with pn junction. *Appl. Phys. Lett.,* **2014**, *105*(12), 122110.
[http://dx.doi.org/10.1063/1.4895661]

[17]   Martinsen, F.A.; Smeltzer, B.K.; Nord, M.; Hawkins, T.; Ballato, J.; Gibson, U.J. Silicon-core glass fibres as microwire radial-junction solar cells. *Sci. Rep.,* **2014**, *4*(1), 6283.
[http://dx.doi.org/10.1038/srep06283] [PMID: 25187060]

[18]   Plentz, J.; Andra, G.; Pliewischkies, T.; Bruckner, U.; Eisenhawer, B.; Falk, F. Amorphous silicon thin-film solar cells on glass fiber textiles. *Mater. Sci. Eng. B,* **2016**, *204*, 34-37.

[http://dx.doi.org/10.1016/j.mseb.2015.11.007]

[19]   Knittel, D.; Kontges, M.; Heinemeyer, F.; Schollmeyer, E. Coatings on textiles for Cu(In,Ga)Se$_2$ photovoltaic cell formation on textile carriers: Preparation of Cu(In,Ga)Se$_2$ solar cells on glassfiber textiles. *J. Appl. Polym. Sci.,* **2010,** *115*(5), 2763-2766.
        [http://dx.doi.org/10.1002/app.30349]

[20]   Peng, M.; Dong, B.; Zou, D. Three dimensional photovoltaic fibers for wearable energy harvesting and conversion. *J. Ener. Chem.,* **2018,** *27*(3), 611-21.
        [http://dx.doi.org/10.1016/j.jechem.2018.01.008]

[21]   Lee, S.; Lee, Y.; Park, J.; Choi, D. Stitchable organic photovoltaic cells with textile electrodes. *Nano Energy,* **2014,** *9,* 88-93.
        [http://dx.doi.org/10.1016/j.nanoen.2014.06.017]

[22]   Ebner, M.; Schennach, R.; Chien, H.T.; Mayrhofer, C.; Zankel, A.; Friedel, B. Regenerated cellulose fiber solar cell. *Flexible and Printed Electronics,* **2017,** *2*(1), 014002.
        [http://dx.doi.org/10.1088/2058-8585/aa5707]

[23]   Liu, J.; Li, Y.; Arumugam, S.; Tudor, J.; Beeby, S. Screen printed dye-sensitized solar cells (DSSCs) on woven polyester cotton fabric for wearable energy harvesting applications. *Mater. Today Proc.,* **2018,** *5*(5), 13753-13758.
        [http://dx.doi.org/10.1016/j.matpr.2018.02.015]

[24]   Arumugam, S.; Li, Y.; Senthilarasu, S.; Torah, R.; Kanibolotsky, A.L.; Inigo, A.R.; Skabara, P.J. Beeby, S.P. Fully spray-coated organic solar cells on woven polyester cotton fabrics for wearable energy harvesting applications. *J. Mater. Chem. A Mater. Energy Sustain.,* **2016,** *4*(15), 5561-5568.
        [http://dx.doi.org/10.1039/C5TA03389F]

[25]   Yun, M.J.; Sim, Y.H.; Cha, S.I.; Seo, S.H.; Lee, D.Y. Three-dimensional textile platform for electrochemical devices and its application to dye-sensitized solar cells. *Sci. Rep.,* **2019,** *9*(1), 2322.
        [http://dx.doi.org/10.1038/s41598-018-38426-1] [PMID: 30787333]

[26]   Song, L.; Wang, T.; Jing, W.; Xie, X.; Du, P.; Xiong, J. High flexibility and electrocatalytic activity MoS2/TiC/carbon nanofibrous film for flexible dye-sensitized solar cell based photovoltaic textile. *Mater. Res. Bull.,* **2019,** *118,* 110522.
        [http://dx.doi.org/10.1016/j.materresbull.2019.110522]

[27]   Zeng, W.; Wang, M.; Li, Y.; Wan, J.; Huang, H.; Tao, H.; Carroll, D.L.; Zhao, X.; Zou, D.; Fang, G. Semi-closed tubular light-trapping geometry dye sensitized solar cells with stable efficiency in wide light intensity range. *J. Power Sources,* **2014,** *261,* 75-85.
        [http://dx.doi.org/10.1016/j.jpowsour.2014.03.050]

[28]   Cai, X.; Zhang, C.; Zhang, S.; Fang, Y.; Zou, D. Application of carbon fibers to flexible, miniaturized wire/fiber-shaped energy conversion and storage devices. *J. Mater. Chem. A Mater. Energy Sustain.,* **2017,** *5*(6), 2444-2459.
        [http://dx.doi.org/10.1039/C6TA07868K]

[29]   Lv, Z.; Fu, Y.; Hou, S.; Wang, D.; Wu, H.; Zhang, C.; Chu, Z.; Zou, D. Large size, high efficiency fiber-shaped dye-sensitized solar cells. *Phys. Chem. Chem. Phys.,* **2011,** *13*(21), 10076-10083.
        [http://dx.doi.org/10.1039/c1cp20543a] [PMID: 21509400]

[30]   Lv, Z.; Yu, J.; Wu, H.; Shang, J.; Wang, D.; Hou, S.; Fu, Y.; Wu, K.; Zou, D. Highly efficient and completely flexible fiber-shaped dye-sensitized solar cell based on TiO2 nanotube array. *Nanoscale,* **2012,** *4*(4), 1248-1253.
        [http://dx.doi.org/10.1039/c2nr11532h] [PMID: 22278314]

[31]   Guo, W.; Xu, C.; Wang, X.; Wang, S.; Pan, C.; Lin, C.; Wang, Z.L. Rectangular bunched rutile TiO2 nanorod arrays grown on carbon fiber for dye-sensitized solar cells. *J. Am. Chem. Soc.,* **2012,** *134*(9), 4437-4441.
        [http://dx.doi.org/10.1021/ja2120585] [PMID: 22300521]

[32]   Wang, W.; Zhao, Q.; Li, H.; Wu, H.; Zou, D.; Yu, D. Transparent, double-sided, ITO-free, flexible dye-sensitized solar cells based on metal wire/ZnO nanowire arrays. *Adv. Funct. Mater.,* **2012**, *22*(13), 2775-2782.
[http://dx.doi.org/10.1002/adfm.201200168]

[33]   Dai, H.; Zhou, Y.; Chen, L.; Guo, B.; Li, A.; Liu, J.; Yu, T.; Zou, Z. Porous ZnO nanosheet arrays constructed on weaved metal wire for flexible dye-sensitized solar cells. *Nanoscale,* **2013**, *5*(11), 5102-5108.
[http://dx.doi.org/10.1039/c3nr34265d] [PMID: 23644717]

[34]   Freitag, M.; Teuscher, J.; Saygili, Y.; Zhang, X.; Giordano, F.; Liska, P.; Hua, J.; Zakeeruddin, S.M.; Moser, J.E.; Grätzel, M.; Hagfeldt, A. Dye-sensitized solar cells for efficient power generation under ambient lighting. *Nat. Photonics,* **2017**, *11*(6), 372-378.
[http://dx.doi.org/10.1038/nphoton.2017.60]

[35]   Lee, M.; Ko, Y.; Jun, Y. Efficient fiber-shaped perovskite photovoltaics using silver nanowires as top electrode. *J. Mater. Chem. A Mater. Energy Sustain.,* **2015**, *3*(38), 19310-19313.
[http://dx.doi.org/10.1039/C5TA02779A]

[36]   Hu, H.; Yan, K.; Peng, M.; Yu, X.; Chen, S.; Chen, B.; Dong, B.; Gao, X.; Zou, D. Fiber-shaped perovskite solar cells with 5.3% efficiency. *J. Mater. Chem. A Mater. Energy Sustain.,* **2016**, *5*(10), 3901-3906.
[http://dx.doi.org/10.1039/C5TA09280A]

[37]   Miao, D.; Hu, H.; Li, A.; Jiang, S.; Shang, S. Fabrication of porous and amorphous TiO2 thin films on flexible textile substrates. *Ceram. Int.,* **2015**, *41*(7), 9177-9182.
[http://dx.doi.org/10.1016/j.ceramint.2015.03.080]

[38]   Arbab, A.A.; Memon, A.A.; Sun, K.C.; Choi, J.Y.; Mengal, N.; Sahito, I.A.; Jeong, S.H. Fabrication of conductive and printable nano carbon ink for wearable electronic and heating fabrics. *J. Colloid Interface Sci.,* **2019**, *539*, 95-106.
[http://dx.doi.org/10.1016/j.jcis.2018.12.050] [PMID: 30576992]

[39]   Borazan, I. A study about lifetime of photovoltaic fibers. *Sol. Energy Mater. Sol. Cells,* **2019**, *192*, 52-56.
[http://dx.doi.org/10.1016/j.solmat.2018.12.003]

[40]   Borazan, I.; Bedeloglu, A.; Demir, A. The effect of MWCNT-PEDOT: PSS layer in organic photovoltaic fiber device. *Optoelectron Adv Mat,* **2015**, *9*(3-4), 347-352.

[41]   Taş, M.; İşlek Cin, Z.; Sam Parmak, E.D.; Çelik Bedeloğlu, A. İşlek.; Cin, Z.; Sam, Parmak. ED.; Çelik, Bedeloğlu. A. Fabrication of unilateral conductive and transparent polymer thin films decorated with nanomaterials for flexible electrodes. *Polym. Compos.,* **2018**, *39*(5), 1771-1778.
[http://dx.doi.org/10.1002/pc.24109]

[42]   Sugino, K.; Ikeda, Y.; Yonezawa, S.; Gennaka, S.; Kimura, M.; Fukawa, T.; Inagaki, S.; Konosu, Y.; Tanioka, A.; Matsumoto, H. Development of fiber and textile-shaped organic solar cells for smart textiles. *J. Fiber Sci. Technol.,* **2017**, *73*(12), 336-342.
[http://dx.doi.org/10.2115/fiberst.2017-0049]

[43]   Hashemi, S.A.; Ramakrishna, S.; Aberle, A.G. Recent progress in flexible–wearable solar cells for self-powered electronic devices. *Energy Environ. Sci.,* **2020**, *13*(3), 685-743.
[http://dx.doi.org/10.1039/C9EE03046H]

SUBJECT INDEX

A

Accumulation 4, 79, 91, 160
  electrostatic 160
Acid 21, 22, 27, 28, 30, 31, 46, 51, 62, 92,
    128, 131, 146, 177, 178, 181, 183, 184,
    199
  ascorbic 28, 92
  boric 128
  caffeic 30
  carboxylic 62
  citric (CA) 27
  electrolyte 46
  formic 22, 178
  heteropoly 181
  Lewis 199
  nucleic 27, 31
  perfluorosulfonic 181
  phosphomolybdic 181
  phosphonic 183
  Phosphoric 177
  picric 28
  stearic 131
  Sulfonic 181
  sulfuric 46, 146
  sulphuric 21
  uric 51
Activated carbon (AC) 41, 45, 153, 169, 170,
    242, 253
Activities 50, 92, 130
  bioelectrocatalytic 92
  bio-electrocatalytic 50
  photocatalytic 130
Advanced 161, 207, 248, 250
  charge storage devices 161
  electro-optical properties 207
  nanomaterials 248, 250
Agglomeration 24, 84, 185, 207, 209
Ag nanowire 166
Alkaline fuel cell (AFC) 177
Amalgamation 78, 166, 244
  reaction 166

Amorphous silicon solar cells 260
Anchoring energy 214, 215, 216
  coefficient 214
Annealing process 44, 251
Anti-ferroelectric mesophases 206
Antigen 22, 52
  carcinoembryonic 22
Applications 62, 78, 79, 83, 130, 250, 251
  energy-related 78, 79
  photonic 130
  photovoltaic 62
  sensor 251
  sustainable 83
  thermoelectric 250
Arc discharge method 7, 8
Aromatic hydrocarbons 181
Asymmetric supercapacitors (ASCs) 127, 152,
    163
Atmosphere 7, 166
  green 166
  helium 7
Atomic force microscope (AFM) 198

B

Balanced charge injection 192
Ball milling technique 128
Band gap semiconductor 110
Base polymer microstructure 182
Battery-type storage behavior 168
Bending conditions 149, 150, 151
Binary metal oxide nanowires 167
Biomolecular interactions 25
Biomolecules 19, 20, 23, 27, 30, 31, 51
  immobilized 20
  redox-active 51
Biosensing platforms 22, 25, 27, 92
  electrochemical 27
  fluorescent 27
  green 92
  transfer-based fluorescent 25
  ultrasensitive 22

www.ingramcontent.com/pod-product-compliance
Lightning Source LLC
Chambersburg PA
CBHW050813220326
41598CB00006B/194